Annual Review of Chinese Architectural Design Works

2010–2011

中 国 建 筑 设 计 作 品 年 鉴

（下）

《中国建筑设计作品年鉴》编委会 编

江苏人民出版社

图书在版编目（CIP）数据

2010～2011中国建筑设计作品年鉴 / 《中国建筑设计作品年鉴》编委会编. -- 南京 : 江苏人民出版社, 2011.12
ISBN 978-7-214-07746-2
Ⅰ. ①2··· Ⅱ. ①中··· Ⅲ. ①建筑设计－作品集－中国－2010～2011 Ⅳ. ①TU206

中国版本图书馆CIP数据核字(2011)第258497号

2010—2011中国建筑设计作品年鉴（上、下册）　　《中国建筑设计作品年鉴》编委会 编

责任编辑：刘 焱 艾 璐
责任监印：彭李君
美术设计：郝泽星 彭 丹 贾 妍
出　　版：江苏人民出版社（南京湖南路1号A楼 邮编：210009）
发　　行：天津凤凰空间文化传媒有限公司
销售电话：022-87893668
网　　址：http://www.ifengspace.cn
集团地址：凤凰出版传媒集团（南京湖南路1号A楼 邮编：210009）
经　　销：全国新华书店
印　　刷：深圳市彩美印刷有限公司
开　　本：889毫米*1194毫米　1/12
印　　张：80（上：41，下：39）
字　　数：1000千字
版　　次：2012年1月第1版
印　　次：2012年1月第1次印刷
书　　号：ISBN 978-7-214-07746-2
定　　价：1058.00元（上、下册）
　　　　（USD 200.00）

（本书若有印装质量问题，请向发行公司调换）

主　　管：中华人民共和国住房和城乡建设部
主　　办：中国建筑文化中心
编　　辑：《中国建筑设计作品年鉴》编委会
　　　　　北京主语空间文化发展有限公司
主　　编：陈建为
执行主编：肖 峰
编辑部主任：王 伟
编　　辑：代 君 郭 晨 金 峰 林 朋 冷娜英 李 悦
　　　　　李 妍 刘笑艳 欧 阳 齐 阳 张 辉 张 澜
编辑部地址：北京市海淀区三里河路13号
　　　　　中国建筑文化中心712室
邮　　编：100037
电　　话：+86-10-88151966/+86-10-88153600/13910120811
传　　真：+86-10-88151958
网　　址：www.archrd.com
E-mail：zgjsmail@126.com

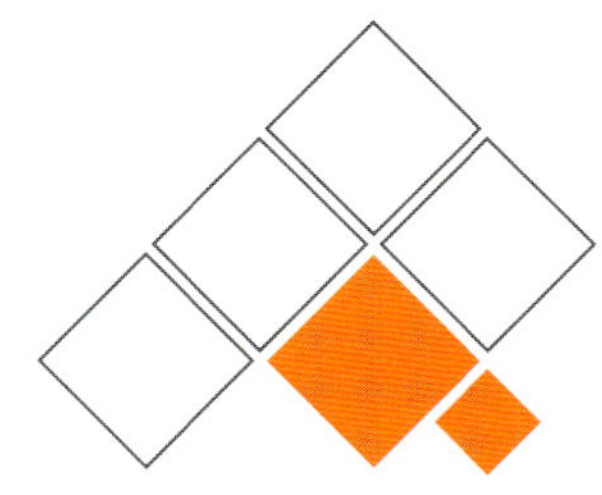

2010—2011
中国建筑设计作品年鉴

谨以此年鉴献给为中国建筑设计行业发展而辛勤付出的设计师们！

《2010—2011中国建筑设计作品年鉴》（以下简称《年鉴》）
收录了国内外180余家设计机构的近2000件优秀建筑设计作品，
这些不同类型、不同设计理念的作品基本反映了当前建筑设计行业多元化的发展状况和实践成果，
从中可以直观地领略到中国建筑设计行业融汇中西、贯通古今的设计思想。
对于各地建设主管部门和从事城市规划、建筑设计、景观设计、工程建设、房地产开发人员以及相关科研教育机构掌握建筑设计行业发展趋向、了解新的设计理念，《年鉴》会有很好的参考、借鉴作用。
《年鉴》在编辑过程中，得到了各地建设主管部门和众多设计院所、设计师的大力支持和帮助，
在此，谨向为《年鉴》提供帮助和支持的单位和个人表示衷心的感谢！
《年鉴》在国内公开发行，并委托中国图书进出口（集团）总公司在国外和国内港、澳、台地区发行。
在《年鉴》的编辑过程中，不免会有疏漏或错误之处，敬请读者指正，并提出宝贵意见，
以便我们不断提高《年鉴》的编辑水平，满足读者的需求。

《中国建筑设计作品年鉴》编辑部
2011年11月

特邀编委
（按姓氏首字母排序）

目录 上卷

目录 下卷

设计作品分类索引 下卷

设计作品分类索引 卷

设计作品分类索引 下卷

设计作品分类索引 下卷

设计作品分类索引 下卷

设计作品分类索引 卷

ECD International Engineering Design Co., Ltd.

卓创国际工程设计有限公司 www.ecd.com.cn

卓创国际工程设计有限公司是具有国际化背景并经国家住房和城乡建设部（以下简称建设部）批准的重庆市建筑设计企业（证书编号：190530–sj），具有独立的法人资格，是一家具有从事民用、工业建筑工程设计及相应的工程咨询甲级资质的设计机构。经过重组兼并，卓创国际已经成为重庆市最大的民营设计机构，并逐步铺向全国。已经成为在上海、深圳、重庆、西安、南宁、昆明、美国等地拥有多个分支机构的国际化设计服务集团公司。

公司现有高素质的各类专业设计人员120余人，其中国家一级注册建筑师5人，国家一级注册结构工程师8人，高级工程师25人，工程师66人，博士、硕士研究生20人。公司以完善的现代企业管理制度、雄厚的人力资源、先进的技术设备、独特的设计理念、丰富的实践经验，始终如一地为客户提供一流的服务。公司下设建筑创作室、项目经理部、总经办、总师办、质管办、人力资源部、财务部等职能部门。公司设备先进，已建成了内部的局域网，使各建筑创作室、项目部及各专业间沟通极为便利，极大地提高了工作效率。多年以来，公司承担各类工程设计百余项，共计完成建筑面积约600万平方米。所完成的项目得到了众多房地产开发商和建设施工单位的好评，并在全国及省市设计评比中多次获奖。21世纪充满着机遇与挑战，建筑设计行业也面临着品牌、技术、人员素质等多方面的竞争，卓创人必将排除万难，发挥聪明智慧去创造一个崭新的未来；将通过提供高品质的产品、高质量的服务，创造新的生活空间。

公司与Dans、Hanganu、STANTEC等国外建筑师事务所建立了良好的紧密合作关系。

愿景：品质卓越，行业领先
核心价值观：尊重、合作、创新、共赢
宗旨：追求卓越，创造价值
人才理念：扬长容短，团队至上；用心专心，共同发展

ECD International Engineering Design Co., Ltd. is an Architectural Design Enterprise with international background and approved by the Ministry of Housing and Urban-Rural Development (MOHURD) (Certificate No. :190530-sj), an independent legal personality, which has engaged in civil, industrial construction and engineering design and the corresponding engineering consulting. After reorganization and merger, ECD International has become the largest international private design institution in Chongqing, and gradually spread to the nation. ECD has become an international design services group with multiple branches of Shanghai, Shenzhen, Chongqing, Xi'an, Nanning, Kunming, the United States of America etc.

The company now has high-quality professional design staff of more than 120 people, including 5 first grade registered architects of China, 8 First Grade registered structural engineers of China, 25 senior engineers, 66 engineers, 20 doctors and masters. The company consistently provides customers with the first-class service in virtue of the perfect modern enterprise management system, strong human resources, advanced technology and equipment, unique design concepts, rich experience. The company has construction workshops, project manager department, General Manager's Office, chief engineer office, quality control office, human resources department, finance department and other functional departments. Advanced equipment, has built an internal LAN, so that communication between construction workshops, project departments and inter-professionals is very convenient, greatly improving work efficiency. Over the years, the company has taken hundreds of all kinds of engineering items, whose total construction area reaches about 600 million square meters. The completed projects have won the praise of numerous real estate developers and construction units, and won numerous awards in design competitions of the nation and provinces. Twenty-first century is full of opportunities and challenges, the architectural design industry is also facing competition in the brand, technology, quality of personnel and other aspects, ECD staff will overcome all troubles, make full use of the wisdom to create a new future, and provide high quality products, quality service, to create a new living space.

The company has established a good close cooperation with Dans, Hanganu, STANTEC and other foreign firms.

Vision: high-quality, industry-leading
Core values: respect, cooperation, innovation, win-win situation
Purpose: the pursuit of excellence, creating value
Talents concept: make best use of the advantages and bypass the disadvantages, team first; diligence, concentration and common development

重庆卓创国际

地址：重庆市渝北区龙塔街道兴盛大道120号天江鼎城博园10幢
电话：+86–23–65479361
网址：www.ecd.com.cn

ECD International in Chongqing
Add: The 10th Building of Tianjiang Ding City Bo Garden, No.120 Xinsheng Avenue, Longta Street, Yubei District, Chongqing City
Tel: +86–23–65479361
Website: www.ecd.com.cn

昆明卓创国际

地址：云南省昆明市关上关兴路288号万裕国际商务大厦15楼CD座
电话：+86–871–7361096

ECD International in Kunming
Add: Tower C and D, No.15 Building, Wanyu International Business Building, No.288 Guanxing Road,Guanshang District, Kunming City, Yunnan
Tel: +86–871–7361096

上海卓创国际

地址：上海市徐汇区漕宝路80号光大会展D座10层
电话：+86–21–64329355

ECD International in Shanghai
Add: The 10th Floor, Tower D, Guangda Exhibition, No.80 Caobao Road, Xuhui District, Shanghai City
Tel: +86–21–64329355

南宁卓创国际

地址：广西壮族自治区南宁市桂春路15号1区1单元303
电话：+86–771–5533388

ECD International in Nanning
Add: Room 303, Unit 1, Zone 1, No.15 Guichun Road, Nanning City, Guangxi
Tel: +86–771–5533388

西安卓创国际

地址：陕西省西安市高新区旺座国际城A座23层
电话：+86–29–88606710

ECD International in Xi'an
Add: The 23th Floor, Tower A, Wangzuo International City, High-tech District, Xi'an City, Shaanxi
Tel: +86–29–88606710

1 天下名宅

建设地点：四川
建筑面积：720 000平方米
设计时间：2011年

2 兰州奥体生态城

建设地点：甘肃 兰州
项目类型：概念规划
设计时间：2011年

3 涪陵水磨滩旅游度假区

建设地点：重庆 涪陵
建筑面积：600 000平方米
设计时间：2011年

4 天津河东美凯龙

建设地点：天津
建筑面积：410 000平方米
设计时间：2010年

5 汇东星城·南宁塔

建设地点：广西 南宁
总建筑面积：730 000平方米（住宅610 000平方米、主塔楼75 000平方米、商业45 000万平方米）
塔楼高度：220米

6 黄埔国际

建设地点：贵州 贵阳
建筑面积：650 000平方米
设计时间：2011年

7 成都融科

建设地点：四川 成都
建筑面积：300 000平方米
设计时间：2010年

8 红星·南平滨江国际城

建设地点：福建 南平
用地面积：14.1公顷
总建筑面积：655 000平方米

9 天地源·御湾雅墅

建设地点：广东 惠州
用地面积：180 000平方米
建筑面积：216 000平方米
设计时间：2009–2011年
竣工时间：预计2013年前竣工
容 积 率：1.2

10 音乐现场

建设地点：重庆
建筑面积：130 000平方米
设计时间：2011年

11

12

13

11 首金美利山

建设地点：重庆
建筑面积：830 000平方米
设计时间：2009年

12 海南酒店概念性方案

建设地点：海南
建筑面积：70 000平方米
设计时间：2011年

13 东和冉家坝广场

建设地点：重庆
建筑面积：300 000平方米
设计时间：2011年

14 南宁规划展览馆

建设地点：广西 南宁
建筑面积：30 000平方米
设计时间：2009年

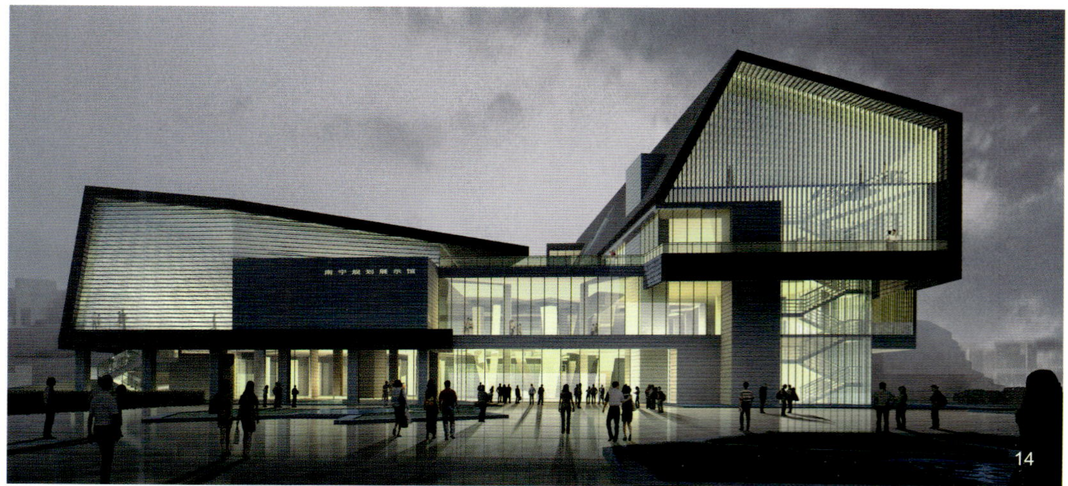

14

重庆同乘工程咨询设计有限责任公司

Chongqing Tongcheng Engineering Consult & Design Co., Ltd.

军工　机械　建筑　景观

军工品质　同乘铸就

重庆同乘工程咨询设计有限责任公司前身为重庆长安设计研究院，具备军工、机械、建筑工程设计甲级及工程咨询甲级资质。2005年，由中国长安汽车（集团）有限责任公司改制而成，经过多年来市场竞争的洗礼，公司由110人发展为230人，产值由5 000万元发展为4.5亿元（含咨询项目），军工出身的"同乘"正以崭新的英姿向前迈进！

业务范围包括：房地产项目的整体策划、各类新建、扩建及技术改造项目的项目咨询（总承包）服务；民用与工业建筑设计、景观规划设计、各类公用动力工程、环保工程、非标设备及机械产品的开发设计（如汽车整车、发动机工程设计和汽车研发试验设施等工程）。

同心合力　精益求精

同，合会也（《说文》）。同乘人深知：诚乃立业之基，质乃企业之本，我们以"迎风雨、见彩虹"的斗志，始终坚持"诚信为先、优质为本、以技术求创新、以管理促发展"的经营理念，秉承公司"优化设计、真诚服务、方寸之间、提升价值"的企业宗旨，同心合力，精益求精，服务每一家客户，"干一项工程、创一项佳绩、树一方信誉、交一方朋友"。

方寸之间　提升价值

寸，十分也；揣度；思量（《说文》）。同乘之作，在于方寸之间为客户提升价值，我们坚持"高品质，低成本，快响应"的服务理念，尊重每一片土地、关怀每一方人文、投入每一份激情、用心凝结每一根轴线、为钢筋混凝土铸造灵魂！

Military / Machinery / Architecture / Landscape

Military quality, casted by Tongcheng

Chongqing Tongcheng Engineering Consult & Design Co., Ltd. formerly the Chang'an Design and Research Institute, has Grade A design qualification and consultation qualification for military industry, machinery industry and architecture project. Tongcheng was reformed from Chang'an Auto (group) Co., Ltd. in 2005. Through many years' marketing experience, the company has got great improvement. Now, Tongcheng is keeping moving forward with a new posture.

Its business scope includes overall planning for real estate project, consultation and general contract for new project, extension project and renovation project; industrial and civil architecture design, landscape planning and design, research and design for public power engineering, environment protection projects, non-standard equipment and mechanic products.

Concerted efforts, keeping improving

"Tong" means union. Tongcheng's staffs know that good faith is the base for starting career, and high quality is the fundamental principle of company. We insist on the management concept of "innovate by technology, develop by management", adhere to "optimal design, sincere service, elaborate space, infinite value", unite to supply the excellent service for every customer. We finish engineering projects, obtain achievement, set up good reputation and make good friends.

Improve value in elaborate space

Inch consists of ten sections. Tongcheng is devoted to improving great value for customer in elaborate space. We keep on "high quality, low cost, rapid response", respect every inch land, care for every humanity, input every passion and cast soul for the reinforced concrete.

长安集团某工厂　CM8汽车焊接生产线　某工厂特种产品办公楼　铜仁江宗仁半岛　重庆美心洋人街A7-1　重庆巴南区政府广场

地址：重庆市江北区建新东路78号
邮编：400023
电话：+86-23-86394658 +86-23-86394600
传真：+86-23-67700054 +86-23-86394699
邮箱：569106136@qq.com
网址：www.tccq.cn

Add: No.78 Jianxin East Road, Jiangbei District, Chongqing
P.C.: 400023
Tel: +86-23-86394658 +86-23-86394600
Fax: +86-23-67700054 +86-23-86394699
E-mail: 569106136@qq.com
Http: //www.tccq.cn

重庆同乘工程咨询设计有限责任公司

Chongqing Tongcheng Engineering Consult & Design Co.,Ltd.

重庆市长生桥中学

设计思路

山径悠然，人文书院

巧妙保留原始地貌高差，山地生长出的有机建筑，现代的造型配以传统的坡顶，加上灰砖竹影，一柱一廊，一砖一瓦，无不散发出现代巴渝人文书院的自然馨香。

四川省通江县人民医院

建筑规模：7万平方米
项目类型：二级甲等综合医院
建设单位：通江县人民医院
建设地点：四川 巴中

重庆市长生桥中学

建筑规模：11万平方米(100班)
项目类型：学校
建设单位：重庆市长生桥中学
建设地点：重庆

重庆正大软件学院

建筑规模：100公顷
项目类型：校园规划及建筑设计
建设单位：重庆正大软件职业技术学院
建设地点：重庆

重庆大渡口中华国学馆

建筑规模：8 000平方米
项目类型：小型观演文化场馆
建设单位：重庆大晟资产经营（集团）有限公司
建设地点：重庆

成都树德外国语学校

建筑规模：7.8万平方米（120班）
项目类型：涉外学校
建设单位：成都树德外国语学校
建设地点：四川 成都

重庆同乘工程咨询设计有限责任公司
Chongqing Tongcheng Engineering Consult & Design Co.,Ltd.

公建类

某公安局办公指挥中心

建筑规模：3.5万平方米
项目类型：综合办公楼
建设地点：重庆

北京天威科技集团研发基地

建筑规模：12万平方米
项目类型：写字楼
建设单位：北京天威科技集团
建设地点：北京

中船重工重庆科技中心

建筑规模：7万平方米
项目类型：甲级写字楼
建设单位：中船重工集团重庆市航舶房地产开发有限公司
建设地点：重庆

大足龙水湖室内滑雪馆

建筑规模：3.5万平方米
项目类型：运动场馆
建设单位：重庆广电集团大足石刻影视文化有限责任公司
建设地点：重庆

梨香溪水上游客接待中心

建筑规模：6万平方米
项目类型：旅游娱乐建筑
建设单位：重庆美心集团房地产开发有限公司
建设地点：重庆

TONGCHENG 重庆同乘工程咨询设计有限责任公司
Chongqing Tongcheng Engineering Consul & Design Co.,Ltd.

吉首锦绣新都

设计思路

现代都市的民族印痕

考虑湘西地域特征，采用大量透空构架和立体绿化以适应当地湿热气候。同时，穿斗、窗格、方洞、菱孔、圆形花饰……又给现代都市留下了土家、苗寨的民族印痕。

吉首锦绣新都

建筑规模：26万平方米
项目类型：大型综合社区
建设单位：吉首福大房地产开发有限公司
建设地点：湖南 吉首

合川德润世家

建筑规模：17万平方米
项目类型：大型综合社区
建设单位：合川德佳集团德泽房地产开发有限公司
建设地点：重庆

遵义上东华城概念性规划

建筑规模：160万平方米
项目类型：大型综合社区
建设单位：重庆祥龙居房地产开发有限公司
建设地点：贵州 遵义

永川山水林涧三期项目

建筑规模：7万平方米
项目类型：类别墅小区
建设单位：重庆宏旭房地产开发有限公司
建设地点：重庆

永川山水林涧三期项目

设计思路

离尘不离城、大隐隐于市。旧时帝王将相府，如今白领桃花园。建筑依山就势，道路沿湖蛇行——道法自然。

首钢金岭华府别墅小区

建筑规模：70万平方米
项目类型：类别墅小区
建设单位：首钢集团重庆首金房地产开发有限公司
建设地点：重庆

住宅类

重庆市巴南区花溪保障性住房项目

建筑规模：21万平方米
项目类型：大型综合社区
建设单位：重庆市渝兴建设投资有限公司
建设地点：重庆

重庆同乘工程咨询设计有限责任公司
Chongqing Tongcheng Engineering Consult & Design Co.,Ltd.

重庆正大软件学院

建筑规模：10万平方米
项目类型：校园景观
建设单位：重庆正大软件职业技术学院
建设地点：重庆

景兰寨项目

建筑规模：3万平方米
项目类型：旅游地产
建设单位：云南海诚投资有限公司
建设地点：云南 西双版纳

景兰寨项目

设计思路

以清新、自然的原味风格让游人耳目一新。以原生的种植序列来体现傣家乡村小院的热带浪漫风情。

重庆正大软件学院

设计思路

山水树人，登高博见！

一心 花溪湖心

花溪湖处于整个校园中央，景色优美，山水兼具，启发心智。

两轴

人文礼仪轴

校风校训轴

三公园

中央公园、人文公园、运动公园。

景观类

中国草海国际养生基地

建筑规模：2万平方米
项目类型：公园
建设单位：重庆伊卓实业有限责任公司
建设地点：贵州 威宁

合道设计集团
HORDOR DESIGN GROUP

厦门合道工程设计集团由有着40多年历史的原厦门市建筑设计院改制发展而来。1960年厦门市建筑设计院成立，2003年经改制成立厦门市建筑设计院有限公司，2007年8月厦门市建筑设计院有限公司及下属十余家参控股公司，本着“道同而融合，循道而成业”的事业愿景，正式成立“厦门合道工程设计集团（下简称‘合道设计集团’）”。

合道设计集团具有建设部颁发的建筑工程设计甲级资质、国家发改委颁发的工程咨询甲级资质以及城市规划、市政工程（景观）、建筑智能化系统工程设计资质等。

合道设计集团拥有规范的现代企业制度、科学有效的管理机制和灵活的市场应变能力，集团立足设计主业，专注打造设计品牌，业务涉及建筑规划、建筑设计、景观设计、施工图审查、智能化设计、工程咨询及经济咨询、房地产开发与策划等领域，客户遍及北京、上海、深圳、福州、武汉、大连、青岛，南昌、昆明、成都、济南等地、迄今已与美国、加拿大、意大利、日本、英国、德国、西班牙、澳大利亚、新加坡、中国香港、中国台湾等国家和地区的设计机构建立了广泛的合作关系。此外，集团始终致力于人才成长平台的研究和打造，培育了一支功底扎实，思想敏锐，敬业守纪的专业技术队伍，其中教授级高工、高级工程师、工程师人数占公司人员总数的44%以上，至今已有百余个工程项目获得部省级优秀设计奖项，赢得中国建筑设计行业分会华东联席会“2001—2005年度改革与发展最具创意奖”、中国勘察设计协会“2006年优秀勘察设计企业”、福建省勘察设计协会“2006年度优秀勘察设计单位”、“2007年中国十大民营建筑设计企业”、“2010年中国最具影响力品牌设计机构”、“2011年中国十大民营建筑设计企业”等荣誉称号。

汗水与荣誉，变革与坚守，50多年风雨历程，合道工程设计集团历久弥新。今天，集团一如既往地关注着您的生活，构筑着城市的蓝图，正以专业的精神打造百年合道品牌。

Xiamen Hordor Engineering Design Group is developed from the original Xiamen Architectural Design Institute which has 40 years of history. In 1960, Xiamen Architectural Design Institute was established; in 2003, Xiamen Architecture Design Institute Co., Ltd. was established by reform. In August 2007, Xiamen Architectural Design Institute Co., Ltd. and its ten more subordinate holding companies, in the career vision of "integrating because of the same target, pursuing the target to establish the career", officially founded "Xiamen Hordor Engineering Design Group (hereinafter referred as 'Hordor Design Group')".

Hordor Design Group possesses Grade A qualification of architectural engineering design issued by the Ministry of Construction, Grade A qualification of engineering consulting issued by the National Development and Reform Commission, as well as the qualifications for urban planning, municipal engineering (landscape), building intelligent systems engineering design and so on.

Hordor Design Group has standardized modern enterprise system, scientific and effective management mechanism and flexible market adaptability. The group takes design as the main business, and focuses on brand building. Its business involves building planning, architectural design, landscape design, construction diagrams checkup, intelligent design, engineering consulting and financial consulting, real estate development and planning and other areas. They have customers in Beijing, Shanghai, Shenzhen, Fuzhou, Wuhan, Dalian, Qingdao, Nanchang, Kunming, Chengdu, Jinan and other places, and so far they have established the broad cooperation with the design institutions in the United States, Canada, Italy, Japan, Britain, Germany, Spain, Australia, Singapore, Hong Kong China, Taiwan China and other countries and regions. In addition, the group is committed to researching and creating the talent development platform, and they have nurtured a disciplined and dedicated professional team with solid foundation and perceptive thought, including professor senior engineers, senior engineers, and engineers representing above 44% of the total member of company. So far there are more than hundred projects received the provincial excellent design awards. The group has been awarded many titles of honor, such as "2001-2005 Most Creative Award for Reform and Development" of China Architectural Design Industry Branch East China Ally, "2006 Best Engineering Design Enterprise" of China Exploration & Design Association, "2006 Best Engineering Design Unit" of Fujian Survey and Design Institute, "2007 China Top Ten Private Construction Design Enterprises", 2010 Annual, China's most influential design agencies, 2011 China Top Ten Private Construction Design Enterprises, and so on. The group ranked the No.24th in 2007-2008 Overall Civil Architectural Design Market in China.

With sweat and honor, reform and insistence, and 50-year ups and downs, Hordor Engineering Design Group stands the test of time. Today, they continue concerning your life and building a blueprint for the city, and they are building a century Hordor brand with the professional spirit.

五缘尊墅

联排别墅：

1．超大面宽，绿化中庭，一宅四院；2．环水而建，与水相依，推窗见景，浪漫气息弥漫四周；3．立面现代简洁又具丰厚的文化底蕴，典雅气度展露无遗。

高层住宅：

1．以短板拼接和点式组合，局部扭转错位体现布局均好性和韵动性；2．三层高超大光景阳台，享受无限海景；3．建筑体量错落有致，富于变化和层次感；灰白素洁的建筑外观给人纯净、清新、欲怜之感。

Wuyuan Respect Villa

Townhouse

1.Super large width, green central hall, a house with four courtyards

2. Build around the water, dependent with the water, window views, and romantic ambience filled the whole atmosphere

3.simple and modern facade with a rich cultural heritage, elegant bearing accentuates

High-rise Residential

1. With a short board splice and dot combinations, partial dislocation reverse display the better layout and dynamic nature of rhyme

2. Three-floor high superb scene balcony, enjoying the unrivaled sea views

3. The body of buildings mass patchwork, full of change and layering; gray and white exterior of the building provides the people pure, fresh, want a sense of pity.

特房・美地雅登商业会所

梦开始的地方

谁还记得那年少时的梦想？谁还记得那曾经的渴望？谁还记得家乡那棵老树的身影？谁还记得村口那湾清澈的池塘？当这一切都已随风远去，只剩下故土的眷念在你的脑海中久久萦绕。

在这个拥挤的小区入口，应以怎样的姿态来面对周围的环境？是温馨还是冷漠？是宁静还是喧嚣？是谦让还是张扬？当你决定以家的气氛来营造这个环境时，温馨、宁静、谦让就成为了建筑的主题。

Special room • Meidi Yadeng Business Club

A Dream Place to Start

Who remembers your childhood dream? Who remembers the desire you had? Who remembers the old tree's shadow in hometown? Who remembers the clear pond at the entrance to the village? When everything is gone with the wind, leaving their homeland fondly lingering in your head.

Entry in a crowded area, what attitude should be to face the surrounding environment? Warm or cold? Quiet or noisy? Humility or publicity? When you decide to create a family atmosphere in this environment, warm, quiet and humility have become the subject of the building.

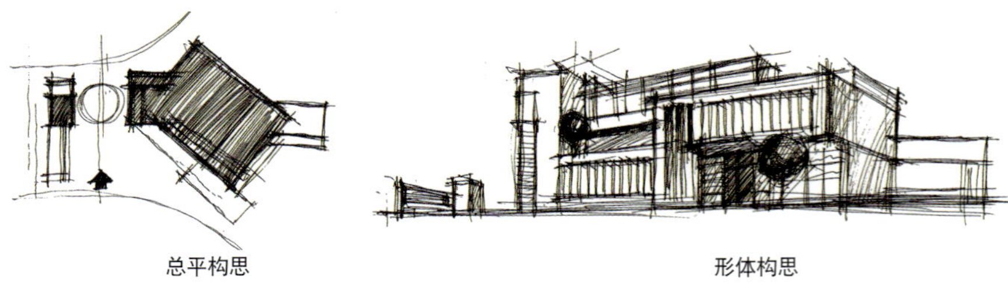

总平构思　　形体构思

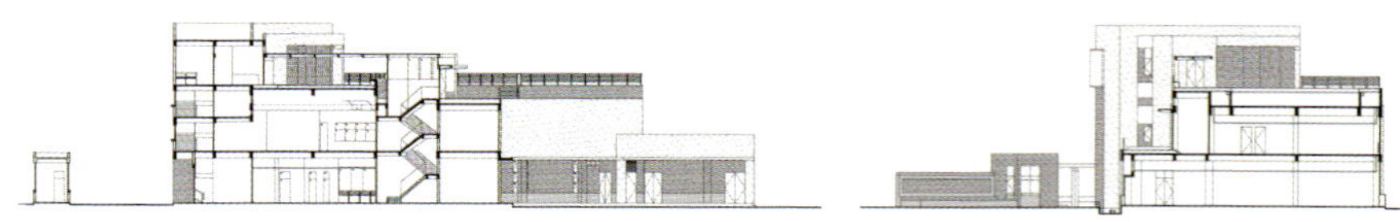

龙岩国际商贸物流园（一号方案）

本案位于龙厦高速路口处，作为城乡形象的窗口，应赋予建筑一种情感特征，使"他"既能展现城市性格特点又能体现出决策者对城市发展的前瞻性及带领龙岩经济腾飞的胸襟和魄力。

Longyan International Trade Logistics Park (No.1 Solution)

The case is in the Longxia high-way intersection. As the window image of the urban and rural areas, construction should be given a kind of emotional characteristics, so that "he" can not only display the city's characteristics, but also reflect the policy-makers foresight to urban development and the mind and courage for leading Longyan economic take-off.

龙岩国际商贸物流园（二号方案）

本项目依托其物流服务功能，搭建龙岩商贸招商引资平台，立足打造闽西地区最大的国内外商品交易、展示采购平台、金融结算平台和电子商务平台，把商贸与物流完美融合。并且辐射闽粤赣，连接全国网络，从更高层次完善商贸物流配套，提升城市形象品位。建筑布局组合形式犹如镶嵌工艺，组合有序，分区明确，并赋予彩色宝石的寓意。建筑以抽象的"茧"的外观形式和"蝶"的平面形式，寓意该项目商贸交流平台的孵化功能，营造"化茧成蝶"的意象之美。

Longyan International Trade Logistics Park (No.2 Solution)

The project relies on the function of logistics services, builds Longyan business investment platform, for the purpose of creating the largest domestic and international commodity trading, showing procurement platform, the financial settlement platform and e-commerce platform in the western Fujian, thus achieving perfect blend of commerce and logistics. Involved in Fujian, Guangdong and Jiangxi, connecting the national network, the business logistics support is improved from a higher level of to enhance the taste of the city's image. Building's layout combinations are like mosaic, with combined order and area clear, giving the meanings of colored gemstones. Buildings with the abstract "cocoon" the appearance of the sense form and the "butterfly" of the plane form, meaning the project commerce communication platform for the incubation function, creating the "emergence into a butterfly" image of beauty.

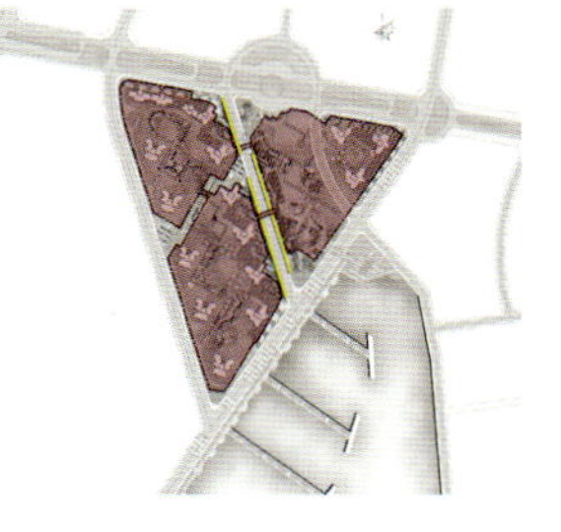

ANALYSIS 1

地下室轮廓与停车

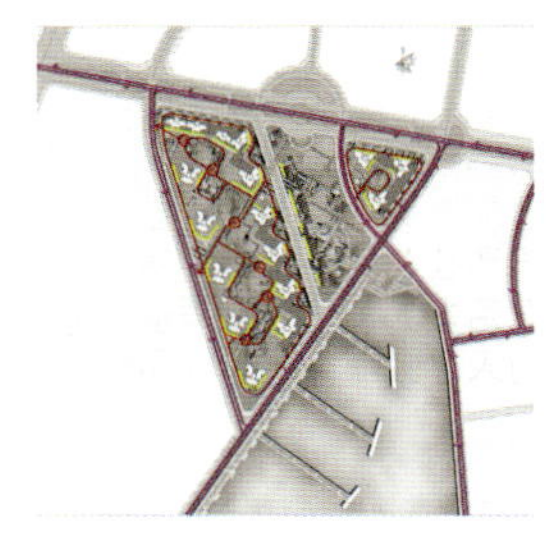

ANALYSIS 2

消防路线分析

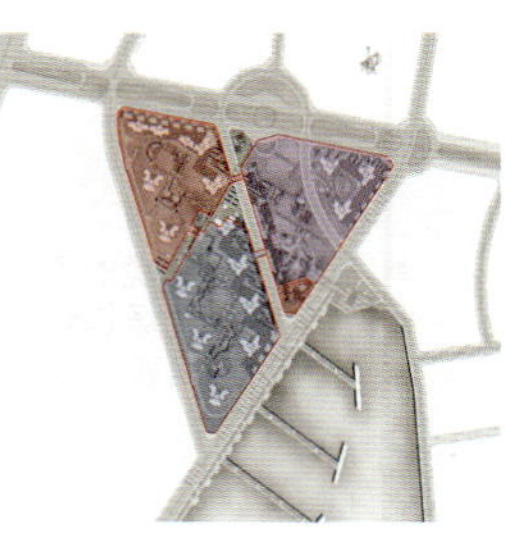

ANALYSIS 3

分期

杏林湾1号花园

1. 设计方案的总平面设计。总体设计充分挖掘基地的海景优势，采用点式布局，保证社区的通透性，同时使城市空间具有视觉穿透力。整体景观的均好性得以实现，中庭空间完整、丰富，为景观设计创造了良好的空间基础，最大限度增加户型的端角单元。

2. 建筑户型设计。户型设计采用全新概念，避免多户一楼的户型缺点，采用全通透式设计，空间方正实用，南北通透，具有单元式住宅的优点。每户均有景观中庭，同时可享受海景，以此提高户型价值，做到户户有卖点，带动社区的经济。强调户型的均好性布局，避免销售盲点，调高销售速率，在最短时间内收回投资成本，实现赢利。

3. 设计合理创新。采用多户设计，减少公摊，减少单元数，大量节省投资，合理控制了造价，又使户内实用面积增加，同时节约用地，使空间更加舒适。建筑群体空间相互错开，使每户观海视线均无遮挡。

Xinglin Bay No. 1 Garden

1. Site plan drawing of the design. Overall design makes full use of the advantages of sea views, takes the dot-style layout, makes the permeability of the community to achieve the best state, while the urban space with a penetrating vision. The overall landscape is the best, the atrium space is integrated and rich, creating a good space-based for the landscape design, maximally increasing the end-angle of unit types.

2. Building type design. Type design takes a new concept to avoid the shortcomings of multi-family in one apartment, using full pass-through design. The space is upright, foursquare and practical, south and north are ventilate, with the advantages of modular homes. Every house has the landscape in the atrium, at the same time enjoys the sea view that can improve the values of types, so every house has the selling point, driving the economic benefits of the community. Emphasizing on the type with the layout is good for less blind spot in sales, increasing the sales rate, recovering investment costs in the shortest time and achieving profitability.

3. Design innovation. Taking the multi-family design to reduce the cost distribution and the number of units, save a lot of investment, reasonably control the cost. Also increasing the usable floor area, at the same time, saving space, so the total space plane is more comfortable. Building space group staggered, so that each house has the boundless sight views of the sea.

厦门邮件处理中心

厦门邮件处理中心的设计总体构思源于邮政的企业文化。中国邮政标志是〝中〞字与邮政网络的形象互相结合，表达了服务千家万户的企业宗旨，以及快捷、准确、安全、无处不达的企业形象。建筑以平行结构为主，稍微向右倾斜的处理，表现了方向与速度感，而这一造型处理正与这种邮政标志设计的构思殊途同归，同时对邮政标志与传输的元素进行充分挖掘与提炼，强调一种现代物流建筑形象，没有过多的装饰、虚实对比，给人深刻的印象。二期与一期结合起来进行整体设计，使得建筑加建前后均能浑然一体。

空间组合方面注重辅助楼与主体的联系，基本层高关系为主体一层、辅助楼二层，辅助楼一、三层与主体一、二层连通，方便功能联系。

Xiamen Mail Processing Center

The overall design concept of Xiamen Mail Processing Center originates from the Post's enterprise culture. China Post sign is "中" word connecting with the image of the postal network, expressing the company aims to serve thousands of families and a fast, accurate, secure, ubiquitous reach of the corporate image. The building is mainly formed by parallel lines, slightly tilted to the right treatment, showing a sense of direction and speed, and this shape shares the same design concept with the postal logo, while the postal mark and transmission elements are refined, emphasizing the image of a modern logistics building, not too much decoration, comparison of the actual situation, giving a deep impression. Phase one and two with overall design considerations make the building to be built around seamless.

The space combination focuses on the link between auxiliary building and the main building. The basic relationship of storey is the first storey of the main building and storey of auxiliary building, the first and third floor of auxiliary building are connected with the first and second floor of the main building, making the functional linkages easy.

下洋体育文化中心

山清水秀、风光旖旎、多山多石是三明的城市特质。即将建成的文体中心，必将成为下洋区的景观中心。自然的形态，棱角分明、极富动感的造型，既呼应了周围优美的自然环境，又很好地体现了作为运动建筑的功能特色，更激发了全民健身的积极性。

- 强调了对环境的尊重。建筑和公共活动场地均对湖敞开，引湖光秀色入其中，形成绿色、生态的健身休闲场所。
- 体育中心的场馆设施采用灵活多变的、可以适应市场变化和发展的多功能原则，保证在可持续发展过程中的最大经济效益。
- 交通流线合理，保证区域内人车分流，确保人员安全疏散。

Xiayang Sports and Cultural Center

Beautiful, natural scenery, mountainous rocky are characteristics of the Sanming city. The sports center will be the landscape center in Xiayang District. Natural, very dynamic angular shape, not only echoes the surrounding beautiful natural environment, but also embodies the function characteristics as a movement building, and increased the affinity of fitness.

- Emphasizing respect for the environment. Buildings and public venues opened to the lake, bringing into the lake sceneries, and forming a green, ecological fitness and recreation venue.
- Sports center stadium facilities use flexible multi-function principle to market changes and the development, which ensure the maximum economic benefits.
- Reasonable traffic line makes the people and vehicles separated and ensures the safety of evacuation.

建发集团长沙总部大楼

本项目运用菱形图案形成疏密相间、错落有致的立面纹理，在加强双塔挺拔感的同时，也构成了高层的外筒结构体系。相互分离的倾斜面在有限的基地范围内为双塔提供了最有利的景观视线；统一而富有变化的形体组合造就了富有自然诗意的超高层城市综合体形象。

Jianfa Group headquarters building in Changsha

The project used the formation of self-similar density and white diamond pattern, patchwork facade texture to enhance the sense of tall towers, and constitute a high-level structure of the outer tube system. Inclined plane separated from each other within a limited base provides the most favorable landscape for the twin towers; unified and innovative form combinations create a rich natural poetic image of high-rise urban complex.

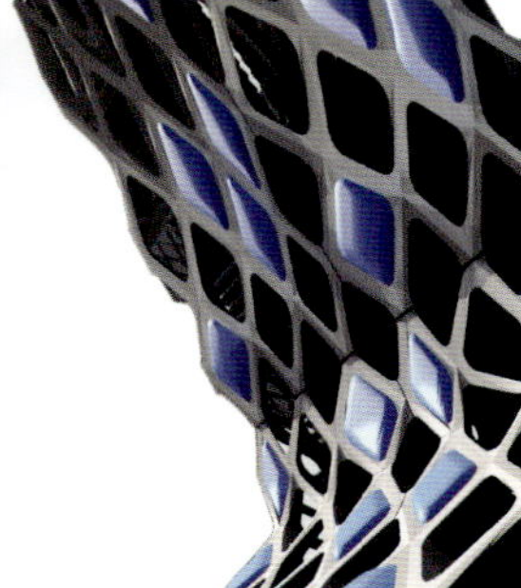

三明市图书馆、图书城、青少年宫

“书山，学海，扬帆起航”广场建筑群分为三个主要的部分：位于抬高广场下的一层书城营业空间，西面的青少年宫，以及东面的图书馆，三个部分紧密结合，连成一个水平伸展的整体。立面采用横向线条、大面积的虚实对比，形成强烈的光影效果和动态的韵律感。水平线条结合架空体量和悬挑平台，使图书馆、青少年宫如同一艘远洋游轮，与粼粼沙溪遥相呼应，在学海中拔锚起航。

Sanming City Library, Bookstore and Youth Palace

"Books, Learning, Sailing" Square building complex is divided into three main parts: bookstore operating space under the elevated plaza, the Youth Palace on the west, and the library on the east, which are closely linked together into a horizontal stretch of the whole body. The facade using horizontal lines, and a large area of contrast, formed strong lighting effects and dynamic sense of rhythm. Horizontal lines with elevated body mass and cantilevered platforms make the library and the youth palace as an ocean cruise, and sparkling sand and stream echoed, surfing the ocean of knowledge.

石狮市图书馆新馆

螺旋上升的变化空间反映在外立面上，形成了富有韵律感的书架结构，同时又隐喻了迂迴书山的路径，建筑造型就是内部空间“书”与“径”的直接反映。玻璃、素混凝土和木纹铺面的材质交替加强了石头的坚实与木头的温暖之间的对比，强烈的虚实对比更形成了丰富的体量感。

Shishi City, New Library

Changes in spiral space are reflected in the facade, forming a structure similar to a bookshelf, also a metaphor for the mountains, winding paths, interior architectural style is "book" and "path" of direct reflection. Glass, plain concrete and wood paving stone alternately strengthen the contrast between warm wood and solid stone, creating a strong presence on the body volume.

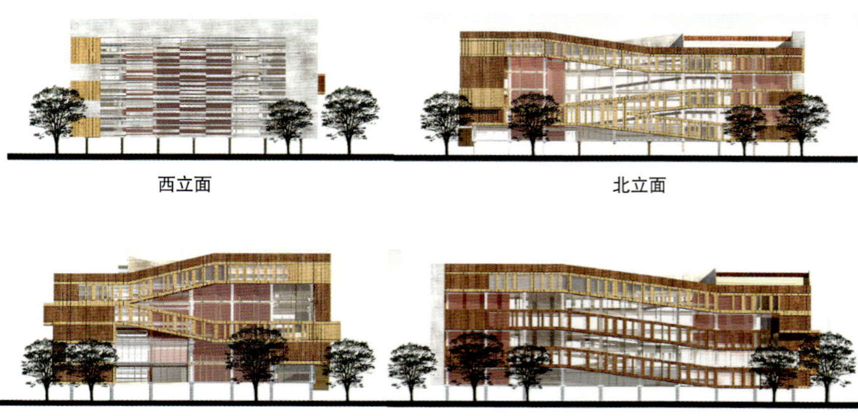

厦门枋湖客运中心

1．延续传统特色：本案将旧厂房改造成了车站，原有工业建筑缺乏表情，与建筑改造后的功能特点相违背。设计中以沿街面的建筑柱廊为主体，附加精致的钢结构支撑，共同创造一个新时代的骑楼作为新车站的过渡空间，以达到外部—灰空间—内部的过渡效果。

2．强调厦门海洋风格：在改造建筑形象的过程中，强调整体的素雅效果，在蓝天的衬托下，犹如海上的朵朵浪花。在侧立面上结合功能需求，以原有厂房结构为基础，加建3米的钢结构骨架，以改变沉闷的盒子形象。"翻腾"的钢结构也暗示出厦门海洋性环境的特点。

3．素雅的整体形象：新车站采用白色亚光漆饰面，使其在区域中卓然不群。

Xiamen Fanghu Passenger Transportation

1. The continuation of the traditional features: the case, renovated the old factory to the station. Original industrial buildings are lack of expression, contrary to function features of the buildings after renovation. Design using colonnade along the street side as the theme, additional fine steel structure supporting, work together to create a new era as a new arcade space station to meet external – gray space – internal transitions.
2. Emphasizing the Xiamen Marine style: in the process of renovation of the building image, the overall effect is elegant, against the backdrop of blue sky, like a sea of blossoming spray. Combined with functional requirements in its side surface and with the original plant structure basis, we added three meters of steel frame, to change the boring image of the box. Billowing out of the steel structure also suggests the characteristics of Xiamen Marine environment.
3. The elegant overall image: the new station is painted with matte white to highlight the extraordinary characteristics in the region.

龙岩云顶大酒店

本项目利用山顶自然地形曲线，形成环抱穿插的并行曲线，水平舒展的形体为绝大部分客房提供了全方位山顶观景角度，建筑尽端超大尺度的悬挑体量令建筑如腾飞于山峦之上。有序错落的阳台造就了富有韵律感的光影，并且为客房提供了深浅不一的室外和半室外空间。

Longyan Yunding Grand Hotel

The project uses the mountain top curve of the natural terrain, interspersed with the formation of the parallel curve surrounded, while the horizontal stretch of the body provides a full range of landscape for the majority of rooms. Large-scale construction makes the cantilevered end of the body look like that building on the mountains. Orderly scattered balcony creates rhythmic sense of light and shadow, and offers different light effects of outdoor and semi-outdoor space for rooms.

天津滨海国际会展中心配套会议中心

Tianjin Binhai International Convention & Exhibition Center Supporting Conference Center

设计充分结合地块周边建筑和环境，同时结合该建筑特有的属性，运用现代建筑的设计手法，采用大跨度结构，巨大且折叠的屋面很自然地形成了建筑主体优雅飘逸的造型，同时在建筑尺度上也体现出了会议中心应有的气势。

建筑整体犹如一只振翅欲飞的雄鹰在海面上跃起，预示着滨海新区也将在渤海湾上腾飞而起，成为一个经济繁荣、环境优美的宜居生态新城区。

The design fully integrated the surrounding buildings and environment of the land, combined with the unique attributes of the building, used the modern building design techniques, taking use of the large-span structures. Large folding roof naturally formed the elegant flowing shape of the main building, while in the building scale also show the magnificent atmosphere of a conference center.
The whole building looks like an eagle flying over the sea, indicates that the Binhai New Area will also fly in the sky on the Bohai Bay, and will be a prosperous economy, harmonious society and beautiful environment of the new eco-livable city.

重庆世贸中心
重庆纽约.纽约
重庆国贸大厦
重庆国贸大厦
重庆邹容广场
重庆时代豪苑
重庆谊德大厦
重庆黄花园立交
重庆万豪酒店
重庆新东方女人广场
重庆邹容广场
重庆奎星楼建筑群
重庆时代豪苑
重庆地王广场
重庆江北滨江路
重庆洲际酒店
重庆帝都广场
重庆群鹰商场
重庆都市广场（重庆大世界）
重庆华侨宾馆
重庆裕轮大厦
重庆建设银行大厦
重庆聚融广场
重庆韩光百货
重庆扬子岛酒店
重庆南国丽景
重庆商业大厦
重庆（和记）商务大厦
重庆东方曼哈顿
重庆御景江山
重庆海逸酒店
长江文具店
重庆美美百货
重庆民族广场
重庆大都会广场

CQADI

ChongQing Architectural Design Institute of China

中国 · 重庆市设计院

重庆市设计院成立于1950年，是国家建设部、国家发改委批准的综合甲级勘察设计和工程咨询单位，拥有建筑工程、市政公用、智能化设计、工程勘察、工程咨询、工程造价咨询、施工图审查等国家甲级资质，经国家外经贸委批准，享有对外经营权，并通过 ISO 9001：2000 质量体系认证，2006 年被评为建设部〝十五〞全国建筑技术创新先进企业，一直以来多次位列全国勘察设计前 100 强。

目前重庆市设计院下设 9 个建筑所、6 个建筑工作室、3 个市政设计所、城市规划研究中心、施工图审定中心等 20 个生产部门和办公室，组织有人事部、财务部、技术质量部、计划经营部等 5 个管理部门，以及后勤服务中心及从事建筑勘察设计延伸业务的迪赛实业（集团）有限公司。设计院现有在职职工 675 人，拥有重庆市设计大师 4 名、享受国务院政府津贴专家 12 名，拥有教授级高级建筑工程师 46 人、高级建筑工程师 159 人、建筑工程师 149 人，以及注册建筑师、注册结构工程师、注册岩土工程师、注册规划师、注册电气工程师、注册公用设备工程师、注册监理工程师、注册造价工程师、注册估价师等注册人员共 186 人。

重庆市设计院始终坚持〝精心设计，求实创新，诚信服务，顾客满意〞的质量方针，以优质的作品和高度的责任心竭诚为广大客户提供满意的服务，建院以来，设计院从事的工程涉及众多行业，遍及国内十多个省、市、自治区，在全国各地设计了众多具有影响力的项目，建院以来共完成勘察设计工程项目 4000 多项，先后有重庆大都会广场、重庆都市广场、四川美术学院综合教学楼等近二百个项目荣获了国家、部、省、市级优秀设计和科技成果奖。

Chongqing Architectural Design Institute ("CQADI"), which was founded in 1950, is an institute with class A certificate approved by the Ministry of Construction and the State Development and Reform Commission in geotechnical survey & design and engineering consultation.It has the national Class A qualification in construction engineering, municipal works, design of intelligent property management, geotechnical survey & design, engineering consultation, project cost consultation,drawing verification etc. After authorized by the National Foreign Economic Relations and Trade Commission, it is also a legal entity to undertake foreign operations and has acquired the ISO9001: 2000 China Quality Certification as well,It was evaluated as the national construction technology innovation advanced enterprise by the Ministry of Construction it and it also has been listed as one of the national top 100 best geotechnical survey & design Institute for several times.

At present, CQADI has 20 production departments and offices in total.which includes 9 construction institutes, 6 architecture studios, 3 municipal design institutes, urban planning and research center, drawings examination center,for more,we have 5 administration departments which consists of personnel department,financial department,technology department,planning department.We also have the logistics service center and Disai business group Co., LTD which is engaged in the extension business of geotechnical survey and design.We have 675 on-the-job staff,4 famous design masters in Chongqing, 12 experts who could enjoy the state council to government subsidies,46 professor-level senior architectural engineers,159 senior architectural engineers,149 architectural engineers and registered architects, registered structural engineer, registered geotechnical engineer, registered planner, registered electrical engineers,certified public equipment engineer,certificated supervision engineers, certified cost engineer and certified appraisers and so on,186 in total.

CQADI always insists on the quality guaranteed guideline of "elaborate design,realistic and innovative approach,faithful services, Customer Satisfaction". It wholeheartedly and dutifully provides the high-quality products to meet customer satisfaction with high sense of responsibility.Since the founding of CQADI,our projects covers many industries throughout the country over 10 provinces, cities, autonomous regions,we've developed many influential projects around the country. CQADI has accomplished more than 4000 geotechnical design projects,and the projects of Chongqing Metropolitan Plaza, the Teaching and Administration Building of Sichuan Fine Arts Institute won the Excellent Design and Scientific & Technological Achievements prizes at national, ministerial, provincial or municipal levels.

ChongQing

National comprehensive A grade prospecting and design institute

Architectural

National Top100 prospecting and design institute lasting 7 years

Design

Experienced over 40000 architecture project

Institute

Gain over 200 prizes

地址：重庆市渝中区人和街 31 号　电话：+86-23-63854124　传真：+86-23-63856935　邮箱：cqadinet@cta.cq.cn　网址：www.cqadi.com.cn

福州市建筑设计院

Fuzhou Architecture Design Institute

福州市建筑设计院成立于1956年3月，是勘察设计综合甲级设计院。具有国家甲级建筑行业建筑工程设计、工程勘察专业类岩土工程、工程咨询及工程监理、工程项目管理和一类施工图审查等资质。技术力量雄厚，是福建省骨干勘察设计单位之一。

我院现有员工200余人，各类专业技术人员近170人，教授级高工、高级职称约占40%；拥有执业注册资格的各类专业技术人才近90人，约占员工总数的45%。其中一级注册建筑师和结构师近40人，注册设备师、岩土师、监理师、造价师近50人，专职、兼职审查师近40人。

我院秉持“顾客为核心，质量是根本”的经营理念，确立“品质成就理想，服务提升价值”的质量方针，倡导“细节决定成败、周到创造感动”的服务理念，精心设计、精心管理，承担的各类公共建筑和住宅小区以及各种大中型工业与民用建筑的设计、勘察、监理、审查及岩土检测与治理工程，得到社会和客户的好评，并有几十项勘察设计成果获得住建部和福建省优秀勘察设计奖或科技进步奖。先后荣获省、市精神文明建设先进单位、思想政治工作先进单位、行业文明单位、先进单位等称号，连续十数年获“重合同、守信用”单位。

Fuzhou Architecture Design Institute was established on March 1956 and is a comprehensive class-A design institute focusing on engineering investigation and design. The Institute has obtained various qualifications, including constructional engineering design qualification of national class-A architectural industry, geotechnical engineering of investigation, engineering consultation and supervision, project management and class-A shop drawing review. It has a strong technical force and is one of the backbone engineering investigation and design enterprise in Fujian Province.

The Institute has over 200 employees, including nearly 170 technical personnel of various disciplines, about 40% of whom are titled professor senior engineer, and nearly 90 staff with registered qualification, 45% of the total staff. There are nearly 40 first-class registered architects and structure engineers, nearly 50 registered equipment engineers, geotechnical engineers, supervision engineers and budgeting engineering, as well as nearly 40 full-time and part-time reviewers.

With the operation principle of "Customer is the core and quality is fundamental", the quality principle of "Quality can fulfill our ideas and services can improve the value", and the service principle of “Detail is the key of the success and consideration can create touching", the Institute are devoted to design, engineering investigation, supervision, review, geotechnical inspection and treatment of various public buildings and residences, as well as all kinds of industrial and civil engineering. It is widely acclaimed by society and customers, and has received dozens of excellent investigation awards, scientific and technological progress awards from the Ministry of Housing and Urban-Rural Development as well as Fujian Provincial authorities. The Institute also wins numerous titles, such as provincial and municipal advanced unit of spiritual civilization construction, advanced unit of ideological and political work, industry civilized unit, advanced unit, as well as unit that observes contracts and keeps promises for ten successive years.

地址：福建省福州市鼓楼区津门路32号　邮编：350001　电话：+86–591–87549980/87556803　传真：+86–591–87616491　邮箱：fzjzsjy@pub2.fz.fj.cn

海峡汽车文化广场一期——汽车超市

项目地点：福建 闽侯
用地面积：124公顷
总建筑面积：15万平方米

福汽研究院及实验中心大楼

项目地点：福建 福州
总占地面积：2 792平方米
总建筑面积：38 183平方米

福建海峡银行

项目地点：福建 福州
用地面积：8 488平方米
总建筑面积：62 211平方米

融汇江山

项目地点：福建 福州
用地面积：3.21公顷
总建筑面积：14万平方米

长乐市青少年校外体育活动中心

项目地点：福建 长乐
用地面积：10 869平方米
总建筑面积：8 058平方米

三盛托斯卡纳

项目地点：福建 福州
用地面积：29.6公顷
总建筑面积：61.65万平方米

福建省司法大楼

项目地点：福建 福州
用地面积：6 078平方米
总建筑面积：15 900平方米

福清海峡商品交易中心

项目地点：福建 福清
用地面积：57 936平方米
总建筑面积：86 900平方米

FUJIAN BOYORK ARCHITECTURAL DESIGN CO., LTD.

福建博宇建筑设计有限公司

福建博宇建筑设计有限公司是一家具有甲级建筑工程设计资质的公司，专业从事建筑工程设计。公司由一批年富力强的建筑师、结构工程师及设备工程师组成，现有员工近150人，其中一级注册建筑师10人、一级注册结构师6人、注册电气工程师3人、注册给排水工程师3人、注册暖通工程师3人、中高级以上职称的技术骨干40余人，有很强的综合技术力量。公司以大型居住区、商业综合体及高级酒店为主要业务类型，在历年各类评奖活动中屡获殊荣，并通过大量的工程实践，积累了丰富的项目设计经验。公司视每一个项目为一次神圣而可贵的机遇，在工作过程中全情投入，不断创新，精益求精，为业主提供完美的解决方案与高质量的专业服务，并以此回报各界的信任和支持。

Fujian Boyu Architectural Design Co., Ltd. is a licensed Class-A architectural design company, committed to architectural engineering design. This Company is composed of a group of young and strong architects, structural engineers and equipment engineers, with nearly 150 employees, including 10 Class-1 Registered Architects, 6 Class-1 Registered Structural Engineers, 3 Registered Electrical Engineers, 3 Registered Water Supply & drainage engineers, 3 registered heating & ventilation engineers, and more than 40 technical experts with titles at mid-high and above levels, of strong comprehensive technical competence. Main businesses include the design of large residential community, commercial complex and high-grade hotel. We have won special honors in many years, and gained rich project management through numerous engineering practices. We deem every project as a sacred and valuable opportunity, dedicate ourselves fully in the works, innovate continuously, keep refining, and provide owners with perfect solutions and professional services of high quality, winning trust and support from all social communities.

福州公司：
地址：福建省福州市铜盘路软件园A区32号楼
电话：+86-591-87508851
传真：+86-591-87508853
邮箱：boyu-2005@163.com

厦门公司：
地址：福建省厦门市云顶中路市政大厦14楼
电话：+86-592-5250203
传真：+86-591-5250183
邮箱：boyu-2005@163.com

Fuzhou Branch:
Add: Building 32, Zone A, Software Park, Tongpan Road, Fuzhou, Fujian
Tel: +86-591-87508851
Fax: +86-591-87508853
E-mail: boyu-2005@163.com

Xiamen Branch:
Add: Floor 14, Municipal Building, Yunding Middle Road, Xiamen, Fujian
Tel: +86-592-5250203
Fax: +86-591-5250183
E-mail: boyu-2005@163.com

福州·中庚城
项目位置：福建 福州
项目规模：41万平方米
项目状况：竣工

Fuzhou Zhonggeng Town
Location: Fuzhou, Fujian
Scale: 410,000 m^2
Project Status: Completed

名城·城市广场
项目位置：福建 福州
项目规模：41万平方米
项目状况：在建

Mingcheng · City Plaza
Location: Fuzhou, Fujian
Scale: 410,000 m^2
Project Status: Under construction

群升国际三期
项目位置：福建 福州
项目规模：20.5万平方米
项目状况：竣工

Qunsheng International Phase III
Location: Fuzhou, Fujian
Scale: 205,000 m^2
Project Status: Completed

徐州科技大厦
项目位置：江苏 徐州
项目规模：3.7万平方米
项目状况：竣工

Xuzhou Science & Technology Building
Location: Xuzhou, Jiangsu
Scale: 37,000 m^2
Project Status: Completed

贵安海峡文化村酒店
项目位置：福建 福州
项目规模：16.8万平方米
项目状况：在建

Guian Strait Culture Village Hotel
Location: Fuzhou, Fujian
Scale: 168,000 m^2
Project Status: Under construction

重庆·中庚城
项目位置：重庆
项目规模：86.8万平方米
项目状况：在建

Chongqing Zhonggeng Town
Location: Beibei, Chongqing
Scale: 868,000 m^2
Project Status: Under construction

中庚办公楼
项目位置：福建 福州
项目规模：3.5万平方米
项目状况：在建

Zhonggeng Office Building
Location: Fuzhou, Fujian
Scale: 35,000 m^2
Project Status: Under construction

名城办公楼
项目位置：福建 福州
项目规模：4.0万平方米
项目状况：在建

Mingcheng Office Building
Location: Fuzhou, Fujian
Scale: 40,000 m^2
Project Status: Under construction

国际华府
项目位置：福建 福州
项目规模：19.6万平方米
项目状况：竣工

International Mansion
Location: Fuzhou, Fujian
Scale: 196,000 m^2
Project Status: Completed

建发·利苑
项目位置：福建 长乐
项目规模：27.3万平方米
项目状况：在建

Jianfa - Liyuan
Location: Fuzhou, Fujian
Scale: 273,000 m^2
Project Status: Under construction

名城・豪生酒店
项目位置：福建 福州
项目规模：4.1万平方米
项目状况：竣工

Mingcheng · Howard Johnson Riverfront Plaza
Location: Fuzhou, Fujian
Scale: 41,000 m^2
Project Status: Completed

三木・家天下
项目位置：福建 福州
项目规模：19.2万平方米
项目状况：竣工

Leisure Realm - Sanmu City
Location: Fuzhou, Fujian
Scale: 192,000 m^2
Project Status: Completed

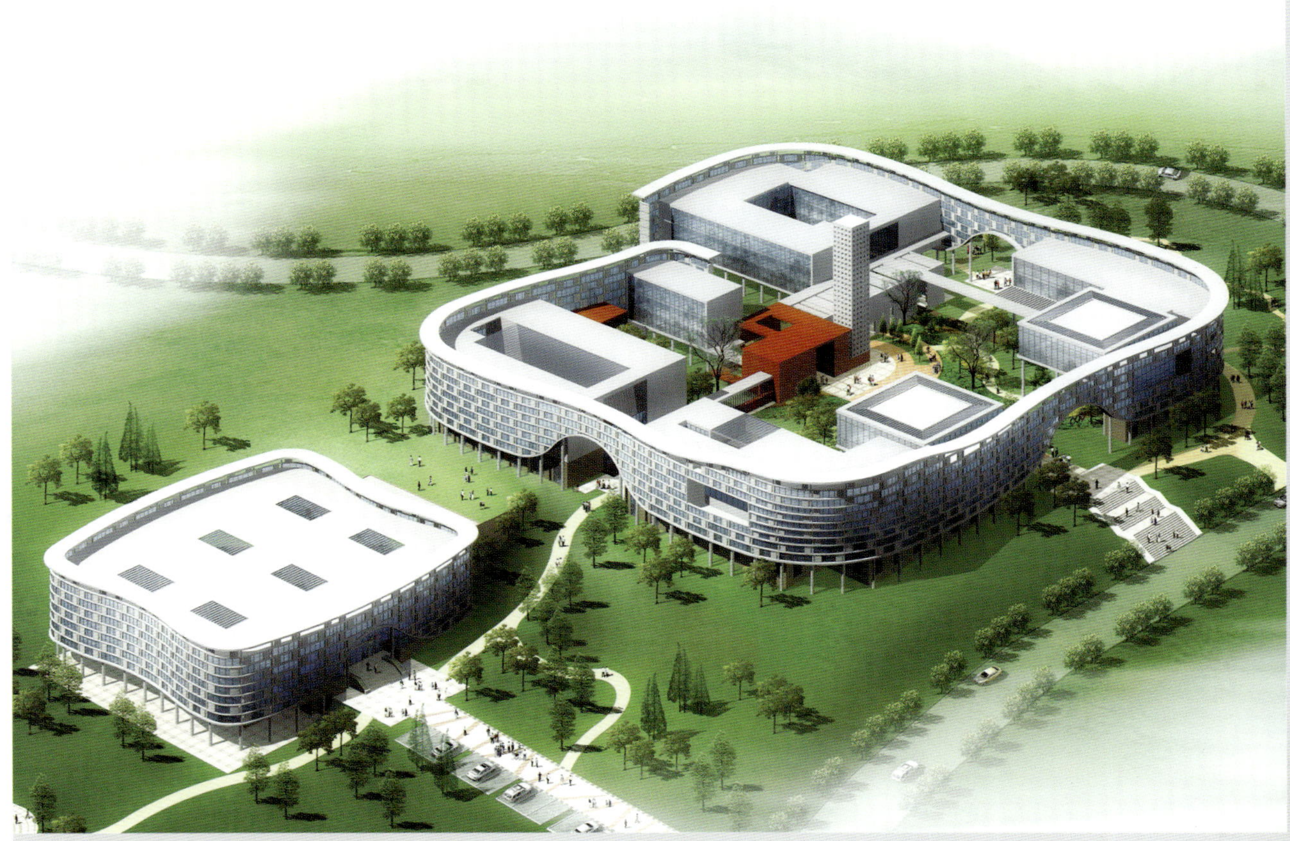

中国举重马江基地
项目位置：福建 福州
项目规模：2.0万平方米
项目状况：在建

China Weightlifting Majiang Base
Location: Fuzhou, Fujian
Scale: 20,000 m^2
Project Status: Under construction

尚城建筑
项目位置：福建 福州
项目规模：7.9万平方米
项目状况：竣工

Shangcheng Architecture
Location: Fuzhou, Fujian
Scale: 79,000 m^2
Project Status: Completed

紫金香山
项目位置：福建 福清
项目规模：25.6万平方米
项目状况：竣工

Zijin Xiangshan
Location: Fuqing, Fujian
Scale: 256,000 m^2
Project Status: Completed

帝国大苑
项目位置：福建 福州
项目规模：8万平方米
项目状况：在建

Empire Garden
Location: Fuzhou, Fujian
Scale: 80,000 m^2
Project Status: Under construction

博宇设计
BYAD
BOYORK ARCHITECTURAL DESIGN

龙域·香醍半岛
项目位置：广西　南宁
项目规模：27.2万平方米
项目状况：在建

Longyu · Xiangti Peninsula
Location: NanNing, Guangxi
Scale: 272,000 m^2
Project Status: Under construction

中恒首府
项目位置：福建　福清
项目规模：23.8万平方米
项目状况：在建

Zhonggeng Capital
Location: Fuqing, Fujian
Scale: 238,000 m^2
Project Status: Under construction

中庚·喜来登酒店
项目位置：福建　福州
项目规模：15.2万平方米
项目状况：在建

Zhonggeng · Sheraton Hotels
Location: Fuzhou, Fujian
Scale: 152,000 m^2
Project Status: Under construction

2010—2011

主要建筑设计作品

Main Architectural Design Works

Annual Review of Chinese Architectural Design Works

广州市纬纶建筑设计有限公司
Win-land Architecture Design Co., Ltd.

····设计无限可能····

广州市纬纶建筑设计顾问有限公司是一个具备雄厚实力的设计团队，旗下的智海建筑设计公司及艺筑建筑设计公司分别拥有建筑行业甲级及乙级资质。纬纶专注于房地产开发建筑设计，在全国完成多项工程，对房地产项目总体规划、市场状况、小区空间环境及住宅户型有着深入的研究，向客户提供商业、零售中心、公建、酒店、住宅、规划等多个领域的建筑设计服务。

纬纶围绕"设计无限可能"为核心理念，以精益求精的态度完成每一项设计作品。创立13年来，累计为客户完成建筑超过2 000万平方米，为60万人提供了优质舒适的居住环境，精品的设计和优质的服务在客户群中树立了极佳的口碑。现已拥有合景泰富、万科集团、新鸿基地产、美林基业、天伦控股、苏宁集团、佛奥集团、保利集团、佛山能兴地产、兴发铝业等知名集团长期客户。项目辐射到佛山、茂名、惠州、湛江、梧州、贵港、南宁、南昌、武汉、南京、郑州、山东、海口等地，其中合景泰富·科汇金谷大型项目、佛奥·星光广场商住项目、佛山宾馆、万科·天景花园、美林湖畔别墅、美林海岸住宅小区、天伦林和村城中村改造、天伦·郑州龙湖小区、苏宁·天润城等重点项目更在业界获得一致好评。

"正·大·光·明"是纬纶出品的最大特色，大气的规划、户型，宽敞的空间，最优的自然采光、明快开阳的建筑感受，令地产商及住户的多层次需求得到充分满足，销售业绩不断创出新高。在建筑业风起云涌的今天，纬纶坚持"无限可能"的设计理念，独特的建筑设计特色，以专业的触觉及市场敏感度，致力打造出符合客户和市场需求的设计精品。

Guangzhou Win-land Architecture Design Co., Ltd. is a design team with abundant strength, Zhihai Architecture Design Co., Ltd. and Yizhu Architecture Design Co., Ltd. belong to it, and respectively owned construction industry Class A and Class B qualification. Win-land focuses on real estate development architectural design, completed various projects in the country, has depth studies in the overall planning of real estate projects, market conditions, residential space environment and residential units, provides commercial, retail centers, public buildings, hotels, housing, planning and other fields of architecture design services.

Win-land takes "design possibilities" as the core concept, with the attitude of excellence to complete each of the design works. Since its foundation for thirteen years, it has complete the cumulative construction area over 20 million square meters for customers, provided 60 million people with high-quality and comfortable living environment, quality of design and excellent service in the customer base and established an excellent reputation. Now it has KWG, Vanke Group, Sun Hung Kai Properties, Mayland Foundation, Tianlun Holdings, Suning Group, Fo Ao Group, Poly Group, Foshan Nengxing real estate, Xingfa aluminum and other well-known group long-term customers. Project radiation to Foshan, Maoming, Huizhou, Zhanjiang, Wuzhou, Guigang, Nanning, Nanchang, Wuhan, Nanjing, Zhengzhou, Shandong, Haikou, etc., including large project KWG • Kehui Golden Valley, the commercial and residential project Fo Ao • Star Plaza, Foshan Hotel, Vanke • Tianjing Garden, Mayland Lake Villa, Mayland coastal residential area, the transformation of Tianlun Linhe village, Tianlun • Zhengzhou Longhu residential area, Suning • Tianrun City and other key projects received more praise in the industry.

"Standard, Great, Glory, Bright" is the most significant feature of Win-land production, the magnificent planning and units, broad space, optimal natural light, bright building experience, so that the developers and tenants of multi-level needs are fully met, the sales achievement kept high record. Surging in the construction industry today, Win-land adheres to the "endless possibilities" design concept, the unique architectural features in a professional sense and market sensitivity, to create the design boutique fit the needs of customer and market.

地址：广东省广州市越秀区中山二路3号粤运大厦3楼A室
电话：+86-20-61181771　　传真：+86-20-37621177
网址：www.win-land.com　　邮箱：winland@win-land.com

Add: Room A, 3F, Yueyun Tower, No. 3 Zhongshan Road 2th, Guangzhou, Guangdong
Tel: +86-20-61181771　　Fax: +86-20-37621177
Http://www.win-land.com　　E-mail: winland@win-land.com

1 凤城商务办公楼

2 佛奥天津金盛购物广场

3 金奥广场

4 佛奥广场

广州市纬纶建筑设计有限公司
Win-land Architecture Design Co.,Ltd.

合景泰富·科技园

湛江·渔人码头

惠州·山水华府

合景泰富・科汇金谷

广州市设计院
GUANGZHOU DESIGN INSTITUTE

广州市设计院组建于1952年，是国内成立最早的甲级勘察设计单位之一。拥有工程设计建筑行业甲级，工程勘察、建筑智能化系统工程甲级，以及工程设计咨询、城市规划编制、市政设计、消防、环保、施工图审查等多项资质，并已通过ISO9001：2008质量管理体系认证，是全国勘察设计综合实力百强单位，在全国民用建筑设计院中一直位居前列，2009年被评为广东省高新技术企业。注重建筑工程精品的设计和技术的创新，自20世纪50年代以来，打造出了一批高质量、富有较强影响力的时代精品。目前已有四百多项（次）工程勘察设计获国家、部、省、市级的奖励。全院员工五百多人，其中"全国工程设计大师"一名，经国务院批准享受政府特殊津贴的专家十多名，各专业的国家注册师一百多名。全院员工将更加热诚地为社会各界提供优质服务，用智慧和汗水创造更多的设计精品。

Guangzhou Design Institute, organized in 1952, is among the first Class-A survey & design organizations in China. It is qualified for engineering design & building industry Class-A, project survey, building intelligence system project Class-A, as well as project design consulting, urban planning & programming, municipal design, firefighting, environmental protection, construction drawing review and others, and has adopted the ISO9001:2008 Quality Management System; it is among the Top 100 National Survey & Design Powers, ranking among the Best National Civil Building Design Institutes, and was appraised as Hi-New Tech Enterprise in Guangdong Province. It pays great attentions to build exquisite design and technical innovation, and since 1950s, it has built up a group of epochal excellencies of high quality and strong influence. Currently, it has won over 400 honors of project surveys & designs at national, ministerial, provincial and municipal levels. This Institute comprises of more than 500 persons, of whom, there is one "National Engineering Design Master", more than 10 experts enjoying national special subsidy granted by State Council, and more than 100 national registered engineers of all specialties. The entirety of this Institute will be more energetic to provide quality services to all social communities, and create more excellent designs with wisdom and great efforts.

地址：广东省广州市体育东路体育东横街3号设计大厦
邮编：510620
电话：+86-20-87544608
传真：+86-20-87544798

Add: Design Mansion, Tiyudong Hengjie, Tiyudong Road, Guangzhou, Guangdong
P.C.: 510620
Tel: +86-20-87544608
Fax: +86-20-87544798

E-mail: bgs@gzdi.com
Http://www.gzdi.com

广州国际体育演艺中心

设计单位：广州市设计院
合作单位：美国Manica建筑师事务所
建设地点：广东 广州
总建筑面积：130 312.8平方米
竣工时间：2010年

Guangzhou International Sports and Performance Art Center

Design Company: Guangzhou Design Institute
Cooperation Company: Manica Architects Inc.
Construction Site: Guangzhou, Guangdong
Overall Floorage: 130,312.8 m^2
Completion Date: 2010

广州国际体育演艺中心作为华南地区首个按NBA标准建设的场馆，能完全体验到原汁原味的美国NBA文化。场馆外观线条充满生命律动感，内部结构设计合理，既能满足2010年广州亚运篮球赛、NBA中国赛等大型国际赛事的要求，同时也为各类演出提供国际一流水准的舞台设计空间和艺术创意享受空间，是一个集篮球及各类体育练习馆、综合娱乐中心和公园于一体的顶级体育演艺中心。

Guangzhou International Sports and Performance Art Center is the first venue which is built according to NBA standard in South China Region, people there can fully experience the genuine American NBA culture. Its rhythmic external line and reasonable interior design can satisfy the requirement of Basketball Game of 2010 Guangzhou Asian Games and large-scale international games such as NBA China Game, etc. as well as offer a space of international first-rate stage design art creation enjoyment for all kinds of performances, it's a top sports and performance and art center with all kinds athletic training venues, comprehensive entertainment center and park.

湖南省人民会堂

设计单位：广州市设计院
建设地点：湖南 长沙
总建筑面积：27 000平方米
竣工时间：2010年

Hunan Provincial People's Hall

Design Company: Guangzhou Design Institute
Site: Changsha, Hunan
Overall Floorage: 27,000 m^2
Completion Date: 2010

（来源：房王网）

广州亚运城运动员村国际区

设计单位：广州市设计院
建设地点：广东 广州
总建筑面积：33 260平方米
竣工时间：2010年

International Zone, Athletes Village, Asian Games Town, Guangzhou

Design Company: Guangzhou Design Institute
Site: Guangzhou, Guangdong
Overall Floorage: 33,260 m^2
Completion Date: 2010

广东省新兴县禅泉酒店

设计单位：广州市设计院
建设地点：广东 云浮
总建筑面积：70 562平方米

Chanquan Hotel, Xinxing County, Guangdong Province

Design Company: Guangzhou Design Institute
Site: Yunfu, Guangdong
Overall Floorage: 70,562 m^2

广州市申派建筑设计顾问有限公司
Simply Arch. Design & Consultant Ltd. Guangzhou

申・国际设计有限公司
SIMPLY ARCH. INTERNATIONAL LIMITED

公司简介 Profile

广州市申派建筑设计顾问有限公司是一个不断成长的空间，对建筑艺术的追求以及对创作作品的高要求使企业员工凝聚在一起，每个人都以不同的方式作出贡献，并对设计享有共同的信念、相同的热情。公司秉承与时俱进的设计理念，积极探索新技术，追求务实、创新、发展，打造有文化、有理想、有灵性的设计作品，为业主提供全方位的优质服务。

Guangzhou Simply Arch Design & Consulting Co., Ltd. is a growing space, pursuing the art of architecture and high quality requirements of the creations, which bring the enterprise employees together. Everyone makes contributes in different ways with a common belief and the same passion on architectural design. Adhering to the design concept of advancing with the times, the company actively explores new technologies, pursues pragmatism, innovation and development, and creates cultural, ideals, spiritual design works, providing a full range of quality services for the owners.

设计理念 Design Philosophy

社会的价值——我们工作和挖掘的核心；
客户的要求——我们进步的助推器；
客户的回应——我们设计的创作的重要思想；
创造性　　——我们工作的主题，并和客户的精神高度相关；
对市场的高度聆听，使我们的创造处于时代发展的前沿。

Social values – the core of our work and excavation;
The requirements of customers – the driving engine of our work and progress;
Customer responses – an important creative, design idea;
Creativity – the subject of our works;
highly following the market keeps our creation at the forefront of the times.

Convention And Exhibition Center of the Center of AKANG Logistics Park, Ordos
鄂尔多斯阿康中心物流园区会议展览中心

地　　点：内蒙古 鄂尔多斯
设计阶段：规划及建筑方案设计
总建筑面积：36 000平方米
合作公司：Molen Associates

Location: Ordos, Inner Mongolia
Design Phase: Planning and Architecture Design Program
Construction Area: 36,000 m^2
Cooperation Company: Molen Associates

Jingdezhen Zhenrutang Pottery Creative Base
景德镇真如堂陶艺创意基地

地　　点：江西 乐平	Location: Leping, Jiangxi
设计阶段：在建	Design Phase: Construction
总建筑面积：10 500平方米	Construction Area: 10,500 m^2

景德镇真如堂项目位于江西省景德镇市东南三宝村，项目占地面积约4万平方米，总建筑面积为22 000平方米左右。设计以"中国瓷都"景德镇为文化依托，形成集艺术陶瓷创作、研发、生产为主体的产业基地，并通过建设陶艺展示中心、艺术家村以及休闲度假酒店，把陶瓷展示、交流文化旅游等功能融为一体。

在建筑造型方面，利用现代建筑语言，表达粉墙灰瓦的传统徽派建筑的意象，通过建筑围合形成错落有致的院落组织，再现田间村落的生活图景。

该项目的功能分区有四大块，其中有：儒释道陶瓷艺术研修中心、中国美术家协会创作中心、日用陶瓷及家具装饰设计与研发中心、香文化产品生产和研发中心及其配套基础设施等。

Jingdezhen Zhenrutang project is located in Sanbao village, southeast of Jingdezhen City, Jiangxi Province. The project covers an area of about 40,000 m^2, total construction area of 22,000 m^2. The design takes the "porcelain capital of China", Jingdezhen as a cultural basis, and forms the main industrial base combining art creation, development and production, and through the construction ceramic art exhibition center, the artist village resort hotels, integrates the ceramics show, exchange of culture, tourism and other functions.

In architecture form, modern architectural language is used to express the gray-brick-wall image of Hui-style architecture, through the building enclosure of courtyard to form patchwork organization, reproducing field picture of village life.

There are four functional areas, including Confucianism-Buddhism-Taoism Ceramic Art Training Center, Chinese Artists Association Creative Center, Ceramics and Furniture Decoration Design and R & D Center, Joss Sticks Culture Production and R & D Center and the Supporting Infrastructure Facilities.

主要经济技术指标

规划用地面积	33 400平方米
总建筑面积	10 500平方米
其中 陶瓷作坊	3 500平方米
生活区	1 450平方米
艺术家区（一区）	4 050平方米
艺术家区（二区）	1 500平方米
首层建筑占地面积	7 250平方米
容积率	0.31
建筑密度	21.7%
绿化率	65%
地面停车位	39辆

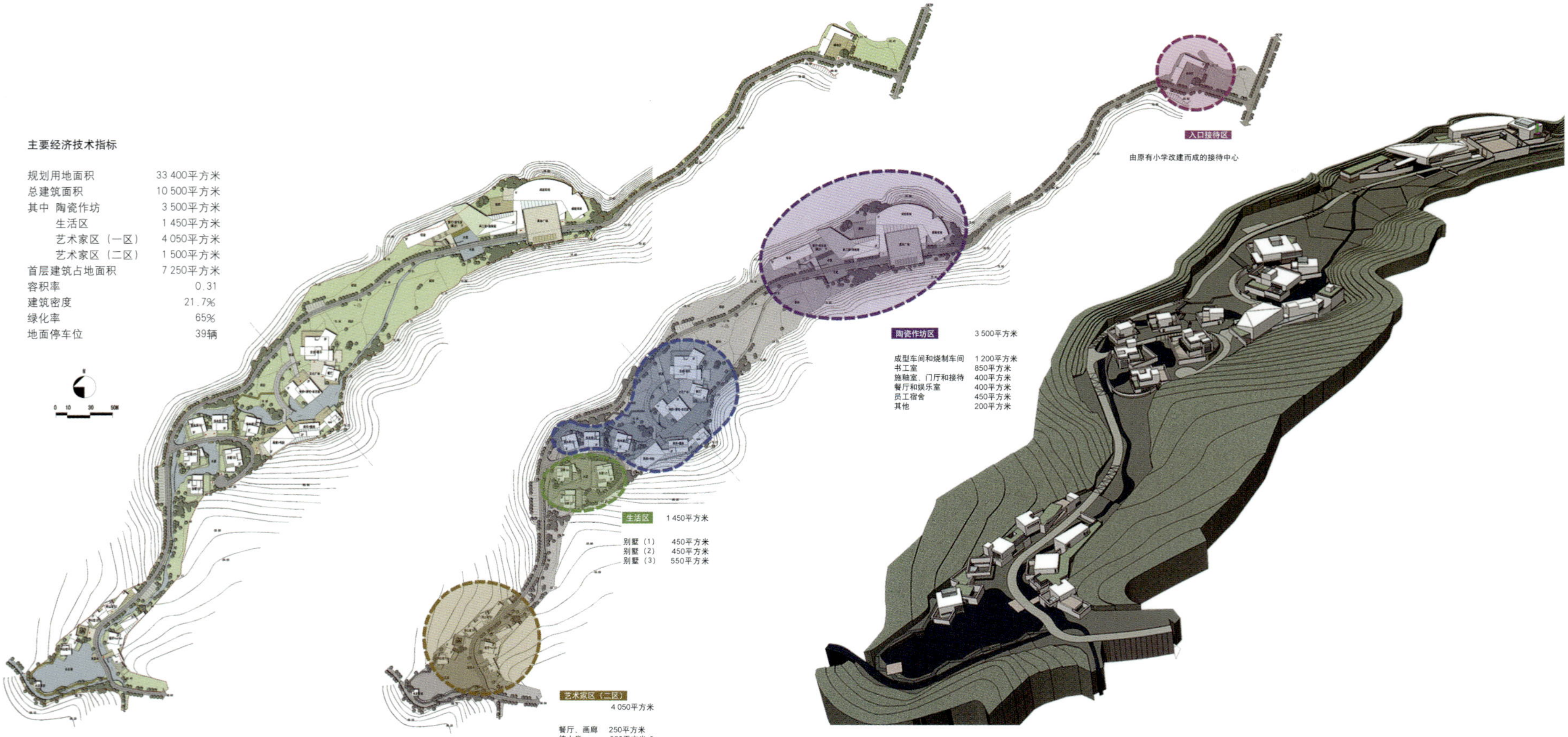

入口接待区

由原有小学改建而成的接待中心

陶瓷作坊区 3 500平方米

成型车间和烧制车间	1 200平方米
书工室	850平方米
施釉室、门厅和接待	400平方米
餐厅和娱乐室	400平方米
员工宿舍	450平方米
其他	200平方米

生活区 1 450平方米

别墅（1）	450平方米
别墅（2）	450平方米
别墅（3）	550平方米

艺术家区（二区） 4 050平方米

餐厅、画廊	250平方米
倚山堂	350平方米·3

Guangzhou Pearl River Foreign Investment Architectural Design Institute

广州珠江外资建筑设计院有限公司

广州珠江外资建筑设计院是华南地区著名的国有综合性建筑工程甲级设计院。

设计院和位于珠江之滨的白天鹅宾馆一起诞生，伴随着祖国改革开放的大潮崛起，记载了南粤大地建设发展的辉煌业绩。

建院以来，设计院坚持技术进步、科学管理、繁荣创作，高质量完成了广州及全国各地面积达上千万平方米的工程设计，项目涵盖公共建筑、居住建筑、景观园林和室内装饰等领域。作品创新意识浓郁，设计风格独特，既有岭南文化的韵味，又具有现代建筑的风采。众多项目荣获国家、部、省、市优秀设计及科技进步奖，赢得了客户及社会各界的广泛赞誉。

具有标志性的作品有：与英国扎哈·哈迪德建筑设计事务所共同设计了广州市标志性文化设施——构思为“圆润双砾”的广州歌剧院；与德国GMP国际建筑设计有限公司合作设计晶莹剔透的珠江边立方体“广州电视台新址”；追求自主创新，大胆采用解构主义象征手法，自主设计了武汉琴台文化艺术中心——“琴台大剧院”、“琴台音乐厅”；完成了具有“舰魂铸碑”之意的武汉“中山舰博物馆”；完全按照澳门当地和欧盟标准设计了“澳门2005年东亚运动会主体育馆”；广州白云国际会议中心及酒店、南沙行政中心、亚运媒体村、汶川县禹羌博物馆及避难广场、长春净月潭保利大剧院、江门演艺中心、柳州市工业博物馆、邯郸汽贸城、辛亥革命纪念馆等也是我们的力作。

各具特色的作品，是我们对新理念、新文化、新技术、新材料以及新的审美观的思考和探索。也是设计师们进行审美体验和诗意追求的心路历程。

铸精品，树品牌，追求原创，赢得市场，是我们的战略理念。面向未来，秉承创新，勇于进取，将更多异彩纷呈的设计作品呈现在世人的面前，谱写灵动七彩的华章。

Guangzhou Pearl River Foreign Investment Architectural Design Institute is a well-known integrated design institute with Class-A License of Architectural Engineering in South China. The institute is built at the same time with the White Swan Hotel at the shore of Pearl River. With the development of Reform and Opening of China, it witnesses the brilliant development of South China.

Since its establishment, adhering to technical progress, scientific management, prosperity creation, the institute has completed engineering design of projects in Guangzhou and all over China, covering an area of ten million square meters. Projects include fields of public buildings, residential buildings, landscape, interior decoration etc. Works of the institute have strong sense of innovation, and an unique design style, with both the cultural flavor of South of the Five Ridges and the elegance of modern architectures. Many projects win national, ministerial, provincial and municipal awards for excellent design and technological progress, as well as good reputation from customers and all walks of life.

Most representative works: iconic cultural facility – Guangzhou Opera Theater with the idea of “Double Round Gravel”, in cooperation with Zaha Hadid Architects (UK); and crystal clear cube near the Pearl River – “New Site of Guangzhou TV Station”, in cooperation with GMP International Gmbh Architects and Designers (Germany); with the pursuit of innovation, boldly use of deconstruction symbolism, the institute designs Qintai Cultural Arts Center, Wuhan – “Qintai Grand Theatre” and “Qintai Concert Hall”; completes “Zhongshan Warship Museum” with the meaning of “Monument of the Soul of Warship”; designs “Main Stadium of Macau 2005 East Asian Games” in full accordance with Macau and EU standards; Baiyun International Convention Center and Hotel, Guangzhou; Nansha Administrative Center; Asian Games Media Village; Yu Qiang Museum and Refuge Square, Wenchuan County; Moon Lake Poly Theatre, Changchun; Performing Arts Center, Jiangmen; Liuzhou Industrial Museum, Automobile Trade Town, Handan, Memorial Hall of the 1911 Revolution etc. are all our masterpieces.

Distinctive works represent our thinking and exploration of new ideas, new cultures, new technologies, new materials and new aesthetic standards. For designers, designing is the mental process of aesthetic experience and poetic pursuit as well.

Casting competitive products, building brand, pursuing of originality and winning market are our strategic concepts. Facing the future, insisting on innovation and enterprising, we will present more colorful design works to the world, and write a smart and colorful chapter.

联系我们
地址：广东省广州市环市东路360号珠江大厦东16楼
邮编：510060
电话：+86-20-83843732
传真：+86-20-83846074
电子邮箱：mail@pearl-river.com/zjtb16@126.com

Contact Us
Add: East Tower 16, Pearl River Building, 360 East Huanshi Road, Guangzhou, Guangdong
P.C.: 510060
Tel: +86-20-83843732
Fax: +86-20-83846074
E-mail: mail@pearl-river.com/zjtb16@126.com

广州歌剧院

设计：英国扎哈·哈迪德建筑师事务所、广州珠江外资建筑设计院有限公司

奖项：广东省2011年度优秀勘察设计一等奖
广州市2010年度优秀勘察设计一等奖

Guangzhou Opera Theater

Design: Zaha Hadid Architects (UK), Guangzhou Pearl River Foreign Investment Architectural Design Institute

Award: First Prize for Excellent Inspection and Design of Guangdong (2011)
First Prize for Excellent Inspection and Design of Guangzhou (2010)

项目说明

1. 完备的观演功能：广州歌剧院以歌剧表演为主，同时兼有芭蕾舞表演、大型交响乐演奏、大型综合文艺演出，实验性话剧、文化艺术交流、研究、培训、新闻发布等辅助功能，是一座个性鲜明，功能齐全的艺术殿堂。两个观演厅分别为1 800座的大剧场和400座的多功能剧场，总占地面积4.3万平方米，建筑面积7.3万平方米。达到国内一流、国际先进的水平。
2. “砾石”形体的塑造；
3. 创新的结构体系：本工程采用了一种新型的钢结构体系，根据其结构构成特点称之为空间折板式三向斜交单层网格结构；
4. 独特的形体创造一流的自然声学效果；
5. 浑然一体的室内空间；
6. 饰面清水混凝土营造的公共“灰空间”；
7. 多方面配套的弱电系统；
8. 预见性强的雨水排水系统；
9. 大空间智能灭火装置；
10. 空调冷源共享；
11. 模拟计算分区气流组织，节约能效。

Introduction

1. Complete watching functions: Guangzhou Opera Theater is mainly used for opera performances, beside of supporting functions such as: ballet performance, large-scale symphony, large-scale integrated theatrical performance, experimental drama, cultural and artistic exchanges, research, training and press releases etc. It is a distinctive, full-featured art palace. Two watching halls are 1,800-seat large theater and 400-seat multi-functional theater, with a total land area of 43,000 m^2 and a total floor area of 73,000 m^2. It is the first-class in China and top of the world.
2. Creation of the shape of “Gravel”;
3. Innovative architecture system: this project uses a new type of steel structure system which is named as space folded-plate three-direction oblique-crossing single-layer grid structure according to its structural characteristics;
4. Unique shape creates first-class natural acoustics;
5. Seamless interior space;
6. Public “gray space” created by fair-faced concrete;
7. Various supporting weak current system;
8. Predictable rainwater drainage system;
9. Large space intelligent fire extinguishing device;
10. Sharing of air-conditioning;
11. Analogue calculate airflow organization of partition, saving energy.

广州市南沙区行政中心一期

设计：广州珠江外资建筑设计院有限公司
奖项：广东省2011年度优秀勘察设计二等奖
广州市2010年度优秀勘察设计一等奖

Administrative Center Phase I, Nansha, Guangzhou

Design: Guangzhou Pearl River Foreign Investment Architectural Design Institute
Award: Second Prize for Excellent Inspection and Design of Guangdong (2011)
First Prize for Excellent Inspection and Design of Guangzhou (2010)

项目说明

广州市南沙区行政中心占地接近12公顷，规划结构采用U形的建筑布局，围绕着中央绿地形成以南北轴为中心的、以主景观空间为重点的建筑群空间组合，建筑主体有公众展示、档案中心、接待中心、党政办公、中介服务（两栋）和市民中心共七栋的建筑，所有的建筑主体之间用架空的连廊联系起来，架空的连廊下还可以通车，既保证了每栋建筑主体的独立性又方便内部的联系。

环境布局上采用绿地、水景及满足功能的广场相互穿插、渗透的手法，使建筑物自然地融入环境中，通过景观要素的有机结合凸显建筑。

Introduction

Administrative Center, Nansha, Guangzhou covers an area of nearly 120,000 m^2. Planning structure adopts the “U-shape” architectural layout surrounding central green land. With a north-south axis, the space combinations of buildings focus on the main landscape space. The main building is divided into seven buildings: public display, archives center, reception center, Party and government office, agent service (two) and civic center. All main buildings are connected by supported galleries, under which, cars can drive through. This design not only ensures the independence of each building, but also makes convenient internal connections.

Environmental layout adopts interspersed green land, waterscape and functional square. Buildings are naturally integrated into environment. Organic combination of landscape elements highlights buildings.

凯云楼
（萝岗中心城区会议及公共服务中心–D1组团）

设计：广州珠江外资建筑设计院有限公司

奖项：**广东省2011年度优秀勘察设计二等奖**

广州市2010年度优秀勘察设计二等奖

项目说明

凯云楼（萝岗中心城区会议及公共服务中心–D1组团）的建筑设计将景观元素渗透到建筑形体和建筑空间当中，以动态的建筑空间和形式、模糊边界的手法形成功能交织，探索使用功能之间的内在关系并使之有机相连，从而实现空间的持续变化和多样杂交。将建筑的内部、外部直至城市空间看成是城市意象的不同但连续的片段，通过切割与联系，形成城市生活和城市景观的融合和共生。

Kaiyun Building
(Convention and Public Service Center – D1 Group of Luogang Central Area)

Design: Guangzhou Pearl River Foreign Investment Architectural Design Institute

Award: Second Prize for Excellent Inspection and Design of Guangdong (2011)

Second Prize for Excellent Inspection and Design of Guangzhou (2010)

Introduction

Architectural design of Kaiyun Building (Convention and Public Service Center – D1 Group of Luogang Central Area) brings landscape elements into shape and space of architecture. Dynamic space and form of architecture and blurring boundary approach form interconnected functions. Design explores internal relationship between functions and connects them organically, so that continuous changes and diversity of space can be realized. Internal and external space of building and urban space are considered as different but continuous segments of city image. By cutting and connecting, urban life and urban landscape can be integrated and coexist.

珠海十字门中央商务区国际会议中心（在建）

设计：罗麦庄马香港有限公司
　　　广州珠江外资建筑设计院有限公司

项目说明

国际会议中心位于一期用地的东南部，地下共两层，主要功能为多功能厅配套及后勤用房、设备用房以及机动车停车库；地上3层，局部4层（部分楼层设有夹层），包括3个多功能厅、公共大厅、各种规模的会议室、办公区、贵宾区、服务配套设施用房、设备用房等。其中多功能厅1为一个1 200座的观演厅，以演出歌剧、舞剧为主，能满足国内外多种歌剧、舞剧、音乐剧、大型歌舞、戏剧、话剧等舞台类演出的使用要求。

International Convention Center of Central Business Area, Zhuhai (under construction)

Design: RMJM Hong Kong Limited
Guangzhou Pearl River Foreign Investment Architectural Design Institute

Introduction

International Conference Center is located in the southeast of phase I construction site. It has two floors underground, which are mainly used for supporting and logistic rooms, equipment storage and vehicle parking garage of multi-function hall; it has 3 floors on the ground, 4 floors in certain areas (some places have interlayer), including three multi-function halls, public hall, various sizes of meeting rooms, office area, VIP area, buildings of supporting service facilities, equipment storage etc. In them, the multi-function hall 1 is a 1,200-seat performance hall, which is mainly used for opera and ballet. It is able to meet requirements of various domestic and international opera, ballet, musical, song-and-dance, drama and other performances of stage drama.

邯郸文化艺术中心

设计：广州珠江外资建筑设计院有限公司

项目说明

邯郸文化艺术中心位于人民路与滏东大街交口东北角，用地东西长度约590米，南北宽约280米，其中包括大剧院（1 567座）、报告厅（622座）、博物馆（3 000人次/天）、图书馆（藏书150万册）和城市规划展览馆、地下车库及配套用房。整体建筑采用曲线造型，一气呵成，雄浑有力，能与周边环境很好协调，使建筑具备视觉连续性，可以从城市内多角度欣赏。利用西侧的博物馆，东侧的图书馆和进入大剧院的大台阶，通过现代的建筑处理手法连接起来，形成如城台般的青铜墙面。大剧院居中，犹如一块无瑕的美玉浮于城台之上。该艺术中心充分体现了邯郸文化的精髓，青铜文化、磁州窑瓷器文化及邯郸建筑传统中高台建筑的特征。它将成为邯郸新世纪城市的标志性建筑。

Culture and Arts Center, Handan

Design: Guangzhou Pearl River Foreign Investment Architectural Design Institute

Introduction

Handan Culture and Arts Center is located in the northeast corner of intersection of Renmin Road and Fudong Street. The land is about 590 m long (from east to west) and 280 m wide (from north to south). It includes grand theatre (1,567-seat), report hall (622-seat), museum (3,000 persons/day), library (collection of 1.5 million books) and urban planning exhibition hall, underground garage and supporting buildings. The whole building is in curve shape, coherent and grand, and can be well coordinated with the surroundings. With visual continuity, we can appreciate the building from multiple perspectives in the city. Museum on the west side, library on the east side and the big step leading into the grand theatre are connected by modern architectural approach, forming a bronze wall like peribolos. The grand theatre is located in the center, like a flawless jade floating on the peribolos. This fully reflects the essence of Handan culture, bronze culture, Cizhoujiao china culture and traditional architectural characteristic of peribolos of Handan. It will become landmark of New Century City of Handan.

京沪高铁苏州北火车站

设计：广州珠江外资建筑设计院有限公司

项目说明：设计项目地处苏州市相城区，车站面积约35 000平方米。

Suzhou North Railway Station of Beijing-Shanghai High-speed Railway

Design: Guangzhou Pearl River Foreign Investment Architectural Design Institute

Introduction: The project is located in Xiangcheng District of Suzhou. The railway station takes up an area of 35,000 m^2.

广州市弘基市政建筑设计院有限公司
Hongji Municipal and Architectural Design Institute Co., Ltd.

广州市弘基市政建筑设计院资质：
建筑行业（建筑工程）甲级、
市政行业（桥梁工程、道路工程）专业乙级

地址：广东省广州市番禺区大龙街富怡路罗家段145号
电话：+86-20-22625666　传真：+86-20-34625803
邮箱：hongji806@126.com　网址：www.hongjisj.cn

清华坊

工程地点：广州 番禺
占地面积：10 000平方米
总建筑面积：80 000平方米
风格：中国传统与现代民居相结合

设计理念

强调天人合一、人与自然和谐相处、创造有灵魂的景观。

宅院格局及特点：住宅格局融合了中国各地具有特色的院落民居风格，前有敦厚沉静的宅门、前庭，中有天井通透天地，后有花木扶疏的围院；古韵悠然、雅致明丽的三层小楼，黑白分明的粉墙黛瓦，透光明亮的塔窗，都充分表现了清华坊融合古今的建筑风貌和独特的风格。

院落艺术品格：充分体现院落式民居艺术风格的质朴无华、随性自然、不矫揉造作；形成一种黑白艺术，具有中国水墨画意趣和质朴的素描风格。

桂林阳朔书童·国际苑↗

桂林阳朔书童·国际苑位于风景冠甲天下的阳朔漓江边，书童山脚，地理位置优越，风光绮丽，一年四季置身于不同风景中。其建筑特色也很显著，建筑外观体为江南水乡民居，也透露出部分徽派建筑风格。建筑面积120 000平方米。

设计理念

1．风格结合建筑本身特色，定位为新中式风格，配以东南亚家具与饰品摆设为中心主题。

2．通过轻装修、重装饰手法及一定数量的字画点缀，着重体现家居文化的品位及高端人士涵养。

3．简洁的户外园林设计，充分体现出山、水、小品景观于一体的水墨诗意及田园情怀。

客天下旅游园↙

工程地点：广东 梅州

建筑面积：70 000平方米

风格：客家文化

设计理念

项目依托梅州特有的客家文化氛围和基地已有的优美自然环境，从地段和环境特点出发，借鉴中国传统山水居所特有的"聚落"模式，充分利用梅州"八山一水一分田"的山地资源优势，秉承以建设新型旅游城镇为主题的城市开发和城市经营理念，按照"总体规划、弹性调整、分步实施"的思路进行开发。

"号召天下客"——高尚化、社区化、人文化、本土化的住居集群是该项目的定位。该项目致力于精品社区的营造，以"客家文化"的传承与发扬为主题，促进各种文化产业共同交流发展；以兼容并蓄的手法来塑造出各具特色又相互联系的社区组团节点，构筑社区整体框架；又凭借基地内部丰富的自然资源和微环境的交互渗透，打造出既具有标志性，又和谐统一的高尚社区整体形象，以此招徕、号召、引领"天下客"。

ATELIER Y 東意建築

广州市东意建筑设计工作室

GUANGZHOU ATELIERY ARCHITECTURAL DESIGN STUDIO

东意建筑设计工作室简介

工作室创建于2004年，2008年更名为东意建筑设计工作室，由多名具有德国留学教育背景的建筑师组成核心设计团队。工作室秉承现代建筑精神，探索本土化设计创新策略；关注研究与创作结合，理念与践行并重。

工作室发展多年，建立了高效而强大的设计团队，确立了明确的设计理念与品格追求，完成了若干具有整体品质的设计作品。

工作室项目涵盖城市规划、城市设计、建筑、景观、室内设计、绿色建筑设计等各个方面，尤其在中小型公共建筑、商业地产、高端房地产及城市公共景观方面颇有建树。同时，积极承担社会责任和职业义务，参与多项社会公共发展项目，并为公益项目提供设计及咨询服务。

东意建筑设计工作室宣言

我们立足于建筑师的专业领域，推动社会文化和技术的进步；
我们关注建筑文化精神的传承，追求建筑学现代价值体现，探索设计发展的创新策略；
我们直面问题，勇于承担，执著坚持；
我们追求从整体到细节的设计品质；
我们致力于建设高水平的设计团队、营造高效而充满活力的工作氛围；
我们乐于交流，积极合作，合力推动行业进步。

Atelier Y - Profile

Established in 2004 and renamed as Atelier Y in 2008, the Atelier Y Architects boast a core architect team with German education background. Atelier Y aims to explore a localized design innovation strategy through the spirit of modernistic architecture, and focuses on combination of research and creation with equal attention to concept and practicality.

Thanks to years of professional practices, Atelier Y has built a highly efficient and strong design team and clearly defined its pursuit for design concept and style. So far, it has delivered dozens of design works of integral quality.

With an extensive portfolio including urban planning, urban design, architecture, landscape, interior, as well as green building design, Atelier Y is especially successful in design of small and medium-sized public buildings, commercial properties, high-class real estate and urban public landscaping. Meanwhile, it also proactively shoulders social responsibilities and professional obligations by getting involved into various public development projects and offering design and consultancy services to public welfare projects.

Atelier Y - Mission Statement

Promote the social, cultural and technological advancement from the professional grounds as architect;
Be aware of inheritance of architecture culture and spirit, endeavor to embody the modernistic value of architecture and explore innovative design strategies;
Face up to problems and be responsible and perseverant;
Pursue design quality both on the whole and on the details;
Build a high-caliber design team and foster efficient and dynamic work atmosphere;
Be ready to communicate and cooperate, and advance the industry through synergized efforts.

地址：广东省广州市天河区五山路五山科技广场C417-C423
邮编：510640
电话：+86-20--85287630
传真：+86-20-85287629
邮箱：ateliery@163.com
网站：www.ateliery.cn

Add: C417-C423, Wushan Technological Plaza, Wushan Road, Tianhe District, Guangzhou, Guangdong
P.C.: 510640
Tel.: +86-20-85287630
Fax: +86-20-85287629
E-mail: ateliery@163.com
Http:// www.ateliery.cn

从化市图书馆新馆（首期）

建设地点：广东 广州
建设单位：从化市人民政府
合作单位：华南理工大学建筑设计研究院
设计时间：2005年
竣工时间：2010年
用地面积：40 000平方米
首期总建筑面积：10 000平方米
项目投资：约2千万元

项目为广州县级市从化市的图书馆新馆。位于城市东北部，临近穿越城市的流溪河及河畔的广东省的休闲绿道。规划新馆规模为2万平方米，首期建设1万平方米。因项目分期建设，设计充分考虑了建筑的使用、建设周期及功能可持续性等因素。将一、二期的建筑形态做了明显的区分。首期建筑包括两部分，分别为图书馆主楼和具有文化活动功能及培训功能的报告厅。

建筑用地紧邻高出的流溪河河堤，设计将建筑主体架空于两层以上，使得建筑的阅览空间得到良好的沿河景观视野。图书馆主楼下部设置为门庭，架空庭院与内庭的屋顶平台形成了一个立体化庭院的布局。主体建筑与报告厅形成一个庭院，较好地处理了不规则的用地关系。图书馆主楼平面集中交通核心和服务用房，使得平面具有良好的灵活性，适应建筑的功能变化需要。

建筑含基本装修造价仅为2 000元/平方米。我们通过设计方式体现建筑的节约和节能。主楼立面采用双层外表面处理，通过水平板和垂直板的组合，形成良好的防热效果；平面围合成内庭，形成良好的建筑自然通风条件。

Conghua Library (New) (Phase I)

Location: Conghua, Guangzhou
Employer: Conghua Municipal People's Government
Partner: Architectural Design Research Institute of South China University of Technology (SCUT)
Design Time: 2005
Completion Time: 2010
Site Area: 40,000 m^2
GFA Phase I: 10,000 m^2
Cost: About RMB 20,000,000 Yuan

The new library building in Conghua, a county city of Guangzhou, is located in the northeast of the city, close to the Liuxi River that runs across the city and the riverside leisure greenway of Guangdong Province. The new library is sized as 20,000 m^2 while Phase I occupies 10,000 m^2. In view of phased project construction, the design fully considers the building's service life and construction period as well as functional sustainability, and differentiates building form of Phase I from that of Phase II. Phase I is composed of two parts, i.e. the main library building and the lecture hall for cultural activity and training purposes.

Since the site is close to the higher level of the Liuxi River, the main building is placed above an open-up structure of two floor height, enabling reading space in the building to enjoy the attractive riverside view. The structure below the main library building is designed into an open-up courtyard, and, together with the roof terrace of inner courtyard, presents a three-dimensional courtyard layout of the building. The main building and the lecture hall form a courtyard, which tactfully handles the irregular site relations. The traffic core and service facilities are centralized in the plan of the main library building to allow for more flexible plan and higher functional adaptability.

The construction cost of the building (including basic finishing) is just 2,000 Yuan/m^2, which is realized through our cost-effective and energy-efficient design solution. The double-skin façade of the main building, coupled with the combination of horizontal and vertical panels, bring about favorable insulation effect; while the inner courtyard enclosed contributes positively to the excellent natural ventilation conditions in the building.

华南理工大学教工活动中心

建设地点：广东 广州
建设单位：华南理工大学
合作单位：华南理工大学建筑设计研究院
设计时间：2002年
竣工时间：2010年
建筑面积：5 000平方米
项目投资：约1千万元

项目用地位于华南理工大学校园东区体育场北侧。建筑北向隔路面对建于20世纪30年代的中国传统建筑风格的大学老体育馆；北面紧邻校园东区主要机动车出入口，下部为下穿隧道的城市干道；南侧为东区体育场。建设场地中央，有一条通往体育场的人行通道，要求被保留。

活动中心主要满足以离退休教职工为主体的教工业余文化娱乐活动需要。提供舞蹈、健身、室内乒乓球、桌球等文化体育活动及老年大学需要的教室、活动室以及校退休协会办公室。主体建筑4层，建筑面积约5 000平方米。

用地经过规划红线退缩布置后，已基本确定了建筑物的外边界。设计通过建筑物的开口分别呼应各个方向的环境。建筑运用简洁的空间与形体关系，处理功能空间。建筑外表面整体式遮阳构件、建筑架空层，竖向庭院与中庭空间的组织，中庭上部开口通风处理，开放式内外廊，以及屋顶绿化等措施综合使用，尝试在低造价建筑中，通过设计的主动思考，营造节能舒适的建筑使用环境，并表现出通透而开放的、富有特色的建筑形象。

SCUT Faculty and Staff Activity Center

Location: Guangzhou, Guangdong
Employer: South China University of Technology (SCUT)
Partner: Architectural Design Research Institute of SCUT
Design Time: 2002
Completion Time: 2010
GFA: 5,000 m^2
Cost: RMB 10,000,000 Yuan

The site is located on the north of the east campus sports ground of SCUT. The building faces a university gymnasium built in 1930's with Chinese traditional architectural style to the north across the road, and neighbors the main vehicular access road leading to the east campus on the north. Under the access road is the underpass section of an urban artery. On the south of the project is the sports ground. In the center of the site, there is a traditional pedestrian passageway leading teachers and students to the sports ground, which is required to be reserved.

The activity center is mainly to meet the demands of the retired faculty and staff for cultural and recreational activity. It offers cultural and sports activities such as dancing, fitting, indoor table tennis and snooker, plus the classrooms and activity rooms needed by the College for the Aged and offices for the university's Association of Retirees. The main building is planned with four floors and a GFA of about 5,000 m².

The perimeter of the building is basically defined by the setback from the site property line. The building openings are designed to echo the environment in various directions respectively, while the functional spaces are handled through neat and simple spaces and massing. With combined uses of various approaches, such as integral sun-shading components on the building skin, the open-up floor of the building, the organization of vertical courtyard and atrium space, ventilation treatment of upper opening of the atrium, the open middle and side corridor and roof greening etc, we intend to, through initiative design thinking, create an energy-efficient and comfortable occupancy environment in a low cost building and present a transparent, open and distinctive building image.

广州市南沙发展电力大厦办公楼

建设单位：广州发展集团、广州控股
合作单位：华南理工大学建筑设计研究院
合作建筑师：周剑云
设计时间：2006年
竣工时间：2010年
建筑面积：12 000平方米
项目投资：约5千万元

Office Building of Guangzhou Power Development Nansha Power Building

Employer: Guangzhou Development Group Co., Ltd.
Guangzhou Development Industry (Holdings) Co., Ltd.
Partner: architectural Design Research Institute of SCUT
Cooperative architect: Zhou Jianyun
Design Time: 2006
Completion Time: 2010
GFA: 12,000 m^2
Cost: About RMB 50,000,000 Yuan

项目为广州控股下属珠江电厂的办公业务大楼。建筑用地面临南沙主要城市干道“环岛路”，周边为南沙区的石化电力产业园区。东向为珠江出海口，建设用地近似方形。

建筑功能按照人流使用的密集程度从下往上布置：分别布置了食堂、会议及培训室、档案、值班室、部门办公以及顶层的领导办公及会议室。建筑地上11层，地下1层。建筑面积12 000平方米。

设计在建筑上部6至11层设置了一个内部中庭，结合在立面上打开的侧庭院，构成立体庭院系统，形成了办公区良好的通风采光空间，庭院系统由一组楼梯串联。建筑立面采用标准化的L形遮阳板，分别在不同的朝向上采用不同的布置方式，与水平板共同形成了复合的外表皮遮阳系统。遮阳板在不同朝向、不同时段形成了遮阳、导光及通风的作用。

The project is an office building of Zhujiang Power Plant under Guangzhou Development Industry (Holdings) Co., Ltd. The site faces Island Ring Road, the main urban artery of Nansha, and is surrounded by petrochemical and power industrial parks of Nansha District. The estuary of the Pearl River lies to the east of the site. The site geometry is similar to a square.

Building functions are arranged in view of occupant density from lower floors to upper floors, including: canteen, conference, training, archives, duty room, department offices, as well as management office/meeting room on the top floor. The building has 11 above-grade floors and 1 basement floor, with a GFA of 12,000 m^2.

An internal atrium is arranged at the upper part of the building from F6 to F11 forming a three-dimensional courtyard system in combination with the side courtyard opened in the façade, thus allowing for favorable spatial ventilation and daylighting for the office area. The courtyard system is connected via a group of stairs. The façade features typical L-shaped sun-shields (with different layout in different directions) in combination with horizontal boards to jointly form a complex building skin sun-shading system. The L-shaped boards could play a role in sun-shading and light/wind guiding at different periods of time in different directions.

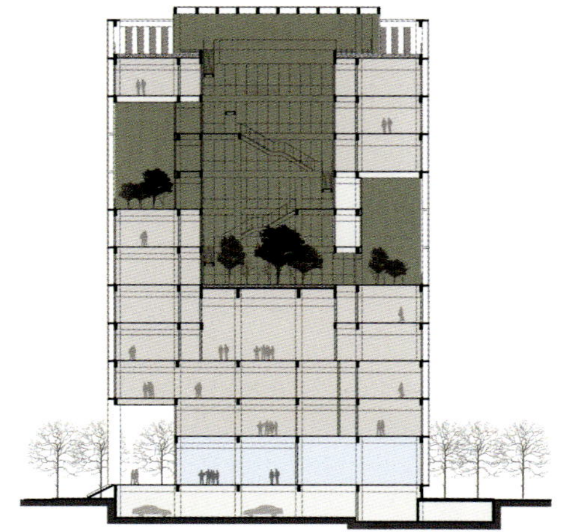

广州南沙广控产业园消防站

建设地点：广东 广州
建设单位：广州发展集团、广州控股
合作单位：华南理工大学建筑设计研究院
合作建筑师：周剑云
设计时间：2006年
竣工时间：2009年
建筑面积：3 650平方米
项目投资：约1 200万元

Fire Station of Guangzhou Holdings Nansha Industrial Park

Location: Guangzhou
Employer: Guangzhou Development Group Co., Ltd.
Guangzhou Development Industry (Holdings) Co., Ltd.
Partner: Architectural Design Research Institute of SCUT
Cooperative Architect: Zhou Jianyun
Design Time: 2006
Completion Time: 2009
GFA: 3,650 m^2
Cost: About RMB12,000,000 Yuan

项目为广州控股下属南沙产业园区配套消防站，同时承担周边城市区的消防服务。

场地紧临城市道路，向东连接南沙区主要城市干道“环岛路”。用地为长方形，消防站面向城市道路展开。建筑首层为消防车库、园区消防中心、修车库及食堂；二层为消防员宿舍；三层为办公、会议室及配套活动用房。

设计结合南方湿热气候特点，采用不封闭的开敞架空车库，结合各层通透及平台处理，形成良好的自然通风和采光建筑环境。连续的公共交通空间将上下层的步行楼梯空间、活动平台及三层的屋顶绿化联系起来，再结合简洁的细节处理和自然光环境，营造了一个竖向的消防员日常生活的“微城市”。

The project is a supporting fire station of the Guangzhou Holdings Nansha Industrial Park. It also offers fire service to peripheral urban areas.

The site is close to the urban road and connects the Island Ring Road, the main urban artery of Nansha District in the east. The site geometry is rectangular while the fire station is developed around the urban road. 1F is planned as fire house, fire center of the park, motor repair shop and canteen, 2F as firemen's dormitory, and 3F as office, conference room and supporting activity facilities.

In view of the humid and hot climate of South China, the design uses open-up garage in combination with floor transparency and platform treatment to create favorable natural ventilation and lighting building environment. Continuous public traffic spaces connect pedestrian stair spaces of upper and lower floors, activity platform, as well as roof greening on F3, which, in combination with simple and elegant details and lighting environment, create a vertical micro-city for the daily life of the firemen.

手牵手计划之儿童早期养育及教育项目

“手牵手计划之儿童早期养育及教育项目”是一项由诺基亚（中国）投资有限公司发起，协同中国青少年社会教育基金会及国际计划、中国光华科技基金会及儿童乐益会共同实施，旨在通过社区、政府、企业、学术机构以及非政府组织的联合努力，为贫困地区弱势儿童群体提供全面的儿童早期发展与养育服务的综合性儿童发展项目。

工作室协同华南理工大学建筑学院作为计划合作的技术支持机构，提供了幼儿园建筑项目的援助设计指导及项目建设管理顾问工作。在两年内为河南、陕西、甘肃的6个国家级贫困县实施建设了9所幼儿园。另外12个幼儿园的设计与建设也正在安徽、河南、湖南、四川、河北、贵州、广西等地的经济欠发达地区展开。

工作室参与了其中三所幼儿园的设计及建设，在设计中通过对场地、儿童活动、功能与需求的关注，结合地方经济及建造水平，很好地达到了儿童友好的、高质量低成本的项目要求。

Early Child Caring and Education Program of the Hand to Hand Project

“The Early Child Caring and Education Program of the Hand to Hand Project”, an integrated children development program initiated by Nokia (China) Investment Co., Ltd. and implemented by Nokia in cooperation with China Youth Social Education Foundation, Plan International, China Guanghua Science and Technology Foundation and Right to Play, aims at providing vulnerable children groups in poverty-stricken areas with comprehensive early child caring and development services through joint efforts of communities, the government, enterprises, academic institutions and non-governmental organizations.

As the cooperative technical supporting institutions of the Program, Atelier Y, in collaboration with the Architectural Design Research Institute of SCUT, have offered design instruction and project construction management consultancy to the kindergarten projects. Within two years, 9 kindergartens were built in 6 state-level poverty-stricken counties in Henan, Shanxi and Gansu. Design and construction of other 12 kindergartens are being implemented in less developed areas in Anhui, Henan, Hunan, Sichuan, Hebei, Guizhou and Guangxi Province.

Atelier Y was engaged in the design and construction of three of the kindergartens. During the design, the studio has successfully reached the child-friendly and high quality/low cost project standard with good awareness of site location, children activities, functions and demands, as well as the local economic and construction level.

陕西省长武县亭口乡西塬幼儿园（6班）　建筑面积：845平方米
Xiyuan Kindergarten (6 classrooms), Tingkou Township, Changwu County, Shanxi Province. Floor Area: 845 m²

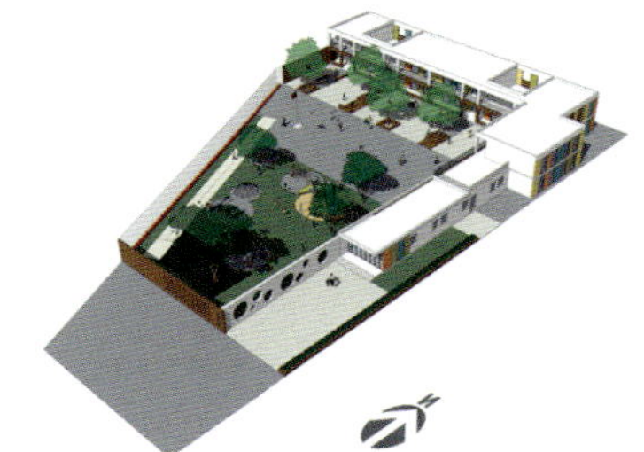

河南省商城县丰集乡幼儿园（7班）　建筑面积：1 443平方米
Fengji Township Kindergarten (7 classrooms), Shangcheng County, Henan Province. Floor Area: 1,143 m²

河南省栾川县叫河乡中心幼儿园（12班）　建筑面积：2 709平方米
Jiaohe Township Central Kindergarten (12 classrooms), Luanchuan County, Henan Province. Floor Area: 2,709 m²

东莞市佛灵湖地区概念性规划国际竞赛

规划地点：广东 东莞
竞赛组织单位：东莞市寮步镇规划管理所
竞赛组织代理单位：广州宏达工程顾问有限公司
合作单位：德国欧博迈亚工程咨询有限公司
设计时间：2008年
规划面积：20.24平方千米

International RFP for Conceptual Planning of Folinghu Area, Dongguan City

Location: Dongguan, Guangdong
Employer: Planning Administration of Liaobu Town, Dongguan City
Bidding Agency: Wang Tat Project Management Consultancy (Guangzhou)
Partner: OBERMEYER Engineering Consulting Co., Ltd. (Germany)
Design Time: 2008
Planning Area: 20.24 m^2

佛灵湖地区位于东莞寮步镇西南侧，毗邻东莞中心城区，与松山湖高科技产业园相邻。本次规划为明确佛灵湖地区的发展目标，深化及完善区域的城市功能与整体形象，提出具有国际视野的建议和思考。

围绕构建大莞城“新生活中心”的目标和理念，在“两轴一环”的规划结构下，重点强调了九个方面的内容：

1. 构建“两轴”中心区功能
2. 营造生态城市带状公园
3. 发展体育产业和健康产业
4. 建立生态健康居住社区
5. 强化区域性商业服务功能
6. 打造创业乐园
7. 深化和提升汽车服务产业
8. 完善文化服务设施
9. 改善交通系统

在规划设计中始终坚持生态自然的理念，注重自然环境的合理利用，通过项目策划与空间的有效组织，完善城市功能，营造全新国际高水平的大东莞新中心。

Folinghu Area is located in the south of Liaobu Town, Dongguan City. It is close to the central urban area of Dongguan and adjacent to the Songshanhu High-tech Industrial Park. The planning proposes suggestions and thinking with international vision for the development objective of Folinghu Area and detailing and perfection of urban functions and overall image of this area.

The conceptual planning centers on the objective and concept of building a new life center in Guancheng District, and, based on the planning structure of Two Axes and One Ring, highlights nine aspects as follows:

1. Build the Two Axes central area function;
2. Create strip-like ecological urban park ;
3. Develop the sports and health industry;
4. Establish ecological and healthy residential communities;
5. Strengthen regional commercial service functions;
6. Create a paradise for entrepreneurship;
7. Deepen and upgrade the motor service industry;
8. Perfect the cultural service facilities;
9. Improve the traffic system.

The concept of nature and ecology has always been incorporated into the planning design while reasonable utilization of natural environment is also highlighted. Through such project planning and effective spatial organization, the urban functions are perfected and a refreshing world-class New Dongguan Center is created.

Guangdong Huafang Architects & Engineers Co., Ltd.

广东华方工程设计有限公司

广东华方工程设计有限公司团队最早成立于1992年4月，成立之初凭借技术服务优势迅速发展壮大，经过多年的发展，已成为一家综合性的设计服务机构。具备国家建设部颁发的建筑甲级（A144014879）、规划乙级（082037）、人防乙级（A244014876）、消防甲级、工程咨询丙级资质。公司成立十多年来，完成了大量优秀工程设计项目，业务范围涉及区域规划、住宅、商业、酒店、写字楼、医院、学校、室内外装修等设计业务，在同行中享有较高的知名度。公司在东莞、广州、中山、成都、南宁均设有分公司，拥有多名具有海外工作经验的内地、香港互认建筑师，高、中级技术人员300多人，专业技术力量雄厚。公司拥有完善的质量管理体系，对设计产品前期策划至建成竣工验收阶段的全过程提供优质的设计咨询服务。

Guangdong Huafang Architects & Engineers Co., Ltd. was founded in April, 1992. The team developed rapidly with the advantage of technology and service at the early time. After tens of years of hard work, it has developed into a comprehensive design service agency. It is approved by the MOHURD, has the Qualification of Grade A in architectural design (A144014879), Grade B in plan design(082037)and civil air defence shelter engineer design (A244014876), Grade A fire protection, engineering consultation. It has accomplished many outstanding engineering projects since its foundation, including area plan, residence, business, hotels, office buildings, hospitals, schools, interior design, landscape design and so on, gained good reputation in the field. It has established branches in Dongguan, Guangzhou, Zhongshan, Chengdu, Nanning. The company has accumulated abundant technical power, owned more than 300 senior and middle engineers and technicians, and several architects approved in Mainland China and Hongkong with overseas work experience. With the perfect quality management system, it is engaged in providing excellent service and effective way to fulfill customer' s needs, from the prophase scheme plan to assessment and acceptance of completed project.

地址： 广东省东莞市南城区元美路华凯广场A座19楼
邮编： 523070
电话： +86-769-22820769
传真： +86-769-22820736
邮箱： winway866@163.net
网址： www.gdhuafang.com

Add: 19F, Tower A, Huakai Plaza, Yuanmei Road, Nancheng District, Dongguan, Guangdong
P.C.: 523070
Tel: +86-769-22820769
Fax: +86-769-22820736
E-mail: winway866@163.net
Http:// www.gdhuafang.com

左庭右院

项目阶段：建设中
项目规模：96 513.016平方米
项目地点：广东 东莞

设计说明

项目以城市区域总规为指导，秉承左庭右院项目的开发理念，将现代的城市规划设计理念与高品质的居住文化相结合。创造一个生态的、有机的，有一定文化内涵和艺术品位的特色小区。

东莞时富花园

项目地点：广东 东莞
项目规模：96 000平方米

项目说明

本项目位于东莞市寮步镇莞樟大道边，北面紧临莞深高速路出口，南临寮城中路，东临东升路，西临东莞市东城汽车总站。本项目的区位特点是毗邻市区中心，交通方便，有一定商业气氛，旺中带静，是规划中大型居住小区的理想地段。根据项目的区位特性，规划为面向中高档市场，以住宅为主，商业为辅的中大型商住小区。因项目基本上是四面临路（三面紧临市区马路，一面与高速路隔绿化带相望），无太多可利用自然景观或绿化资源，因此在规划上努力营造小区内部绿化环境，增加楼距和绿化率，改善居住环境。

国际公馆四期

项目地点：广东 东莞
项目规模：122 017.263平方米

项目说明

本项目树立了“生态—自然”的整体环境形象，有效地利用自然资源。创建了一个更注重环境的生活空间，内部以溪涧式水景为主，营造了“小桥流水”的生活意境。各个户型内采用全明布置，南北均有开窗，具有良好的自然采光通风，创造了非常生态的居住空间。内部道路实行便捷，畅通，合流与分流的不同处理，保证了交通安全。绿化层次分明，系统结合了空间关系。使业主能够共享自然景观资源。小区交通现象可分为动态与静态两类，同时道路等级设置清楚，保障了交通的安全，小区主干道呈不规则的曲线环形。设专门的步行通道穿行于各组团之间，创造了步移景异的宜人生活环境。建筑的形体与内部空间有机结合，以简洁雅致的艺术装饰风格为设计基调，造型简约，刚中带柔，垂直的柱式与横向线脚形成对比，相映成趣，外墙形式丰富多样，室外空间形成室内空间的延续。

华凯帝庭园

项目地点：广东 东莞
项目规模：52 389.55平方米

项目说明

华凯帝庭园位于东莞市东城区东城大道与东纵大道之间，金月湾旁。周围商业及生活配套设施齐全，交通便利，环境优越，南面能眺望旗峰山自然景观，是高档居住小区。以人为本的设计原则，以整体社会效益、经济效益与环境效益三者统一为基准点，着重创建优质办公环境，突出健康绿色的生活理念，塑造环境优美、舒适便捷、和谐交流的办公环境和生活体系。

桃源一品

项目地点：山东 烟台
项目规模：85 468.69平方米

项目说明

本项目位于烟台市中心区红旗路魁星楼隧道东口北侧，基地地貌呈盆状。规划总用地面积约为3.7公顷，规划总建筑面积85 468.69平方米，其中，地上建筑面积62 025.28平方米，地下建筑面积23 443.41平方米，容积率1.71。规划布局力求最大限度地利用该地块既能眺望海景，又能享受到自然绿化环境的特性，积极地展现高档生态社区的独特魅力。

中惠香樟绿洲二期

项目名称：中惠香樟绿洲二期
项目规模：302 544.438平方米

项目说明

项目注重生活中每一环节，每一时段，均感受自然的宁静、身心的放松，这方是栖居美境的本源含义。规划设计的核心，在于如何带来更多的观景单位，以及观景单位的均好性。院落是中国传统空间形态，院落住宅则是中国人千百年来的理想住所。在当今高密度的城市中，人们已难以感受到"明月时至，清风自来，行无所牵，止无所枙"的生活乐趣。本项目采用中国传统的居住空间形态——院落，让居住的人能感受到传统空间带来的文化内涵，大大提高居民的安全感、舒适感和认同感。居住社区的意义本身，在于营造富有情境的生活氛围，服务于大众。庭园空间，建筑造型，环境景观等要素共同构筑生活的氛围。

大连中拥蓝天下

项目地点：大连 旅顺
项目规模：65 151.67平方米
项目阶段：已竣工

项目说明

本项目位于旅顺区老城区模珠街尽端，用地两端面向黄海，东西两侧有日坛、月坛两个山体公园。用地西邻4栋在建高层住宅，东临小区中心景观区及酒店会馆，周边自然条件极佳。基地形态基本呈梯形，场地西北部略高，高程为27.00米；南部沿海略低于北段，高程为6.5米。基于对自然环境的尊重，方案的规划设计充分结合地形，因地制宜，提出组团概念。在整体规划考虑与C区结合，形成一个集高层、低层为一体的综合住宅。本项目考虑经过与朝向因素，结合现状地形，形成有序、梯级的组团规划布局。

广东弘业建筑设计有限公司 HY GUANGDONG ARCHITECTS & ENGINEERS LTD.

地址：广东省佛山市南海桂城天佑六路九号(邮编：528200)
电话：+86-757-86231578 86326105(业务咨询)
传真：+86-757-86390768
邮箱：info@gd-hy.cn
网址：www.gd-hy.cn

A: 9 Tianyou 6th Rd., Nanhai, Foshan, Guangdong, 528200, P.R.China
T: +86-757-86231578 86326105 (Business Consulting)
F: +86-757-86390768
E: info@gd-hy.cn
H: www.gd-hy.cn

公司简介

广东弘业建筑设计有限公司创立于1987年，是对建筑工程的建造与使用全过程提供策划、规划、设计、分析与施工管理等专业服务的综合性设计机构。

弘业设计师一贯坚持以独立研发为原则，不断地开发整合新理念与新技术，公司一直坚持走可持续与稳健的发展路线 。至2010年底，公司已设计并完工的项目有近3300个，涵盖行政办公、综合商业、文化教育、医疗保健、商品住宅、工业厂区等多个领域，其中多个项目获省、部级优秀设计奖。作为拥有近百名专业设计师的职业化团队，公司非常重视树立共同的价值观并保持公司整体的协调发展，逐渐形成了成熟的组织模式和适应变革的体制，为公司未来不断为客户提供优秀的服务提供了坚实的基础。

我们期待着通过服务客户来实现职业价值，并通过与客户共同的协作来实现更大的社会价值。

ABOUT US

HY Guangdong Architects & Engineers Ltd. was originally founded in 1987, and our services occupy the whole stage of project construction including development consulting, city planning, architectural design, analysis and project management etc.

HY is committed to providing professional services based on innovation and also sticks to the development of sustainability and stability. By 2010 we have completed more than 3,300 projects including various project types specializing in large scale mixed use commercial center, residential developments, educational facilities, administrative and industrial buildings, and lots of our projects earned ministerial or provincial awards excellent design. Today we've got the ability of providing high quality services by matured model of team-work and adaptable management systems which are derived from shared value and corporate culture.

We are looking forward to have opportunity to materialize our creative ideas and to realize our social value through serving customers.

Kingdom Global Center

景兴环球中心大厦

设计时间　2009年
竣工时间　2011年
广东 · 佛山
佛山市汇智实业发展有限公司
33 375平方米
钢筋混凝土框架结构/钢结构/石材及玻璃幕墙

Kingdom Global Center

Design Time　2009
Complete Time　2011
Foshan, Guangdong
Foshan City Wisdom Property Development Co., Ltd.
33,375 m^2
Reinforced concrete / steel frame / Glass and metal curtain wall

华南国际金融中心

设计时间　2010年
竣工时间　2014年
广东 · 佛山
广东新天鸿物业发展有限公司
385 540平方米
钢筋混凝土框筒结构（钢管柱）/钢结构/ 石材及玻璃幕墙
此项目由本公司和凯达环球合作设计

South China International Financial Centre

Design Time　2010
Complete Time　2014
Foshan, Guangdong
Guangdong Sun Tien Hung Property Development Company Limited
385,540 m^2
Reinforced concrete / steel frame / Glass and metal curtain wall
Joint venture with Adeas Ltd.

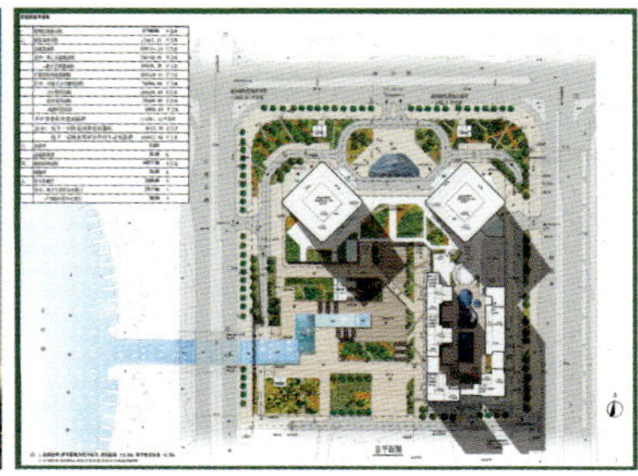

South China International Financial Centre

深圳市水木清建築設計有限公司
THINKER ARCHITECTS & ENGINEERS
广东省深圳市金田路3037号金中环商务大厦42层
42F Golden Central Tower, No.3037, Jintian Rd,
Shenzhen, Guangdong P.R.C
Tel: +86-755-83690555 Fax: +86-755-83690777
E-mail: sz@authinker.com
Http://www.authinker.com

思考创造价值 认真铸造品质

水木清设计的历史可以追溯到1992年成立的“林怀文建筑师事务所”。1992年，在时任惠州市人民政府副市长的庄礼祥先生的大力支持下，林怀文和陈怡姝在惠州创立了以林怀文的名字注册的“林怀文建筑师事务所”。事务所没有资质，只是拥有合法的市场身份。这在当时的历史条件下，已属难能可贵了。

1995年的春天，“林怀文建筑师事务所”整体加入了清华大学建筑设计研究院深圳分院，开始了其在清华分院的12年辉煌历程。

公司架构

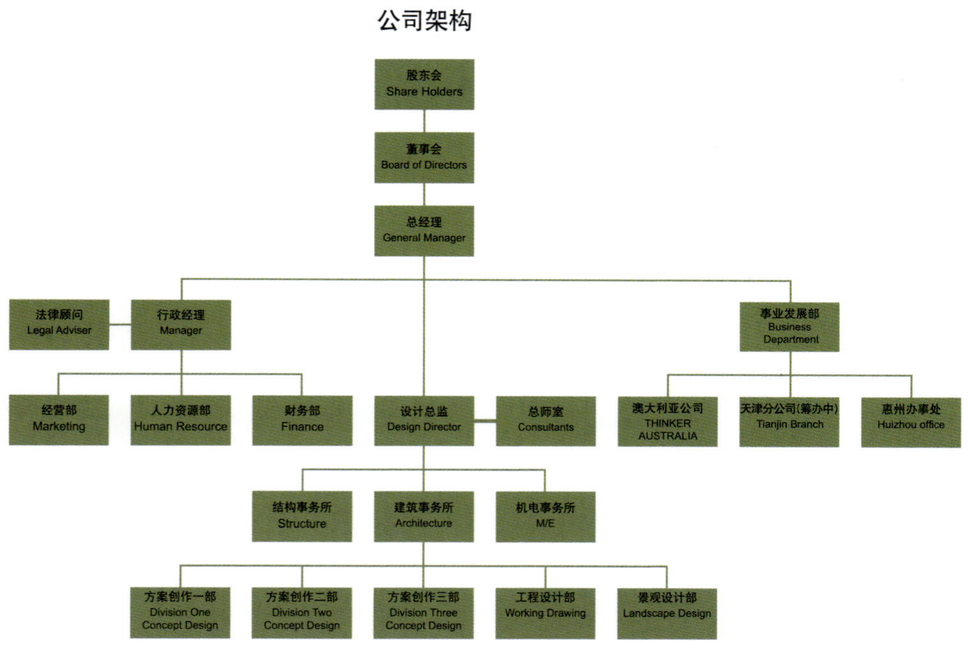

2007年的春天，水木清设计开始创立自己的设计品牌，并在澳大利亚注册成立了“THINKER DESIGN & CONSULT PTY LTD”(澳大利亚水木清设计顾问有限公司)。目前，我们拥有国家建设部建筑设计甲级资质（资质证书编号：A144010821）、建设部结构设计甲级资质（资质证书编号：A144010839），以及建设部机电设计甲级资质（资质证书编号：A144018473）。

在市场经济的大潮中，水木清设计形成了自己独特的设计文化。我们充分尊重合作各方的多元需求，非常重视在设计过程中与客户的“互动设计”，我们提倡以适度超前的理念、适度领先的技术，寻求建筑功能、建筑技术、建筑艺术与市场需求有效结合的“适度设计”；我们专注于挖掘项目的内在价值，尽最大可能地提高建筑的使用效率，最大限度地为社会创造财富，为我们的客户“创造价值”。

我们提供范围广泛的建筑设计服务，包括办公建筑、校园建筑、公共建筑、商业建筑、高层住宅与住宅小区等等。设计服务内容涵盖了建筑设计、室内设计与景观设计。而在超高层建筑设计方面，我们更是拥有丰富的经验。我们在北京、深圳设计，并已经建成投入使用的超高层项目如下（按建成时间排序）：

1.	深圳万科俊园	已建成	公寓	168米（46层）
2.	深圳港丽豪园	已建成	住宅	126米（40层）
3.	北京富尔大厦	已建成	办公	120米（25层）
4.	深圳新世界中心	已建成	办公	238米（55层）
5.	深圳兰亭国际公寓	已建成	住宅	150米（50层）

公司已经完成阶段设计，或正在设计中的超高层项目如下（按设计时间先后排序）：

1.	广东移动通信枢纽大厦（广州）	投标中标	办公	180米（40层）
2.	北京世纪城市	方案设计	城市综合体	200米（50层）
3.	广州涛景湾三期	方案设计	住宅	180米（50层）
4.	深圳港丰大厦	初步设计	商业、公寓	130米（36层）
5.	福州中旅城	立面设计	城市综合体	180米（40层）
6.	福州海西广场	方案设计	城市综合体	250米（60层）
7.	珠海蓝天大酒店	方案设计	酒店、公寓	170米（45层）
8.	佛山坚美展贸中心	方案设计	办公	130米（32层）

公司的作品能够得到社会的持续认可与嘉奖，在于我们对品质的不懈追求与重视。我们认真对待每一个项目，认真分析每一个项目的特定因素，包括但不限于地理环境、气候条件、建筑文脉、社会心理、消费习惯、经济基础、建造技术、流行时尚、业主特征、业主期望等影响因素，寻求一个最好的也是唯一的可行方案。我们的认真、执著，贯彻整个工程设计的始终。

公司能够得到长足的发展，在于拥有优秀的管理团队。公司创办者、现任董事长林怀文先生是我国知名建筑师。林怀文先生1984年毕业于清华大学建筑系，1994年参与创办了“清华大学建筑设计研究院深圳分院”，1995年获得中国建筑学会“青年建筑师奖”。作为公司的代表，林怀文先生入编建筑界权威杂志《世界建筑》主编的《青年建筑师·中国》（2001），并获邀参加在北京中国美术馆举办的首届“中国国际建筑艺术双年展”（北京·2004）。

公司另一位创办者陈怡姝女士是我国知名建筑师。陈怡姝女士1987年毕业于清华大学建筑系，2000年毕业于哈佛大学建筑与设计研究院，师从哈佛大学建筑与设计研究院院长彼德·G·洛先生（Mr. Peter G·Rowe），获得哈佛大学设计研究硕士，是学有所成的海归派建筑师。

ISO 9001
质量管理体系认证证书
深圳市水木清建筑设计事务所

ISO 9001
The Certification Certificate
Of Quality Management System

2010年8月，公司通过ISO9001质量管理体系认证

佛山坚美展贸中心

建设单位：广东坚美铝型材厂有限公司
建设地点：广东 佛山
设计时间：2010年

总建筑面积：9.58万平方米
用地面积：1.2万平方米
容积率：5.3
覆盖率：40%
绿地率：30%
层数：地下3层，地上31层
建筑高度：130米
结构类型：钢筋混凝土
方案设计：庄绮琴、刘天庆、林 郴
毛铁勇、何健新、梁应恒、欧阳华

本项目是著名铝业公司坚美铝业的展贸中心。设计中，注重经济适用，合理处理办公、展贸、金融和商务等多项功能的空间关系，空间布局简洁流畅，通行便捷。布局上，与南侧地块共同安排，共同围合成一个大尺度中庭，使两个地块在满足自身单独运行的前提下，更可紧密结合，形成大规模商业空间，以提升价值。塔楼布置在临街位置，突出主楼体量，利于塑造公司形象。外立面设计现代、简洁，经久耐看。外形既充分体现“坚美铝业”的企业标志性，又与城市环境相协调，庄重、大气并富有创新性；用材方面精心选择，充分运用坚美铝材产品，使整个项目也成为企业自身产品的一大展示空间。

（文：庄绮琴 田东明）

总平面图 1：500

惠州仲恺高新区总部经济大厦

建设单位：惠州市仲恺高新区产业开发区管理委员会
建设地点：广东 惠州
设计时间：2009年

总建筑面积：17.5万平方米
用地面积：5.42万平方米
容积率：3.0
覆盖率：29.7%
层数：地下2层，地上28层
建筑高度：110米
结构类型：浇钢筋混凝土结构
方案设计：朱鸿晶、庄绮琴、赵　佳、毛铁勇、何健新、林　郴、梁应恒

折板式的建筑轮廓，蕴涵着丰富流畅的信息数码的时代气息，以新型的建筑手法，打造高新区品牌，提升区域形象并丰富新区的天际线。退让东南角，形成汇集人流的四层通高灰空间，预示对外服务功能的开放性、透明性和主动性，并有良好的展示和包容的效果。半围合的开放性中心广场，提供给外来及内部使用人员以展示、集散、礼仪的多功能空间。（文：朱鸿晶　田东明）

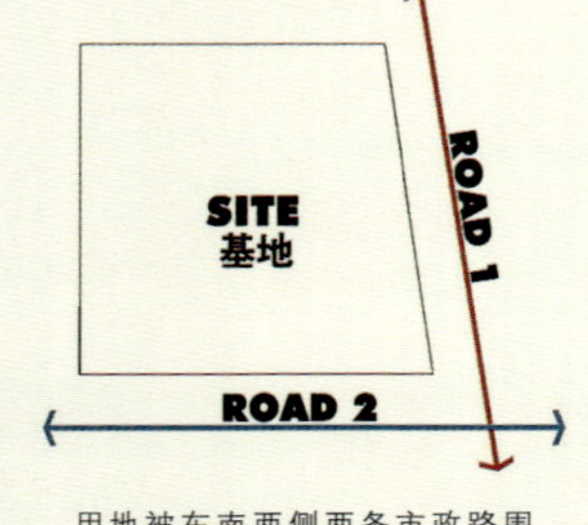

用地被东南两侧两条市政路围绕，南侧为仲恺大道，东侧为和畅六路。

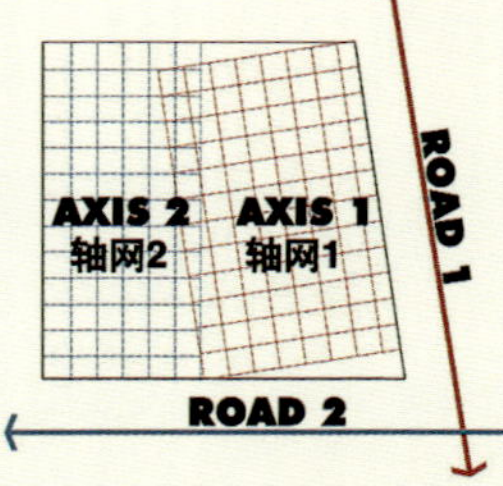

依据两条道路的走向，引入两套斜交轴网。

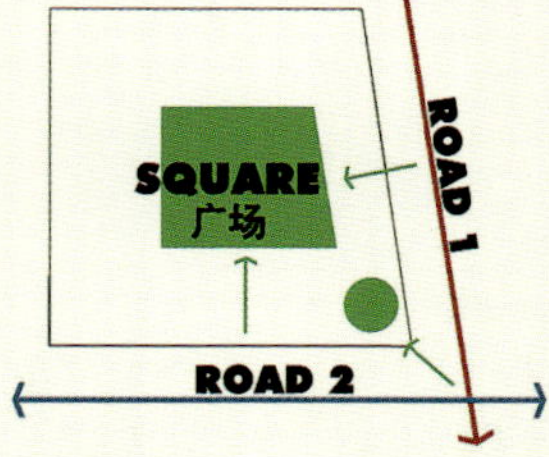

设计要求的中心广场的设置，考虑与两条主干道的有机联系；东南角退让市民广场空间，减轻高层建筑对道路的压抑感。

数字化高科技元素的引入。

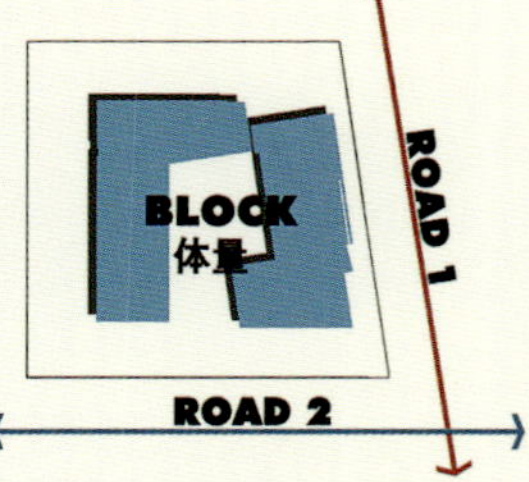

在两套轴网基础上，生长出一二期的建筑体量。

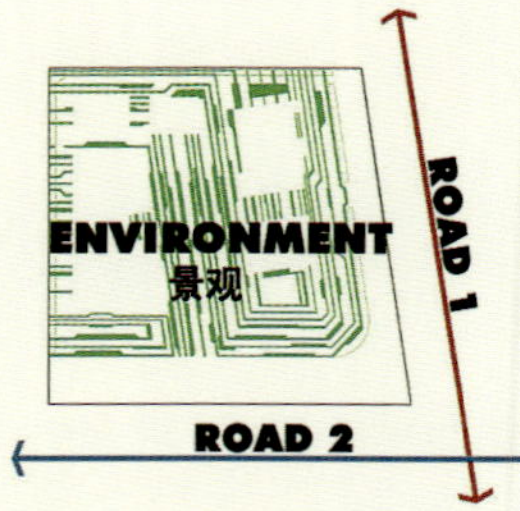

环境的意象与高科技语素的融合。

华大基因产业发展基地

建设单位：深圳华大基因建筑发展中心
建设地点：广东 深圳
总建筑面积：366 000平方米
用地面积：103 000平方米
容 积 率：2.0
绿 地 率：60%
设计时间：2010年

规划总平面图

地下一层平面图 1:1500

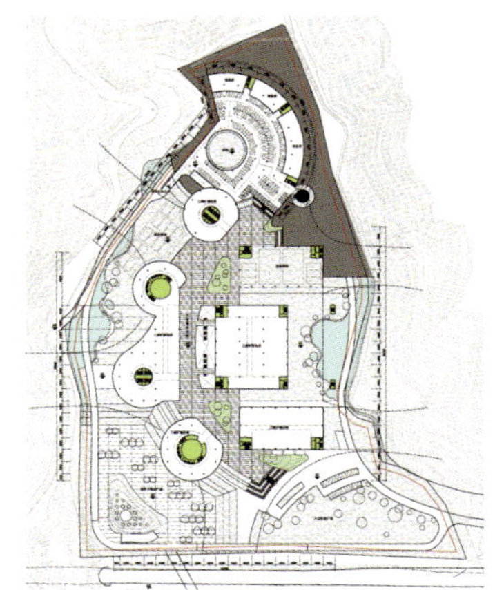
一层平面图 1:1500

三、四层平面图 1:1500
（含二期扩建区城示意）

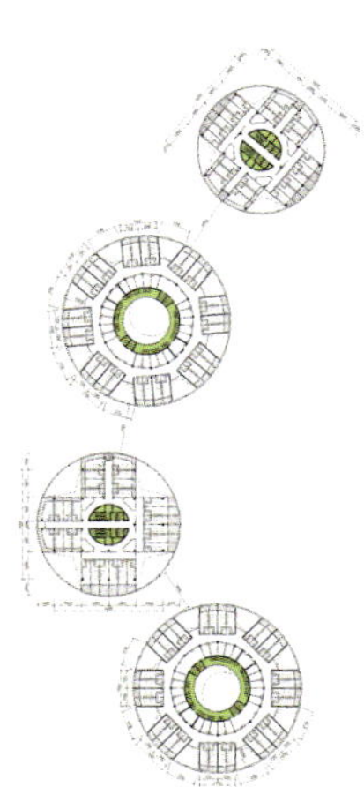
塔楼病房区标准层平面图 1:750

东立面图 1:2000

西立面图 1:2000

福州恒力创富中心

建设单位：福建恒力房地产发展有限公司
建设地点：福建 福州
设计时间：2010年

总建筑面积：51 799.80平方米
用地面积：6 875.28平方米
容 积 率：6.0
绿 地 率：20%
建筑高度：83.1

惠州半山一号

建设单位：惠州市瑞亨企业集团有限公司
建设地点：广东 惠州
设计时间：2009年

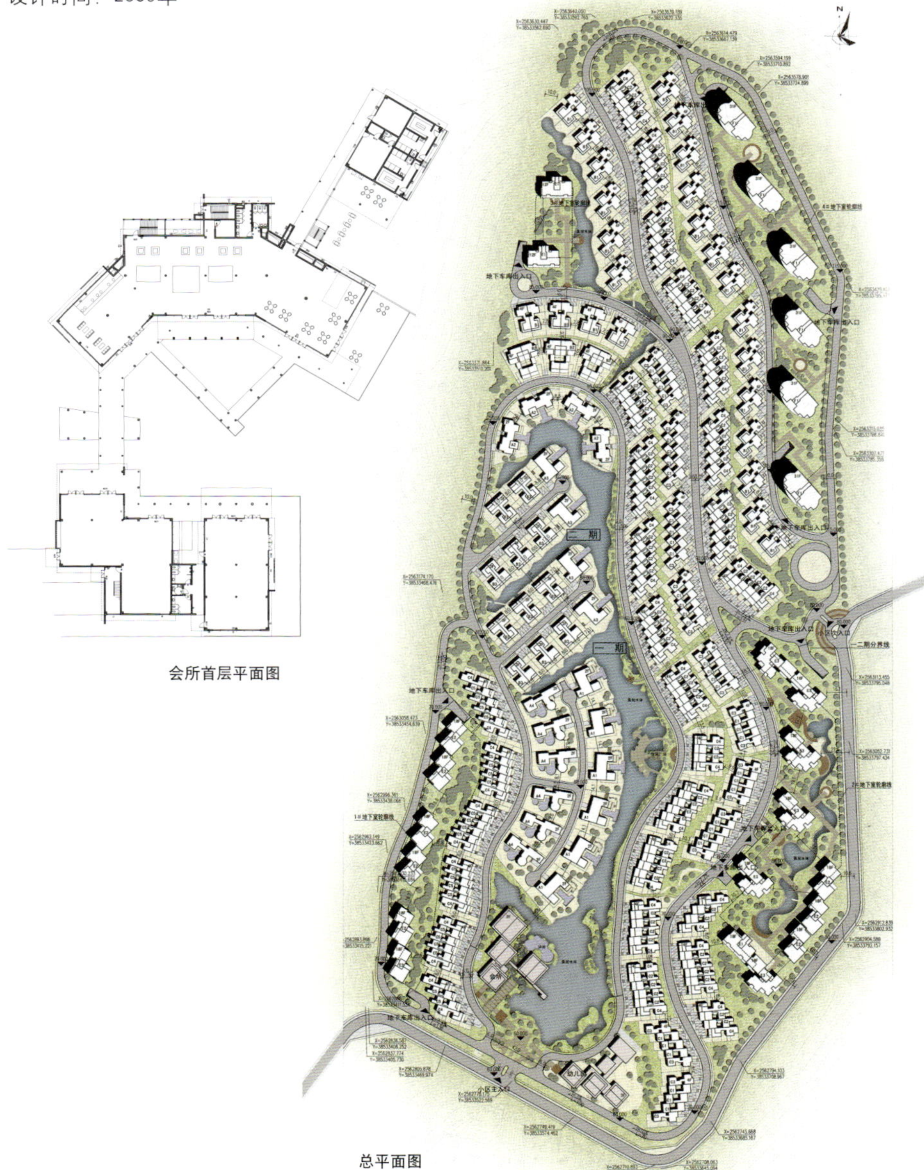

会所首层平面图

总平面图

总建筑面积：39.1万平方米
用地面积：25.7万平方米
层　　数：地下2层，地上3至31层
建筑高度：70.48至99米

方案设计：张维昭、林怀文
建　　筑：张维昭、杨　勇、刘　佳、陈学忠、邱睿敏、黄华添、倪海生
结　　构：关柏岩、张浩远、张英敏、何汝珍、祝　蓉
设　　备：倪达峰、李　雪、石章杆、刘　庆

该项目位于惠州西部，东江以北的山谷中，定位为高档住宅区，小区环境资源利用达到最大化。将高层建筑布置在周边四处台地上，在综合解决土方难题的基础上，占据制高点，将小区以及周边城市山水景观尽收眼底，提升高层住宅品质。整理后的坡地分解成有高差之地块，用于布置联排或双拼别墅，确保各住户景观视线无遮挡。部分双拼别墅与独栋别墅坐落在山水之谷的核心地带，拥有私家水岸。这是山谷与水岸的市镇，闹市中诗意静谧的居第。　（文：张维昭）

深业广州水岸金沙

建设单位：深业南方地产（集团）有限公司
建设地点：广东 广州
总建筑面积：338 650平方米
用地面积：94 519平方米
容 积 率：2.8
绿 地 率：38.2%
设计时间：2010年

户型分布　消防流线　车行流线　人行流线

惠州市金泓升项目

建设单位：深业市金泓升投资集团
总建筑面积：1 378 000平方米
用地面积：226 952平方米
建设地点：广东 惠州
设计时间：2010年

承构建筑

美国承构建筑师事务所
深圳市承构建筑咨询有限公司
上海承构建筑设计咨询有限公司
Made & Make Architects

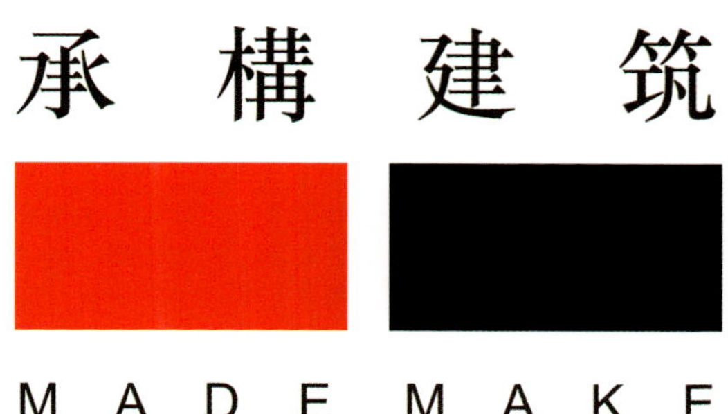

承构建筑2008年成立于美国纽约，现有建筑设计师100余人。在中国深圳与上海都拥有自己的设计机构。作为一个有实力的设计机构，承构建筑的作品覆盖全国，服务于不同类型的客户。

承构建筑自成立以来，已完成近千万平方米的建筑设计，建筑类型包括住宅类建筑、教育类建筑、商业建筑、办公综合体建筑、城市设计等，客户包括香港置地、龙湖地产、中海地产、万科地产、招商地产、华润置地、金地地产、保利地产、中冶集团、创维集团、泰达建设、大唐电力集团、奥林匹克花园、佳兆业、协信地产等众多知名企业，并完成了许多学术价值及市场口碑俱佳的建筑作品。承构的作品主要集中在京津地区、长三角、珠三角以及西南地区（重庆与成都）这四个中国最受瞩目的区域和武汉、沈阳、西安等重要城市。

承构建筑的实践包括建筑单体设计、城市设计、城市规划等相关领域。我们认为建筑实践是一种承接的行为，建筑实践与现存的城市地理环境，社会经济条件息息相关。成功的建筑实践不仅满足建筑功能及美学的需求，还应根植于市场，并且通过构建新的空间，延续城市和文化的整体记忆，提供新的生活体验。

承构的设计宗旨
关注市场——在产品设计上注意结合市场和领先市场。
关注客户——保证与客户的密切配合和沟通。
关注设计——每一个成功的建筑都应是有灵魂的；我们始终追求创意与激情、理想与现实、艺术与市场统一的建筑作品。
关注建造——在设计阶段注重建筑所在地的技术及材料条件，避免设计上的浮夸和不切实际；在施工阶段密切配合甲方，全面保证质量。

Made & Make Architects is founded in 2008 in New York, USA and now it has over 100 architectural designers. Shenzhen and Shanghai Made & Make Architects are the branch offices of Made & Make Architects in China. Made & Make Architects provides service to various clients, and our projects spread across whole country.

Since its establishment, Made & Make Architects has completed numerous projects, and the overall area accounts to 10 million square meters. Our project types vary from residential, academic, commercial complex to urban design. Made & Make Architects serves clients with different background, such as Hong Kong Land Holdings Limited, Longfor Properties Co., Ltd., China Overseas Land & Investment Ltd., Vanke Estate, China Merchants Property Development Co., Ltd, China Resources Land Ltd., Gemdale Group, Poly Group, China Metallurgical Group Corp, Skyworth Group, Teda Construction, China Datang Corporation, Olympic Garden, Kaisa Group, Sincere Group. Made & Make Architects has designed many buildings with market success as well as academic value. The projects of Made & Make are located mainly in four areas with most economic growth in China, such as Beijing and Tianjin Area, Yangtze River Delta region, Pearl River Delta region and Southwest region (Chongqing and Chengdu).

Our practice covers architectural design, urban design, landscape planning, and other relevant areas. We consider architectural practice as a way of sustentation; architectural practice is closely related to our physical environment and social economical conditions. Successful architectural design practice should not only meet the functional and market requirement, but also carry aesthetic consideration. By the construction of physical space, we help to sustain the collective memory of city and culture, and provide new living experience as well.

Design Approach
Focus on market – to be the leader of the market by integrating the project design and the market demands.
Focus on clients – guarantee close cooperation and communication with the clients.
Focus on design – each successful architecture has its own soul; what we pursue is the integration of creativity and enthusiasm, ideal and reality, art and market.
Focus on construction – pay great attention to construction conditions at design stage, avoid anything grandiose and unpractical, closely cooperate with clients during construction stage to guarantee best quality.

深圳市承构建筑咨询有限公司
地址：广东省深圳市福田区深南大道2008号中国凤凰大厦1号楼20C
电话：+86-755-33067800
传真：+86-755-33067801
邮箱：work@mademake.com

Shenzhen Address: Phoenix Building No.1, 20C, 2008 Shennan Road, Futian District, Shenzhen, Guangdong
Tel: +86-755-33067800
Fax: +86-755-33067801
Email: work@mademake.com

上海承构建筑设计咨询有限公司
地址：上海市杨浦区淞沪路388号创智天地广场7号楼202室
电话：+86-21-61437001
传真：+86-21-33067021
邮箱：work@mademake.com

Shanghai Address: KIT Village 7# Building, Unit 02, 2/F, 388, Songhu Road, Yangpu District, Shanghai
Tel: +86-21-61437001
Fax: +86-21-33067021
Email: work@mademake.com

公司网址: www.mademake.com

创维总部大厦

项目地点：广东 深圳
开发单位：创维集团
设 计 人：柴 晟、邓 卿、胡金涛、张倩倩、陈 林

在去年的年鉴上，该项目已经出现过一次。经过一年的深化设计，建筑的各个方面已日趋成熟。此次我们希望将创维大厦的设计发展呈现给大家。

在创维总部的方案中，吸入式的音箱式入口既是方案中最精彩的部分，也是工程中难度最大的地方。我们的设计着重针对这个入口空间，在保持原方案的魅力的基础上，对多种工程实现的方法做了比较。在摒除了热弯玻璃（造价太高）、三角平板玻璃组成三维曲面（削弱了弧形入口的流畅感）等多种方法后，我们选择了“鳞甲”式的方案——由十余种不同规格的平板玻璃沿各自的圆周按照不同的角度拼贴，最终形成整个弧形入口。

Skyworth Headquarters Building

Address: Shenzhen, Guangdong
Developer: Skyworth Group
Designers: Chai Sheng, Deng Qing, Hu Jintao, Zhang Qianqian, Chen Lin

This project was listed in last yearbook. Over a year's further design, the building is being improved in all respects. In this phase, we want to present the design developments of Skyworth Building to you. In the plan of Skyworth Headquarters, adoptive sound-box-style entrance is the most beautiful component, which is the most difficult part of the construction. With special attentions to this entrance space, our design has compared many approaches to realize the project, while maintaining the charms of original plan. After excluding hot bending glass (too expensive), triangular flat glass to form 3D curved surface (detrimental to the fluency of curved entrance) among many other schemes, we choose "scales" pattern – to use more than 10 specifications of flat glass to paste in different angles along their respective circumference, to finally form the entire curved entrance.

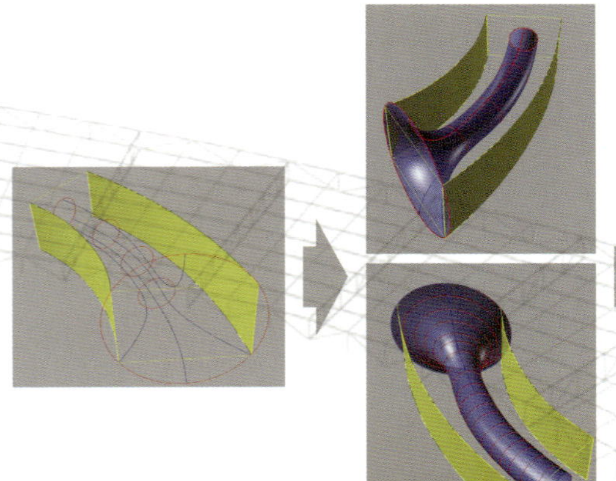

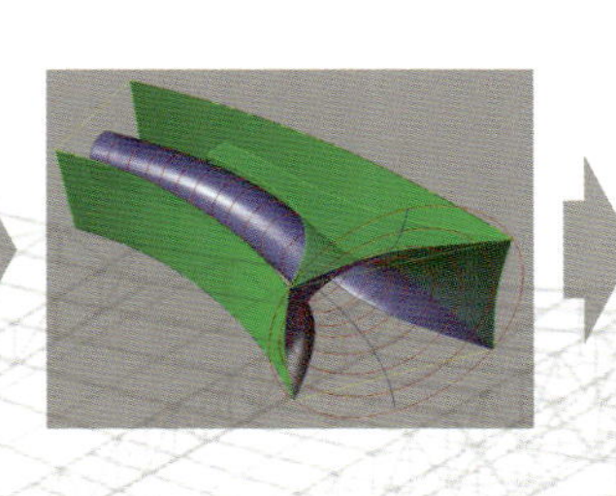

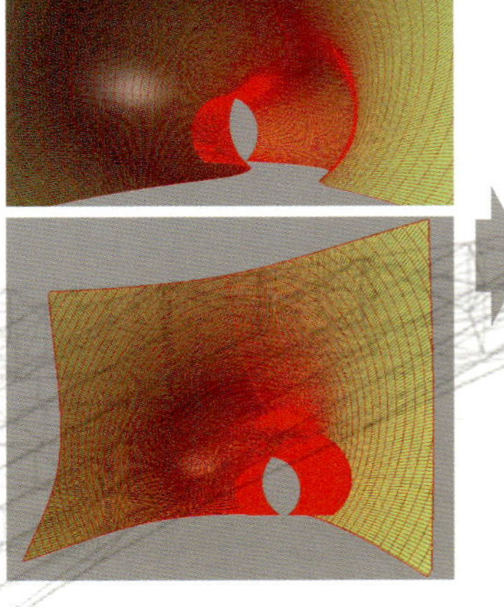

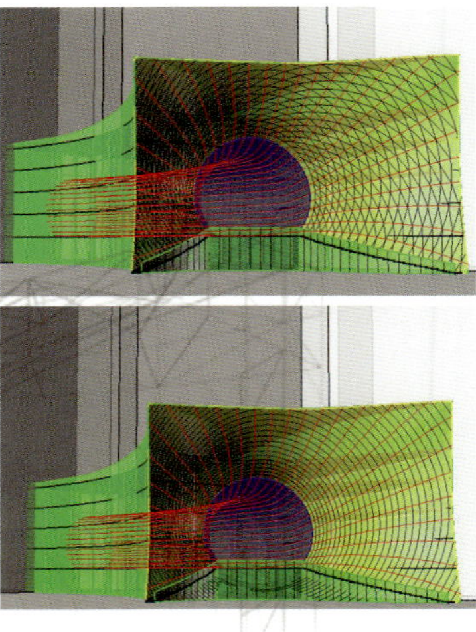

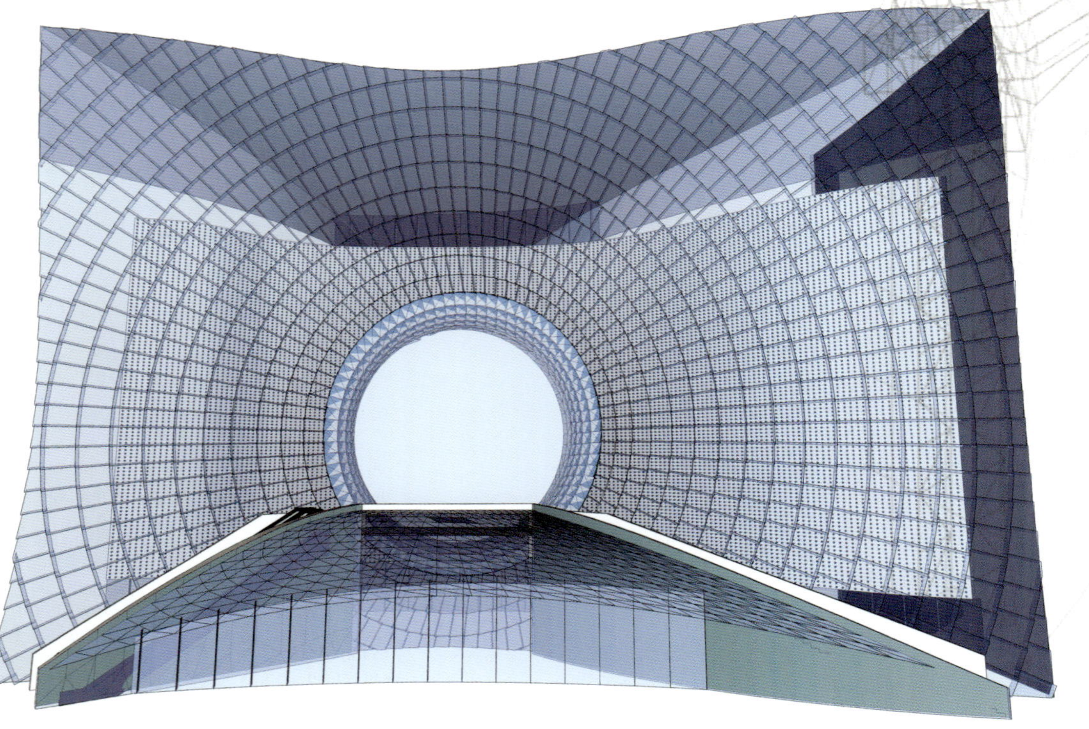

在解决了入口的幕墙方案之后，我们还遇到了一个较大的难题——原来的音箱式入口是由一个巨大的曲面LED显示屏构成的。设计过程中我们与厂家合作做了多种方案的实验，包括（1）直接用LED箱体形成入口。这种方法即使解决了散热问题，也无法解决入口处的物理环境问题，另外LED也无法供人近距离观看。（2）入口玻璃幕墙与LED的结合也有影像被玻璃削弱或影像变形的问题。一系列实验后，我们不得不放弃了直接用LED形成入口空间的做法，转而达成另一个解决方案——让入口的三维喇叭式玻璃幕墙独立形成一个空间，而使原本与幕墙相接的背墙退后8米，在墙体与三维玻璃幕墙之间形成一个可用的空间。在退后的背墙上我们设计了以矩阵形式排列的灯具组合，使得具有奇幻色彩的入口空间在这面墙上得以体现。

After solving the problem of curtain wall plan, we encounter another big difficulty: the former sound-box-style entrance is a huge curved LED display screen, so in the design process, we have tried many schemes in cooperation with the manufacturer, for example: (1) LED box is used to form the entrance, which cannot solve the problem of physical environment at the entrance, even if it solved the problem of heat dissipation; furthermore, LED is not visible in a short distance; (2) The combination of entrance glass curtain wall and LED also has the problem of image weakening or distortion by the glass. After a group of experiments, we have to give up this practice that using LED to form the entrance space, instead, we enter into another solution: we make the 3D trumpet-style glass curtain wall form a separate space, and the back wall next to curtain wall retreat 8 m away, so as to form a usable space between the wall and the 3D glass curtain wall. On the retreated back wall, we design lighting unit arranged in matrix form, so the magic entrance space will be shown on this wall.

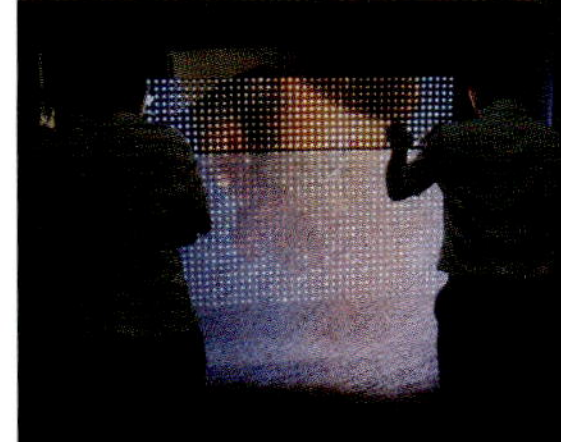

LED型入口效果图

鳞状音箱式入口效果图

LED型入口试验

保利电厂城市改造项目

Poly Power Plant Urban Reconstruction Project

项目地点：贵州　贵阳
开发公司：保利贵州置业集团有限公司
设计时间：2011年
设 计 人：柴 晟、赵伟、严小兰、何运祥、孔庆峰、吴 姮

该项目位于贵阳市南明区太慈桥，属于老城区核缘区域，是连接贵阳市区中心到南部小河及花溪的交通咽喉，原属于贵阳电厂用地。随着城市的更新改造，2011年由保利贵州置业主导这一片区的更新设计，改造后建筑面积达250万平方米，包含住宅、公寓、风情商业街、水上购物中心、超高层办公等复合型业态组合，形成贵阳的又一城市副中心。规划设计对贯穿基地内的南明河上的城市生活、高密度的城市体量与凤凰山的山体关系予以高度关注，让保利电厂改造项目成为一个现代都市中的山水之作。

Location: Guiyang, Guizhou
Developer: Guizhou Poly Property Group Co., Ltd.
Design Time: 2011
Designer: Chai Sheng, Zhao Wei, Yan Xiaolan, He Yunxiang, Kong Qingfeng, Wu Heng

This project is located at Taiciqiao, Nanming District, Guiyang City, in the peripheral old city zone, connecting the downtown to the traffic hub of southern Xiaohe and Huaxi, the former land of Guiyang Power Plant. During the urban reconstruction, in 2011, Guizhou Poly Property leads the renewal design for this area, and the floor area after reconstruction will be 2,500,000 m^2, including residential buildings, apartments, exotic shopping streets, on-water shopping center, super high-rise office buildings and other complex, forming another sub-center of Guiyang City. The planning design pays high attention to the urban life on Nanming River, high density urban capacity and Mt. Phoenix massif relations throughout the base, to make Poly Power Plant Reconstruction Project a landscape masterpiece of modern metropolis.

大中华国际交易广场投标方案

项目地点：广东 深圳
设 计 人：柴 晟、阮 洋、高逸嘉

该项目位于深圳市罗湖区火车站前广场东，是一个大都市中典型的高密度复合城市综合体（容积率达16.0）。项目由一栋360米高的办公及五星级酒店塔楼，和一个购物商场组成。

塔楼下部的办公部分平面呈正方形，向上随着功能的变化逐渐过渡成为椭球形（适于客房分布），最后以一个玻璃（叶子）所覆盖的空中大堂作为结束，向东南侧可俯瞰香港新界之美景。随着塔楼体型变幻，其在空间上产生的优美曲线与下部购物商场的天顶酒吧街的采光天棚连为一体，一气呵成。购物商场则以一个旋涡式的中庭吸引着来往的行人。

Bidding Plan of Great China International Trade Plaza

Location: Shenzhen, Guangdong
Designer: Chai Sheng, Ruan Yang, Gao Yijia

This project, located at east of the Square in front of Luohu Railway Station, is a typical high density urban complex (floor area ratio 16.0) of metropolis. This project is composed of a 360 m high office & 5-star hotel tower, and a shopping mall.

The lower part of the tower is cubic, grows to be spheroid in the upper part as its functions changing (suitable for guesthouse distribution), and finally ends with an aerial lobby covered with glass leaf-shaped lock, from the southeast of which we can have a bird's view of New Territories, Hong Kong. As the tower pattern is changing, the beautiful curves that are produced in space integrate with the daylight sunshade of the rooftop aerial bar street at the lower shopping mall, to form a consistent building cluster. The shopping mall attracts people by a swirling atrium.

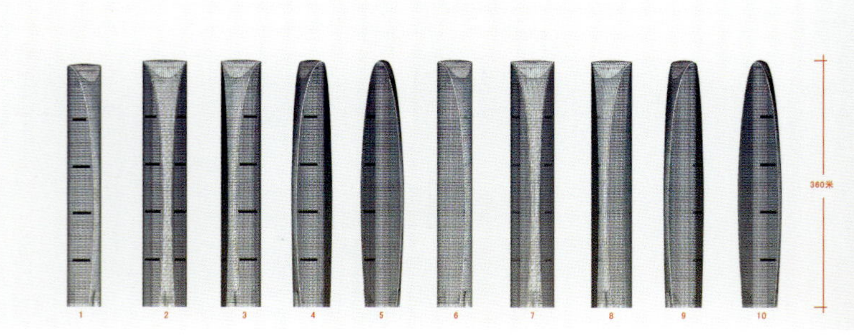

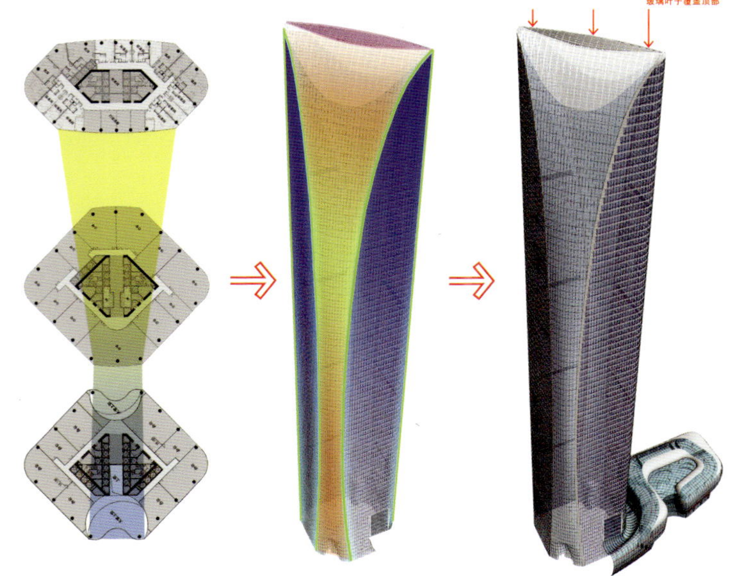

福建泉州新天城市广场

项目地点：福建　泉州
开发公司：新宇（泉州）置业有限公司
设计时间：2011年
设 计 人：许 峰、刘叮咚、刘晓丹、张学森

该项目位于泉州鲤城区，总体规划为以超高层住宅为主，共享城市公园、地块内部景观的花园式住宅区，同时具备大型购物中心的城市综合体，成为该区域的新标志。

其中的购物中心设计定位为区域级中心，规模8万平方米，是涵盖超市、普通购物、情景步行街、文化休息、餐饮娱乐等功能的大型一站式购物中心。建筑设计着重创造富于雕塑感的立面造型，沿街立面如同波浪般起伏变化、极富动态感；结合东西两侧主入口和步行街设计了丰富的室外灰空间，为购物中心提供了良好的室外商业氛围。

Xintian Urban Plaza, Quanzhou, Fujian

Location: Quanzhou, Fujian
Developer: Xinyu (Quanzhou) Property Co., Ltd.
Design Time: 2011
Designers: Xu Feng, Liu Dingdong, Liu Xiaodan, Zhang Xuesen

This project is located in Licheng District, Quanzhou. The general plan is mainly super high-rise residential building, that is, garden-style residential community sharing City Park and the landscape inside the plot; meanwhile, it should be an urban complex with large-sized shopping center, to become a new landmark of this region.

Its shopping center is designed to be a regional level center, 80,000 m^2, that is, a large-sized one-stop shopping center covering supermarket, ordinary shopping, scenario walking street, cultural & recreational, catering and entertaining functions. Its architectural design pays attentions to vertical shape full of sculpture sense, where the roadside facades move like waves, full of dynamics; it also designs rich gray space outdoor in addition to the main entrance at either of east and west sides and the walking street, providing good outdoor business atmosphere for the shopping center.

天津万科东丽湖镇中心商业

开发公司：天津万科发展有限公司
设计时间：2010年
设 计 人：许 峰、李世华、尹铭令、杨冰峰

该项目位于天津市中心城区的东侧，虽隶属于滨海新区的东丽湖生态居住组团，仅2.5万平方米，却是滨水社区的生活核心。其主要功能是综合商业和体育馆场，供万科东丽湖社区居民使用。项目希望根植于天津的建筑传统，采用天津五大道风格，混合了西方与中方建筑特点的风格创造一个适合于休闲、并具有建筑传统的识别性的镇中心项目。

Vanke Donglihu Town CBD Project, Tianjin

Developer: Tianjin Vanke Development Co., Ltd.
Design Time: 2010
Designers: Xu Feng, Li Shihua, Yin Mingling, Yang Bingfeng
This project is located at the east side of central city zone of Tianjin. It is only 25,000 m^2, belonging to Donglihu Ecological Habitat Group in Binhai New District, however, it is the living core of waterside communities. Its main functions are commercial complex and gymnasium, available for residents in Vanke Donglihu Community. The project, in the hope of being rooted in Tianjin building traditions, adopts the styles of the Five Avenues in Tianjin, mixing with western and Chinese architectural features, to create a town center project suitable for leisure purpose, with traditional building identity.

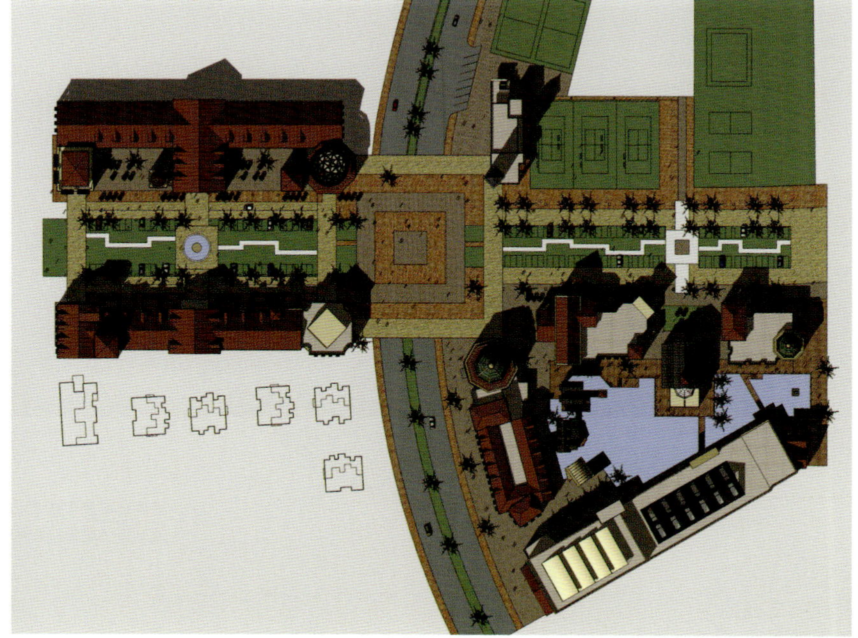

成都龙湖时代天街与北城天街

项目地点：四川 成都
开发公司：龙湖集团
设 计 人：柴 晟、唐 聃、左小冬、高逸嘉、王 远
杨冰峰、何显鳌、黄珏磊、吉成钢

今年承构在成都完成了两个重要的商业综合体，即位于成都北城五块石的北城天街和位于成都电子科技大学一侧的时代天街。两个项目均包含了面积达40万平方米以上的大型商业建筑，其中时代天街还包括了高达20万平方米的销售式商业——商铺。时代天街的商铺以大型综合商业为依托行程环状商业街。因为容积率的关系，商业街高达五至六层。用什么办法将人流向高楼层引入呢？我们给出的方案是一个长约60米的商业单元体。

（一）对于位于地下首层和二层的商业，我们将他们定义为“随机逛街”的商业载体，可以运用于任何业态，如时装店、药房等；

（二）三至六层的商业都有自己的室外拓展空间，我们将其定义为目的式商业，适于做餐饮、咖啡酒吧等。

通过这样的分类，不仅解决了容积率的问题，更解决了商业上总划分的问题。同时由于它是以空中四合院的形式组合商业，使三层以上每个商铺都有花园，为枯燥乏味的简单行列式的商业街铺带来全新的空间感受，被甲方称为具有成都传统特色的商业街——空中的“宽窄巷子”。

Longfor Times Plaza and Beicheng Plaza, Chengdu

Location: Chengdu, Sichuan
Developer: Longfor Group
Designers: Chai Sheng, Tang Dan, Zuo Xiaodong, Gao Yijia, Wang Yuan, Yang Bingfeng, He Xian'ao, Huang Yulei, Ji Chenggang

This year, Chenggou has finished two important commercial complexes in Chengdu, that is, Beicheng Plaza in Wukuaishi of Beicheng District, and Times Plaza beside Chengdu University of Electronics Science and Technology of China. Both works include large-sized commercial buildings held up to 400,000 m^2, of which, Times Plaza also includes stores up to 200,000 m^2 for sale. The stores at Times Plaza based on large-sized commercial complex form a circular shopping street. Due to floor area ratio, the shopping street is 5 or 6 storeys high. Then, how can we direct the flow of visitors into higher floors? Our answer is a commercial unit about 60 m long.

(I) For the commercial buildings in the underground floor-1 and floor-2, we will define them as "strolling random" commercial carriers, which may be in any form of operation, such as fashion store, pharmacy;

(II) The commercial buildings, 3-6 storeys, have their own outdoor expansive space, which will be defined as purposeful commercial buildings, suitable for catering, café and bar etc.

Through such classification, we not only solve the problem of floor area ratio, but also resolve the problem of general commercial division. Meanwhile, since it is a commercial complex in the form of aerial courtyard house, where every store above the third floor has a garden, bringing brand new space sense to simple linear shopping street, which is called by Party A as aerial "wide and narrow lanes", a shopping street featured by Chengdu traditions.

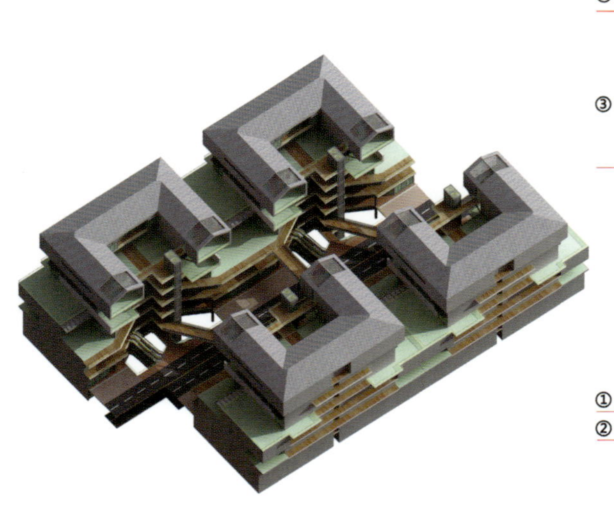

⑦ 六层的商业也拥有自己的入户花园。
③ 二层的商铺由商业街走廊进入，如果一层和二层商铺合并，则可减少二层的商业廊道。
① 首层商铺有车辆可以直达。
② 地下层的商铺也车辆可以直达。
⑧ 每隔60米的交通体可以保证人员上下自如。
⑥ 四层，五层的商铺都有自己的入户花园。
⑤ 三层店铺成为其他店铺的花园。
④ 位于三层的无顶天街由每隔60米的垂直交通和位于商业街入口处的扶梯进入，再通过入户花园进入到每个店铺。

重庆龙湖源筑一期

项目地点：重庆
开发单位：龙湖集团
设 计 人：柴　晟、唐　聃、张钟方、刘晓丹

该项目位于江北区体育公园一侧，包含了三项主要功能：（一）位于基地西侧的超高层写字楼及大型商业综合体；（二）高密度住宅产品——花园洋房；（三）面向体育公园的高层住宅。花园洋房的立面设计采用现代法式风格。由于成本的原因，立面装饰材料不能采用最适合法式建筑的石材，我们在设计中尝试使用真石漆及GRC材料来还原法式立面及浅脚，取得了良好的效果。高层住宅的立面则采用了与法式建筑协调的装饰艺术风格。

Longfor Yuanzhu Project, Chongqing

Location: Chongqing
Developer: Longfor Group
Designers: Chai Sheng, Tang Dan, Zhang Zhongfang, Liu Xiaodan

This project is located beside the Sports Park of Jiangbei District, comprising three major functions: (I) super high-rise office buildings and large-sized commercial complex on the west side of the base; (II) high density residential products – garden houses; (III) high-rise residential buildings facing towards the Sports Park. The façade design of garden houses is modern French style. For cost reason, the façade decorative materials are not stone materials that are the most suitable for French style buildings, instead, we try to use natural stone coating and GRC materials to restore French style façade and architrave, and gain good effects; while the façade of high-rise residential buildings is of Art Deco style in harmony with French style buildings.

上海旭辉玫瑰苑二期

项目地点：上海
开发公司：上海旭辉地产
设计时间：2011年
设 计 人：康 俊、赵灿彬、彭俊玲、黄鸿文、李 媛

该项目以法国古典主义建筑为蓝本，结合现代居住者功能需求，重新捕捉那些美好、高贵具有长久生命力的生活方式。

CIFI Rose Garden Phase II, Shanghai

Location: Shanghai
Developer: Shanghai CIFI Land
Design Time: 2011
Designers: Tang Jun, Zhao Canbin, Peng Junling, Huang Hongwen, Li Yuan

This project takes French classical architecture as blueprint, combines modern inhabitants' living demands, and recaptures those beautiful, noble and long-living lifestyles.

重庆虎溪别墅区

项目地点：重庆
开发公司：龙湖地产
设 计 人：魏 壮、左小冬、赵丹丹

重庆龙湖虎溪项目位于重庆市西部沙坪坝区虎溪镇重庆大学城规划区内。本项目规划用地约135 924平方米，容积率2.75，其中别墅区计容面积约为90 000平方米。

"虎溪"别墅立面设计具有一定"草原风格"。主材选用现在同类型别墅中比较少见的咖啡色横向条形面砖，使立面更具质感，也更加体现设计意图。同时搭配有荔枝面浅色石材与真石漆。坡屋顶设计有1米挑檐，以深灰色槽钢收口，各层挑檐穿插出现。细部设计相对来说并不复杂，通过适当的细节刻画来平衡及实现之前提到的设计原则。

"虎溪"别墅并非想完成一个纯复古的作品，而是通过很多细节的刻画使整个建筑既不失尊贵又透露出少许现代感。

Tiger Brook Villa Zone, Chongqing

Location: Chongqing
Developer: Longfor Property
Designers: Wei Zhuang, Zuo Xiaodong, Zhao Dandan

This project is located in the planning area of Chongqing Campus City in Huxi Town, Shapingba District of western Chongqing City. Its planning land area is 135,924 m^2, at FAR 2.75, including villa FAR area 90,000 m^2.

"Tiger Brook" villa façade design is of somewhat "prairie style". Its main material is coffee transverse strip tile, enriching the texture of façade, and enhancing our design purpose. Meanwhile, it also uses bush-hammered finish of tint stone materials and natural stone coating. The pitched roof is designed with 1m corbel table, closing by dark gray beam channel, with corbel tables of all floors appear alternatively. Detailed design is not so complicated, where we use appropriate detail depicting to balance and realize the aforesaid design principles.

"Tiger Brook" Villa Zone is not going to be a purely antiquated artwork, instead, through many details depicting, the entire building does not lose any nobility, and adds a few modern sense.

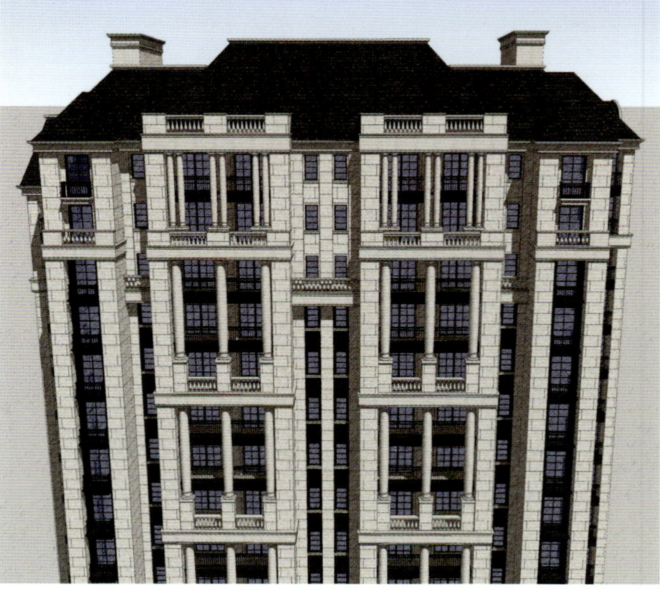

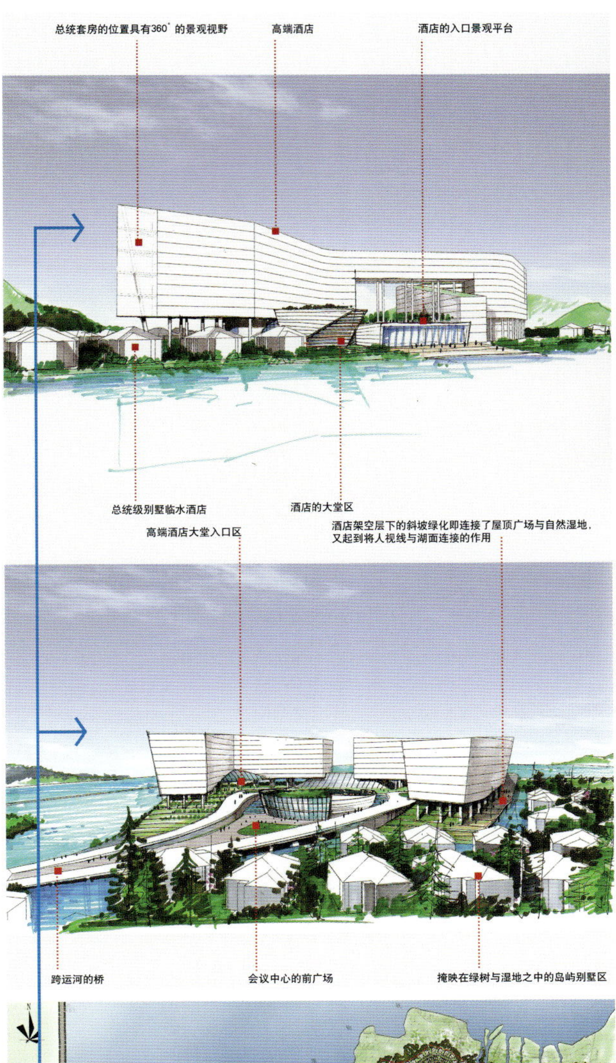

云南仙湖锦绣项目

项目地点：云南　江川
开发公司：云南龙湖地产
设计时间：2011年
设 计 人：柴 晟、赵 伟、何运祥、王宇熙、徐中华

Xianhu Splendor Project, Yunan

Location: Jiangchuan, Yunnan
Developer: Yunan Longfor Property
Design Time: 2011
Designers: Chai Sheng, Zhao Wei, He Yunxiang, Wang Yuxi, Xu Zhonghua

该项目是龙湖集团开发的第一个旅游地产项目，地块紧邻中国水体最大的抚仙湖，重峦叠嶂，风景优美。项目总面积250万平方米，包含一个13公顷的高尔夫球场。围绕球场与抚仙湖，龙湖集团计划修建三座五星级酒店和会所，面积达30万平方米的滇南风情商业街，以及可用做小型酒店经营的度假别墅和公寓，成为世界上最大的度假酒店群。

This project is the first tourist property project developed by Longfor Group, which plot is close to the biggest water body of China, the Fuxian Lake, where chains of hills surround, forming a beautiful landscape. The project total area is 2,500,000 m^2, including a golf course 133,340 m^2 large. Longfor Group plans to build three 5-star hotels and clubs around the golf course and Fuxian Lake, in addition to South Yunnan Folks Shopping Street 300,000 m^2 large, as well as resort villas and apartments operating as mini hotels, which will become the largest resort hotel groups of the world.

華藝設計顧問有限公司
HUAYI DESIGN CONSULTANTS LTD.

公司简介

华艺设计顾问有限公司1986年在香港注册，同年在深圳独资设立香港华艺设计顾问（深圳）有限公司，设有上海、南京、武汉、北京、重庆和广州分公司，是具有甲级工程设计资质的工程设计咨询企业（证书号为：AW144017173）。2009年成立的北京中海华艺城市规划设计有限公司持有"城乡规划编制甲级资质证书"[建]城规编第（101205）。

华艺是拥有500多人的高质素专业设计团队，设有规划、建筑、结构、强电、弱电、给排水、暖通空调、总图、概预算、室内设计等专业，主要承接各类公共与民用建筑工程设计、城市设计、居住区规划与住宅设计、室内设计及前期顾问和建筑策划研究等业务。公司凭借先进的管理，国际视野的底蕴，具有强大的竞争力。发展足迹已遍布全国。

华艺多年来坚持"创意为先、质量为本"的质量方针，赢得了客户、政府、合作伙伴等社会各界的广泛赞誉和尊重。先后共有110多个项目获得260余次国家、省部级及深圳市优秀设计奖。

Company Profile

HUAYI Design Consultants Ltd. (HUAYI) was registered in Hong Kong in 1986, and in the same year the wholly-owned company in Shenzhen was established, named Hong Kong HUAYI Design Consultants (Shenzhen) Ltd.
HUAYI today is an engineering consultant enterprise with Grade A Construction and Engineering Design License, with branches in Shanghai, Nanjing, Wuhan, Beijing, Chongqing and Guangzhou.
Beijing China Overseas HUAYI Urban Planning & Design Ltd. (HUAYI Planning) was established in 2009, with the national Class-A Qualification Certificate for Urban Planning.

HUAYI has more than 500 highly-educated, and experienced personnel, the design team has various professions of planning, building, structure, heavy-current, light-current, water supply and drainage, HVAC, general layout, engineering budget and interior design, mainly undertaking businesses of public and civil construction design, urban design, residential planning and design, interior design, pre-consultancy and construction strategy researches. HUAYI showed the outstanding competence with advanced organization and international sights.

HUAYI sticks to its quality guideline, which is Creativity First, Quality Oriented, and gained acclaims and respects from customers, governments and cooperative partners. Over 110 projects from HUAYI have won more than 260 national rewards, provincial or Shenzhen rewards.

25

1986-2011

www.huayidesign.com

南京栖园
CHI GARDEN / NANJING

莱蒙水榭春天
TOPSPRING THE SPRING LAND / SHENZHEN

中海金沙熙岸
COHL BLOSSOM RIVERIME / FOSHAN

招商华侨城曦城
CMOCT BUENA VISTA / SHENZHEN

中海阳光玫瑰园
COHL ROSARY / SHENZHEN

东莞松山湖长城世家
SONGSHAN LAKE RESIDENCE COMMUNITY / DONGGUAN

深圳长虹研发大厦
RD MANSION OF CHANGHONG / SHENZHEN

深圳安联大厦
ANLIAN BUILDING / SHENZHEN

中国海洋石油大厦
CHODC MANSION / SHENZHEN

深圳赛格大厦
SEG PLAZA / SHENZHEN

海口行政中心
ADMINISTRATIVE CENTER OF HAIKOU / HAINAN

成都南湖国际酒店
NANHU INTERNATIONAL HOTEL / CHENGDU

中海龙岗圣廷苑酒店
COHL LONGGANG PEVILION HOTEL / SHENZHEN

三亚阳光大酒店
GRAAN SOLUXE HOTEL & RESORT / SANYA

安庆市明珠广场
PEARL PLAZA / ANQING

济南中海广场
COHL PLAZA / JINAN

贵阳金融中心
FINANCIAL CENTER / GUIYANG

济南恒隆广场
HENG LONG PLAZA / JINAN

深圳保荷医院
BAO HE HOSPITAL / SHENZHEN

龙岗妇儿保健院
LONGGANG MATERNAL & CHILD CARE SERVICE / SHENZHEN

深圳新安医院
XIN AN HOSPITAL / SHENZHEN

龙岗区人民医院住院大楼
LONGGANG PEOPLE HOSPITAL / SHENZHEN

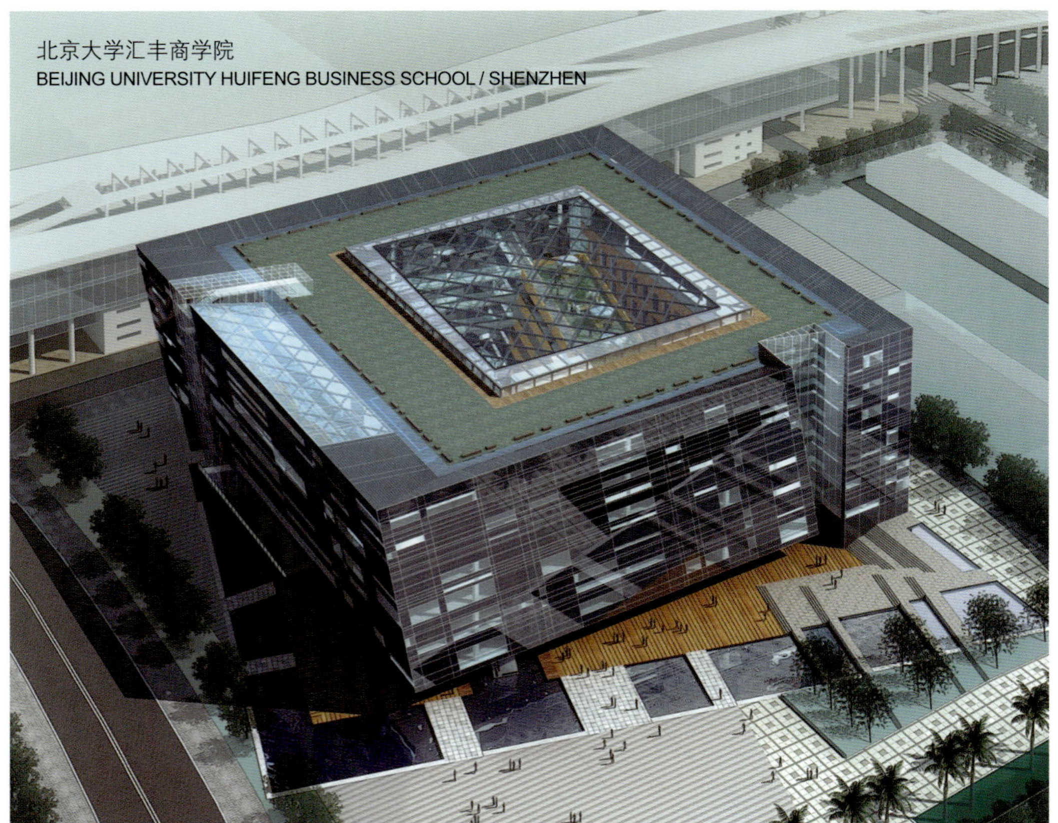
北京大学汇丰商学院
BEIJING UNIVERSITY HUIFENG BUSINESS SCHOOL / SHENZHEN

大鹏地质博物馆
DAPENG GEOGRAPHY MUSEUM / SHENZHEN

长春市城市规划展览馆
CHANGCHUN CITY URBAN EXHIBITION HALL / CHANGCHUN

深圳市福田区图书馆
FUTIAN LIBRARY / SHENZHEN

深圳大学基础实验室一期
THE BASIC LABORATORY OF SHENZHEN UNIVERSITY / SHENZHEN

北京中海华艺城市规划设计有限公司
Beijing China Overseas HuaYi Urban Planning and Design Ltd.

公司简介

北京中海华艺城市规划设计有限公司是中国海外发展有限公司（港股代码HK0688）旗下以城市规划设计为主的国有设计企业，具有国家住房与城乡建设部颁发的城乡规划编制甲级资质（证书号101205）。

依托香港华艺设计顾问（深圳）有限公司25年的技术积累与社会资源，秉承国际化城市规划的先进理念，高起点介入城市规划设计行业，承担了大量的规划设计项目，涵盖了城市发展战略研究、城镇体系规划、城市总体规划、近期建设规划、分区规划、控制性详细规划、修建性详细规划、城市设计、概念规划以及城市规划管理方面的专题研究等领域，取得了可喜的成果，赢得了业界的中肯评价。

伴随着业务的发展，公司继承发扬华艺设计的传统与作风，在组织建设、管理制度、技术研究方面下工夫，努力建设创新型、研究型、可持续发展的规划设计平台，组建了一支作风扎实、技术精湛、务实创新的规划设计队伍。

华艺规划将一如既往，坚持新概念、高起点，以需求为导向的研究路线、以规划与建筑整合的设计特色，为社会提供符合时代发展趋势并具操作性的专业规划技术服务。

Company Profile

Beijing China Overseas HUAYI Urban Planning and Design Ltd is a professional design firm under China Overseas Group (HK Share No.: HK0688). It has been awarded the Chinese National Class-A Qualification Certificate for Urban Planning (Certificate No. 101205).

Since its establishment, HUAYI Planning has been affiliated to HongKong HUAYI Design Consultants (Shenzhen) Ltd. Combining with HUAYI Design over 25 years of successful experience; HUAYI Planning has accomplished many highly recognized projects. HUAYI Planning has a wide range of projects, ranging from development strategy consultant, master planning, system planning, immediate plan, district planning, regulatory detailed planning, to urban design, site plan, conceptual planning and urban planning management.

The traditional philosophy of HUAYI Design has put a strong emphasis on design management and talent resources. It gives HUAYI Planning a starting point to establish a dynamic, professional and creative design team, and continues a customs to encourage skills improvement and progression.

HUAYI Planning will take full advantage of our comprehensive planning and architecture design ability and market familiarity, and provide professional services to all our clients.

Shenzhen University Institute of Architectural Design & Research

深圳大学建筑设计研究院

深圳大学建筑设计研究院成立于1984年，是建设部批准的全国建筑工程甲级设计单位，主要从事城镇规划、居住区规划设计、建筑工程设计和施工图设计文件审查、室内外装修设计和园林景观设计，涉及建设前期咨询、规划研究、社区评价和建筑技术咨询等研究课题，完成的项目遍布全国十数个大中城市。

深圳大学建筑设计研究院依托于深圳大学广博的学术、科研和教学资源，集设计、教学、科研为一体，以实力雄厚的建筑系和规划系为后盾，并作为建筑系、规划系的教学、科研和实践相结合的基地，有多学科的最新科研成果作为支持，走出一条生产、教学、科研三赢的局面，其学术研究、科技成果转化和建筑设计的水平在深圳市名列前茅。

我院尤其重视人才的选拔、培养和使用，吸收了一大批博士、硕士和国外学成归来的优秀人才，形成了一支结构合理、专业配套的设计队伍，在广东省内同行业中具有较强的竞争力。我院现有各类专业技术及管理人员245人，国家一级注册建筑师35人，一级注册结构工程师12人；教授级高工12人，高级工程师55人；博士后、博士、硕士研究生42人，留学归来的设计人员10多人。

我院坚持与深圳大学建筑系、规划系紧密结合，设计实践和学术研究并行，至今我院超过百人次出国专业考察和参加国际学术会议，参与完成了多项国家、省、市规范，规程，地方标准和设计资料集的编制工作，包括《高层办公综合建筑设计》、《实用高层建筑结构设计》、《建筑设计资料集》、《高层混凝土建筑结构设计规程》、《高层建筑结构资料集》、《建筑抗震设计规程》、《深圳建筑防水构造图集》、《地下工程防水技术规范》、《建筑防水工程技术规程》等，承担了国家、省、部级科研课题数十项，担任了市政府多项大型工程的顾问工作，并与SOM、ARQ、HOK等在内的多个国际著名建筑事务所和包括万科、中海、金地、卓越集团、佳兆业、振业等在内的许多著名房地产开发商有全方位的合作关系。

深圳大学建筑设计研究院坚持树立创新创优的精品意识，争创名牌设计院，建立了一整套完善的科学管理制度和严格的质量保证体系，1995年我院被深圳市评为"勘察设计单位综合实力50强"第三名，2000年通过ISO9001质量管理体系认证，2002年和2004年被深圳市企业评价协会评为中国深圳行业十强企业。

The Shenzhen University Institute of Architectural Design & Research (SUIADR) was established in 1984 and it is authorized A design level by the national ministry of construction. The main services provided by the institute are town planning, residential planning, architectural design, construction planning, interior design, landscape design, planning and design consulting, etc. Projects finished are located in many cities in China, and the types include public, commercial, sports, academic and education. Projects we have finished are located through the whole country included fields from public architecture, commercial architecture, sports architecture, educational architecture and residential architecture.

After twenty year's growth, we have a strong design team composed by design experts. Currently we have 245 staffs, and among them, 35 people are registered architects, 12 people are registered structure engineers, 12 people are professors, 55 people are senior engineers, 42 people own master or higher degree, and 10 people have overseas education background. We have been emphasizing on combining research, academic and practice, and created a good base. Many of our staffs have been to other countries for international conference and anticipated planning and design projects there. We also published many academic books and they became good reference for the design field in China. Besides above, we have established good cooperation relationship with many real estate developers and architect firms, and many of our staff have visited other countries for conference or research.

We have won many awards, including 80 from the city to the national level. Some of the residential design projects won awards. We have become one of the most famous design institutes in China, especially in south east region. We were awarded No. three in the "top 50 design institutes" in Shenzhen in 1995, and in 2000, we passed ISO9001 qualification management system. In 2002 and 2004 we were awarded one of the top ten companies in Shenzhen.

单位地址：深圳大学建筑与城市规划学院院馆B座
资质范围：建筑行业（建筑工程）专业甲级
联系电话：+86-755-26732800
网　　址：www.suiadr.com

Add: B tower, office building of Architecture and Urban Planning College, Shenzhen University
Qualification Range: First Grade of Architecture and Engineering in Architecture Industry
Tel: +86-755-26732800
http://www.suiadr.com

深圳市东江环保环境科学研究与工程开发中心

建设地点：广东 深圳
用地面积：4 999.98平方米
总建筑面积：20 404平方米
设计时间：2006年
竣工时间：2009年
主　　创：高 青

Shenzhen Dongjiang Environmental Science and Engineering R&D Center

Construction Site: Shenzhen, Guangdong
Land Area: 4,999.98 m^2
Total Floor Area: 20,404 m^2
Design Time: 2006
Completion Time: 2009
Major Creator: Gao Qing

设计创意

力求将城市空间及用地特征、相邻建筑与环境特点，与业主的功能需求、企业文化特色结合起来，通过建筑造型语言，展示作为科技型环保企业研发建筑的文化内涵。

Design Idea

Making effort to combine the owners of the functional requirements corporate culture with urban space and land features, adjacent buildings and environmental character, through architecture modeling language, show the culture connotation of environmental science and technology R & D buildings.

深圳外国语学校国际部

Shenzhen International Department of Foreign Language School

建设地点：广东 深圳
建筑面积：41 971平方米
占地面积：24 103平方米
设计时间：2009年
竣工时间：2011年
主　　创：刘尔明

Construction Site: Shenzhen, Guangdong
Floor Area: 41,971 m^2
Land Area: 24,103 m^2
Design Time: 2009
Completion Time: 2011
Major Creator: Liu Erming

项目介绍

深圳外国语学校国际部项目是一所接受外籍人士子女就读、采用美国课程体系的国际学校。办学规模为54个班、1 080个学生、教职工142人，包括幼儿部、小学部、初中部三个教学区块。 为了创造一个能够在玩耍中学习的多层次和有趣的空间，运用了庭院概念来组合各种功能，如教室、体育馆、剧场、食堂以及教师公寓。用动态的连廊将庭院、室内平台和屋顶花园联系起来，使建筑空间不仅反映了孩子们灵活的学习方式，同时应于当地气候。

Project Introduction

Shenzhen Foreign Language School International Department is a international school, accepts expatriate children attended, using the American curriculum. There are 54 classes, capacity of 1080 students, 142 teacher staff, including kindergarten, elementary, junior School three teaching blocks. In order to create a multi-layered and interesting space where students can study through playing, it used the courtyard concept to combine a variety of functions, such as classrooms, gymnasium, theater, cafeteria and teacher apartments. Corridor to the courtyard with a dynamic, platform and a roof garden room to link the building space not only reflects the children's flexible approach to learning, but also adapt in the local climate.

深圳市老干部活动中心

建设地点：广东 深圳
建筑面积：25 988.85平方米
占地面积：30 190.2平方米
设计时间：2007年
竣工时间：2010年
主　　创：杨文焱
设计方案助理：李蔚波
施工图设计助理：俞培柱

Shenzhen Veteran Cadres Activity Center

Construction Site: Shenzhen, Guangdong
Floor Area: 25,988.85 m^2
Land Area: 30,190.2 m^2
Design Time: 2007
Completion Time: 2010
Major Creator: Yang Wenyan
Design Assistant: Li Weibo
Construction Design Assistant: Yu Peizhu

项目介绍

深圳市老干部活动中心位于深圳市中心繁华地段，用地地块呈规整的矩形，南北长于东西，80%是绿化水面，为成片数十年树龄的荔枝树覆盖。设计根据项目和用地特点，确立尽量保留园内绿化，营造符合老年人生理、心理的宁静、平实和安详的现代建筑空间氛围的设计目标。因此，建筑的布局沿原建筑基址布置，东、北、西三面围合，同时形成三个带状活动分区。由于地处城市中心区，内外气氛迥异，建筑处理遵循内外有别的原则。建筑面向园外的面做封闭处理，尽量减少墙面开洞面积，以屏蔽来自城市干道的噪声和不利朝向的影响，形态上也以简单的直线构成。面向园内则在控制遮阳的前提下，尽量开敞，并根据具体情况向外凸出，以争取尽可能多的与园林绿化的接触面，并在转角处加入曲面元素，力求跃动的感觉，使建筑更好地融入绿化环境中。建筑以暖色为基调，处理上延续"双重原则"，软黄色面砖与白色涂料交错：临城市的外界面，白色涂料点状镶嵌在大面积的面砖上，厚实而不失活跃；临园区的内界面则以白色涂料的线、面体与面砖交错编织，以获得更强的动感。

Project Introduction

Shenzhen Veteran Cadres Activity Center located in Shenzhen downtown area, was a neat rectangular plots of land, the distance from north to south is longer than that from east to west, 80% of which is water, covered by Litchi tree of decades age. According to features of the project and the land, the design established to preserve the green of park to create a line with the elderly physical and psychological quiet, modest and serene atmosphere of the modern architectural space design goals. Therefore, the layout of new buildings along the base layout of the original buildings, east, north and west sides of the enclosure, while the formation of three band active partition. Because it is urban central area, both inside and outside the atmosphere are different, the construction process to follow the principle of differentiated. Surface outside the building for the park closed deal to do to minimize the area of wall openings, in order to shield the noise from the city roads and adverse effects of orientation, is also a simple form of straight lines. The premise control for the park in the shade, as open and protruding outward depending on the circumstances, to gain as much contact with the landscape, and adding surface element in the corner, and strive to feel vibrant, so building can better financial into the green environment. Built of warm tone, the handling of the continuation of "dual principle", the soft yellow brick and white paint staggered. the surface outside near the city, the white paint dots embedded in a large area of brick, solid but active; while interface within the park places the line of white paint, body and face brick are woven in order to get more dynamic.

深圳中海世纪建筑设计有限公司

Shenzhen CHC Architectural Design Co., Ltd.

深圳中海世纪建筑设计有限公司(CHC)成立于1999年，其前身为深圳市中雅图设计有限公司，经过多年的辛勤耕耘和市场磨合，已经发展成为一个综合性的建筑设计公司，并已取得国家建筑行业（建筑工程）甲级设计资质、城市规划乙级资质。CHC公司由一批年富力强的资深设计师组成，拥有良好的专业素质和优秀的教育背景，富有创造力。所涉及的领域包括城市设计、区域规划及各类公共建筑、住宅建筑、商业建筑、文化建筑的设计，取得了一批丰硕的设计成果，获得了良好的社会效益。

近期，我们将集中力量与一批高等院校合作，致力于政府安居房、城市更新等领域的专项研究，贴近市场、服务社会，公司于2010年通过了ISO9001国际质量体系认证。

CHC公司业务范围已覆盖了中国境内十余省、市、自治区的重点城市，并且致力于建立一个开放的平台，与各地区的优秀设计团队合作，发扬"创新、质量、服务"的中海世纪精神，为客户提供一流的设计服务。因为我们深信，设计优化生活。

2010年，深圳中海世纪建筑设计有限公司（CHC）站在了新的起点，迎来了新的挑战，CHC将致力于打造一个开放式的发展平台，在建筑设计的领域里置换一种全新的理念，创建一种全新的模式，以技术和管理占领市场，开拓创新，再创辉煌。

Shenzhen CHC Architectural Design Co., Ltd. was established in 1999, the former is Shenzhen Zhongyatu Design Co., Ltd. Now, Shenzhen CHC Architectural Design Co., Ltd. has become a comprehensive architectural design company by many years' market research and hard working. In addition, it has achieved the A grade certification on national architectural industry (construction engineering), Qualification B in City Planning. CHC has been composed by a group of experienced and passionate designers, who have good specialized quality, outstanding educational background and great creativity capability. We are programming on urban designing, region planning and public, housing, business and civilization construction design, in which has achieved great design results, and good society reputation.

Recently, we are focusing on projects through cooperating with universities, which details on government economically affordable housing, urban re-construction and so on. To be market orientation, serve for society, the corporation achieved the ISO9001 International Quality Management System Certification in 2010.

The business of CHC has covered important cities of over ten provinces in China, also we has established an open platform, strength the cooperation with other great design teams, carry forward "innovation, quality, service" spirit of CHC, provide the higher class design service, because we believe design optimizes life.

In 2010, Shenzhen CHC Architectural Design Co., Ltd. stepped ped the new start, faced the new challenge, aims to establish an open developing platform, and create new model instead the old one in the architectural design area, guide the market as technology and management, development and innovation to achieve greater performance.

单位名称：深圳中海世纪建筑设计有限公司
地址：广东省深圳市福田区上步北路1028号多彩城16楼
电话：+86-755-22274118
传真：+86-755-22274289
邮箱：chc755@163.com
网址：www.chcsz.com
Name: Shenzhen CHC Architectural Design Co.,Ltd.
Add: Colorful city, No.16 Floor, Shangbu North Road Futian District, Shenzhen, Guangdong
Tel: +86-755-22274118
Fax: +86-755-22274289
E-mail: chc755@163.com
Http://www.chcsz.com

■ 余錾经
教授级高级建筑师
中国特许一级注册建筑师
国务院政府特殊津贴专家

■ 张文华
高级建筑师
公司总经理

■ 吴科峰
国家一级注册建筑师
南昌大学建筑学学士

■ 赵献忠
国家一级注册建筑师
清华大学建筑学学士

■ 段玮娜
建筑工程师中级职称
深圳大学建筑学硕士

■ 姚金凌
教授级高级建筑师
中国建筑学会建筑创作大奖获得者

■ 时芳萍
国家一级注册建筑师
教授级高级建筑师

■ 梁　呐
同济大学建筑学博士
国家一级注册建筑师

■ 万　强
国家一级注册建筑师
高级建筑师

■ 宋兴彦
国家一级注册建筑师
高级建筑师

■ 胡振中
国家一级注册建筑师
教授级高级建筑师

■ 周志仪
南昌大学建筑学硕士
高级建筑师

■ 蒋雪枫
国家一级注册建筑师

■ 陈　松
国家一级注册结构师工程师

■ 熊　冲
国家一级注册结构工程师

■ 戴征志
国家一级注册结构师高级工程师

■ 舒克强
国家一级注册结构师、高级工程师

■ 潘桂平
国家注册电气高级工程师

■ 丁东成
国家注册电气高级工程师

■ 章飞昔
国家注册给排水工程师

■ 项智宇
重庆大学城市规划与设计硕士
高级规划师

■ 余　华
英国Northumbria University项目管理学硕士
高级经济师

上海公司　深圳公司　烟台公司
南昌公司　CHC中海世纪　昆明公司
东莞公司　成都公司　重庆公司

JIANGXI ROAD TRAINING CENTER
江西公路培训中心

业　　主：江西公路开发总公司
功能性质：办公、酒店、公寓
建筑面积：120 000平方米
建设地点：江西 南昌
设计时间：2005年
建成时间：2010年

和平的使者应当是花园而不是鸽子，我们谋求把自然引入城市，而且力求把自然引入建筑内部。通过"场所精神"的建立，找回人对城市空间的拥有权和以往失落的人情味。人与自然共生，人与文化共享，人与科技共荣，营造高品质的办公、科研环境。

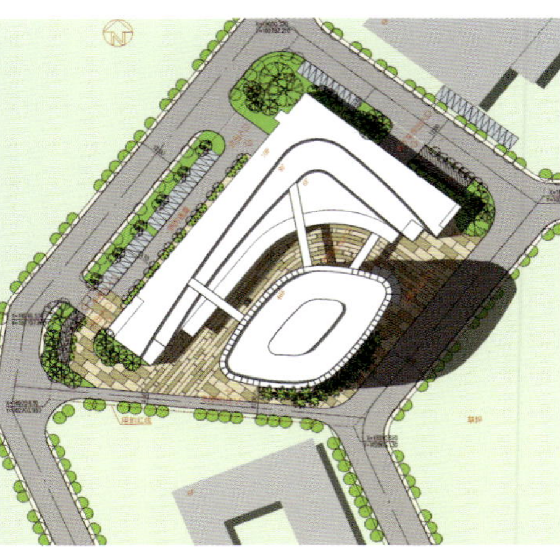

DASHI BUILDING
达实大厦

功能性质：办公
建筑面积：97 400平方米
建设地点：广东 深圳
设计时间：2011年

空中花园通过镶嵌，一组大小不同的"盒子"，如同一个个的橱窗，层叠而上，直通天际，在进与退之间形成一个比较粗糙的面，与光滑的玻璃墙形成鲜明的对比。

在建筑的底部保留原有的建筑曲线，新建建筑与之呼应，并且在底六层局部设计六层高的大堂，形成比较通透的视觉空间，在两组建筑之间保留室外的通廊空间。

在建筑物的顶部设计一处精美的企业会所，如一颗熠熠生辉的明珠，喷薄而出，预示着企业像一颗冉冉升起的新星，光辉照耀大地，达到兼善天下。

LAN TIAN JUN

蓝天郡

业　　主：江西民航置业
功能性质：商业住宅公寓
建筑面积：220 000平方米
建设地点：江西　南昌
设计时间：2006年
建成时间：2009年

项目以环境为中心，商业、广场、居住区三者完美的结合，多种建筑形式，使该项目成为当地最成功的开发项目之一。

TIME SQUARE

时代广场

业　　主：江西丰威置业有限公司
功能性质：办公、商业
建筑面积：50 000平方米
建设地点：江西 南昌
设计时间：2005年
建成时间：2008年

时代广场项目用地位于南昌市胜利路步行街和民德路交汇处，妇儿商店和市公安局原址，属于旧城改造的范畴，地理位置处于城市的中心区域，是商业旺区。

该项目用地对于城市景观要求较高，胜利路两侧建筑都具有一定的历史价值，为西式仿古建筑，因此本项目既要有时代的特点又要继承部分历史的文脉。

该项目地处黄金旺地，它的形象无论是对项目的本身还是对整个城市的景观都起到至关重要的作用。由于地处繁华的胜利路，其中骑楼、钟楼和欧洲古典建筑的一些符号形成胜利路的独特风景线，该项目本着对历史文脉的尊重，采用竖向三段式的构图和一些哥特式的符号：直指天空的尖攒、拱门、钟楼和壁柱等构成自己独有的气质——给人一种向上澎湃的气势，古朴的石材和精细玻璃的对比，金碧辉煌的钟塔和整个建筑群的高低起伏构成优美的天际线，显示皇家的气派。项目的落成给胜利路增添了亮丽的一笔。

LINGXIU ERA
领秀时代

业　　主：株洲市上景地产
功能性质：住宅
建筑面积：300 000平方米
建设地点：湖南 株洲
设计时间：2010年

通过设计建成更美好、更舒适的居住建筑，尊重原有环境，保留原有山势，人性化的交通组织，溪谷森林，利用原有峡谷、山脉、森林营造偏自然的景观，会呼吸的建筑空间，层层叠叠的建筑体系。

JINGUI GARDEN

金桂园

业　　主：金桂置业房地产开发有限公司
功能性质：商业、住宅
建筑面积：100 000平方米
建设地点：河南 驻马店
设计时间：2009年

时尚现代的气质，使一块狭小的空间得到完美的展现，成就了城市一道亮丽的风景。

TIANDU STAR GARDEN

天都星城

业　　主：江西民生集团
功能性质：住宅
建筑面积：800 000平方米
建设地点：河南 驻马店
设计时间：2007年

SUZHOU GUANGCAI TOWN
宿州光彩城

业　　主：宿州生生置业有限公司（江西民生集团）
功能性质：商业、住宅
建筑面积：1 020 000平方米
建设地点：安徽 宿州
设计时间：2005年
建成时间：2010年

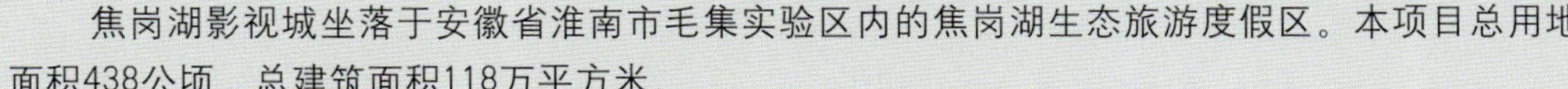

焦岗湖影视城坐落于安徽省淮南市毛集实验区内的焦岗湖生态旅游度假区。本项目总用地面积438公顷，总建筑面积118万平方米。

项目共分十个区域：

1. 影视拍摄区
2. 饮食文化博览区
3. 嘉年华旅游区
4. 花博景观区
5. 声光电娱乐区
6. 宗教文化区
7. 皖省地域文化艺术区
8. 健身休闲区
9. 生态民居园区
10. 文化旅游产品购物区

JIAOGANG LAKE STUDIO

焦岗湖影视城

功能性质：影视基地、酒店、公寓
建筑面积：1 180 000平方米
建设地点：安徽 淮南
设计时间：2010年
设计成员：余鎏经、项智宇、刘志平等

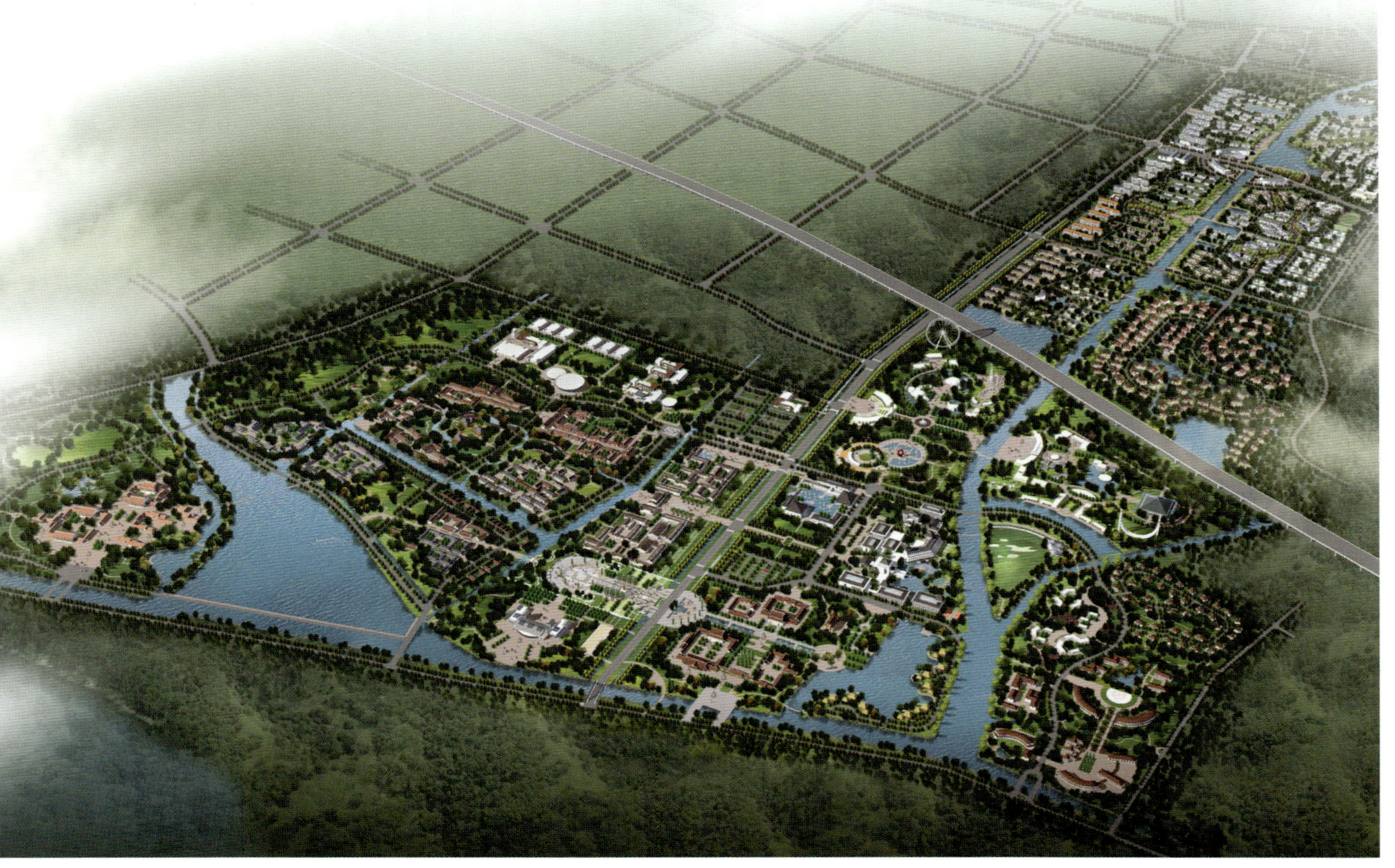

BCCI

BEIJING CCI
ARCHITECTURAL DESIGN

北京中外建建筑设计有限公司深圳分公司

北京中外建建筑设计有限公司深圳分公司2006年成立，坐落于风景宜人的深圳华侨城北部——中航沙河LOFT创意园，拥有3 000平方米的办公区域，与众多知名文化产业公司毗邻，目前在册员工170人。

北京中外建建筑设计有限公司深圳分公司拥有稳定的、经验丰富的设计队伍，先后与万科集团、金地集团、华侨城地产、花样年集团、合正地产、佳兆业地产、卓越集团、中海地产、新世界集团、承翰地产等大型知名地产开发公司合作。

与社会、与环境共生是我们的经营理念，在创作上，我们不简单追求冲击人感官的“奇观景象”，而是将建筑编制进环境的肌理当中。

我们以建筑实践活动参与文化语境和专业价值体系的建设。

Try to build a dialogue platform for the people, express the real and pure pursuit for sports spirits.

建设地点：深圳 南山区
设计时间：2006–2007年
用地面积：20 190平方米
建筑面积：5 130平方米
业　　主：深圳市华侨城房地产开发有限公司

建设地点：山东 潍坊
设计时间：2010年
用地面积：1 994 400平方米
建筑面积：1 114 082平方米
业　　主：山东新建业集团有限公司

International YACHT Club Weifang Shandong

山东潍坊国际游艇会

Combination of the "energetic and passionate" of city life and natural ecology.

贯穿南北东西向的城市主干道便捷地将都市人引入自然的宁静，消除生活的压力。本案将会极大地提升海滨城市的形象，将成为低碳生态、功能完善、循环经济、宜居宜业的现代化高效生态海滨新城的重要组成部分。

本项目位于临沂市解放路与通达路的交汇处，是集商业、办公、住宅和停车楼于一体的商业综合体。设计灵感来自奔腾向前的沂水的鸟瞰自然姿态，水蓝色的办公大厦有如水自天上来的源泉，使得建筑造型既予人亲切自然之感，更营造出热诚和谐的商业氛围，给文化古城增添新的地标景观。

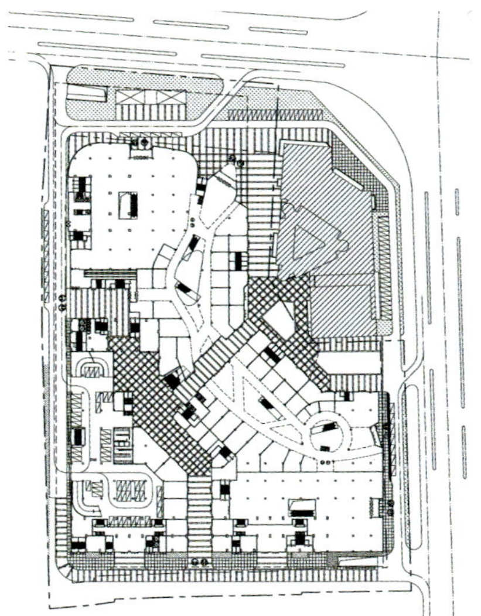

Yinzuo Shopping Centre of Shandong

山东银座购物中心

建设地点：山东 临沂
设计时间：2006–2009年
用地面积：40 404平方米
建筑面积：247 712平方米
业　　主：山东银座地产有限公司

建设地点：山东 临沂
设计时间：2009年
用地面积：155 398平方米
建筑面积：901 298平方米
业　　主：临沂银泰投资有限公司

Organic integration of four function divisions to create a complete business leisure shopping ecological chain.

Linyi Intime Center

临沂银泰中心

项目位于山东省临沂市南坊新区市政府行政中心北侧，处于该片区城市主景观轴上。设计充分考虑整体城市规划，强调城市轴线。在城市总体规划的基础上，将用地划分为商业、酒店、办公、住宅四个功能分区，通过各个功能分区的有机整合，加入体验式的城市商业步行街，打造一个完整的商业休闲购物生态链。通过混合功能的规划设计达到经济效益最大化。而简洁大气的现代主义摩天楼成为现代化都市的一张名片，创造了临沂标志性的建筑群，使项目的品质与众不同。

Nanning Hangyang Xinhe Mansion

南宁航洋信和大厦

塔楼看似随意的横切及水平错动将建筑分隔成数个轻盈的小体块，通过表面肌理变化带来灵动的生机，露天之处形成了令人惊喜的空中花园，催生了一种全新的空中花园式生态办公模式，营造出独特的空间体验。

建设地点：广西 南宁
设计时间：2010年
用地面积：24 310平方米
建筑面积：320 817平方米
业　　主：广西航洋信和置业有限公司

建设地点：广东 深圳
设计时间：2006–2007年
用地面积：20 934平方米
建筑面积：2 698平方米
业　　主：深圳华侨城房地产有限公司

Shenzhen OCT Tianlu Ninth District

深圳华侨城天麓九区

在规划设计中，结合基地坡地地形的特点，我们尝试探讨一种现代的合院生活方式：建筑在最大限度上尊重自然，并融入自然中。建筑本体采用多层次庭院的组合方式，创造一个和自然充分交融的界面，力图让居住在其中的人们感受到中国传统的高品质院落生活。在规划中保留了坡地厚生植被，居住者根据需要改造景观，真正实现了天人合一的居住梦想。

Jinxiu Pan King Valley

金秀盘王谷

建设地点：广西 来宾
设计时间：2010年
用地面积：19 817平方米
建筑面积：26 408平方米
业　　主：金秀瑶族自治县圣景旅游产业开发有限公司

项目以瑶文化为主题，由一期五星级度假酒店，二期公寓、高尚别墅区共同组成。规划设计、建筑形态以及景观的营造都使社区在不同界面呈现一种自然发展的生态感和层次感。设计强调私密性，在充分争取良好景观视野的前提下，避免视线、交通干扰，营造惬意的度假环境。

Image and landscape arrangement make the block with natural ecological and layer sense in different interfaces.

iNgAmE Office Ltd. in Shenzhen

深圳市局内设计咨询有限公司

“局内”源于英文“in game”，直译为：“在游戏之中”。我们的宗旨：理解建筑设计的实质性“游戏规则”。首先，在设计方法上，我们除了用专业知识与经验为业主创造价值之外，更习惯与业主建立畅通充分的交流渠道，确保完全了解业主直接或潜在的需求，为其提供最适当、最完美的解答方案；其次，我们认为，设计的价值不应该仅仅是设计师的个体艺术修养及美学创造力的表达，我们更感兴趣的是探索、挖掘每个设计任务中的独特元素，并将其潜在的机会用设计汇成价值，更令人兴奋的是创造性地转化“不利因素”或“限制条件”为作品独特设计的引爆点。

我们不相信流行与时尚，只相信恰当的，精明的，才是智慧的设计。局内已经并将持续进行独立的、内部的学术交流，技术知识系统的建立、梳理与更新，以确保我们的专业服务保持最前沿的领先性。我们绝不满足于已经收获的成功，而是从超越自我中获得满足。我们喜欢用批判的眼光审视所有已经实现的作品，包括自己的与他人的。

设计是一种生活态度乃至哲学，而不是一种职业或技能，这只不过是实践设计哲学的手段而已，而好的设计应该是这种态度的载体。

局内iNgAmE设计于2004年在美国波士顿成立，主要设计师来自哈佛大学、耶鲁大学、多伦多大学、清华大学等著名高校，有良好的教育背景和专业能力，有着丰富的建筑实践水平和宽阔的国际视野。公司曾先后与多家国内外知名建筑事务所合作，如：荷兰大都会建筑师事务所（OMA）、日本隈研吾建筑师事务所、美国EDAW/AECOM、美国CITYMARK/AECOM、CCDI中建国际、英国AA/GROUNDLAB等。为了更好地服务于中国客户，2007年底正式在中国深圳成立局内设计咨询有限公司。

“iNgAmE” can be understood as “in the game”, meaning we happily play by the rules that are embedded in the field of architecture. Our design philosophy is to identify the “rules” that are inherent in architectural design. These rules are in no way restrictions but merely mark the starting point.

We first approach design by establishing good communication with our clients to fully understand their needs, apparent and latent, in order to provide the most suitable solution. Secondly, we believe that the most valuable part of a design is not necessarily personal style, or personal expression of the architect, rather, we spend the time to discover and explore the unique elements to each project, and transform what might be perceived as disadvantages to the highlights of a project.

We don’t believe in what’s fashionable, but what is appropriate, smart, and wise. iNgAmE has and continue to work internally to pursuit academic interest in architecture. It ensures our firm’s fresh perspective of the industry with a critical eye to our and other’s work. The only challenges are how to outdo ourselves in the next round.

Design is an attitude or life philosophy even. Don’t dare to think for a second that it’s just an occupation or a way to make a living. Good design should reflect that attitude.

iNgAmE was established in Boston U.S, 2004. Major designers were graduated from Harvard, Yale, Toronto University, Tsinghua University, etc. With well-educated background and global professional experiences, iNgAmE has collaborated with many outstanding professional firms, such as OMA, Kengo Kuma, EDAW/AECOM, CITYMARK/AECOM, CCDI, AA/GROUNDLAB etc. iNgAmE Shenzhen office, China, was founded in 2007.

地址：广东省深圳市南山区华侨城东部工业区B-1栋606-608
邮编：518053
法定代表人：张人铤
电话：+86–755–86096263
传真：+86–755–86096263
邮箱：ingameoffice@gmail.com
网址：www.ingameoffice.com

Add: Rm 606 - 608, Building B - 1, Eastern Industrial Area, OCT, Nanshan District, Shenzhen, Guangdong
P.C.: 518053
Legal Representative: Zhao Renting
Tel: +86-755-86096263
Fax: +86-755-86096263
E-mail: ingameoffice@gmail.com
Http:// www.ingameoffice.com

安徽置地·栢悦中心（办公综合体）

建筑面积：360 000平方米
项目地点：安徽 合肥
项目状态：概念设计

项目地处合肥中央政务区，是文化金融中心，是未来城市最重要的地标性建筑。整个项目由一栋250米高及4栋80到160米高的办公楼围合成一个尺度超大的中心花园，并且将主花园依照其功能进行合理的划分，整个基地内将营造出尺度从大到小，功能从动到静并具有丰富竖向变化的、多层次多功能的、丰富的城市空间序列。它们包括精品时尚广场、购物商业街、城市商务广场、文化艺术环以及屋顶商务花园等等。

北京CBD某办公楼

建筑面积：213 600平方米
项目地点：北京
项目状态：投标方案

该项目位于北京朝阳核心区，该区为北京市商业金融区，也是朝阳区政治、经济、文化中心。

在建国路上形成独特的“城市舞台”门户形象，整个建筑的基座是由两个35米高的立方体围合成一个高达70米震撼气派的底层大堂，承托三个相互垒叠的高度分别为40米、60米、130米的“立方环”。顶层130米的“立方环”围合了一个高80米、宽18米、深70米贯通东西的金色“办公社区共享空间”，建立了办公塔楼与中心绿地互动对望的窗口；中区60米的“立方环”内植入了一个高20米、宽18米、深70米贯通南北的“总部展示大厅”，适合进行大型的展示以及集会活动。南侧面对建国路的开口使展厅变成了一个面对城市的舞台；低区40米高的“立方环”朝下的开口与基座围合的空间共同联系形成了一个高达75米的入口大堂空间，惊人尺度与开放性奠定了总部机构无与伦比的大气与尊贵。

佛山市三水区规划档案图书馆

建筑面积：36 300平方米
项目地点：广东 广州
项目状态：投标入围方案

目标定位及设计概念

该档案图书馆位于行政中心轴的南端，是行政中心区中轴广场向南的延伸段。为保证自中央行政大楼向江边视线的通畅，将档案图书馆中间部分压低与行政中心区地面基本持平，并通过步行天桥平台相连，使其形成行政轴南端的终点。

一方面档案图书馆是行政中心轴规划的重要组成部分，另一方面，也与文化馆共同形成以三水大桥为入口的临江标志性门户。与文化馆简洁、阳刚、强烈的形式形成对比。为了突出行政中心区行政建筑群的主导地位，区档案图书馆则采取舒展、水平叠加的柔和的地景式建筑形式，并与南侧的滨江景观公园形成形式上的连续与统一。与文化馆形成刚柔并济、阴阳互补的和谐江岸天际线。

自由退台式的建筑形式可以最大限度地在建筑的各个楼层为使用者提供便利直接的户外活动空间，尽享美丽的江景。

南油购物公园

建筑面积：300 000平方米
建设地点：广州
项目状态：竞赛获奖方案

游憩和商业的本质没有差别，寻求的都是人在漫无目的、放松状态下的愉悦体验。而南油购物公园的项目为游憩和商业的融合提供了一个机会。这个机会要求我们把方盒子式的购物中心塞进碎纸机。南山的商业困局已经宣判了已有商业模式的过饱和——再进行重复就是死亡。设计运用强烈的当代性建筑语言创造一种地标型的，大胆的空间体验，同时对亚热带季风气候和环境友好策略做出积极的回应。

深圳后海中心某办公楼

建筑面积：13 551平方米
项目地点：广东 深圳
项目状态：概念设计

塔楼朝北设计，让出北侧开放公共空间；塔楼占据东西两角，让出城市视觉通廊；塔楼体量做退台处理，增加室外活动场地；用扶梯朝上引导人流，在屋顶设置开放绿地，给后海商务中心增加立体公共花园，丰富城市功能；

大厦南侧的裙房是商业的集中展示面，通过在竖纹石材的完整大面上开凿出一条充满活力的"裂缝"。"裂缝"内侧热烈通透的橘红色与外围严整冷静的立面形成鲜明对比。中央以一条跨越三层的超长自动扶梯为这一视觉的焦点增加动感。在"裂缝"的顶部镶嵌一颗"红宝石"——巨大的LED广告显示屏。

校友大厦

建筑面积：80 000平方米
项目地点：广东 深圳
项目状态：投标方案

现代高层建筑的尴尬在于它们不断创造人工构筑的高度记录，却剥夺了像自然山体那样为人们提供的健康、低碳、环保的户外健身运动平台。不论是垂直电梯，还是消防楼梯，都是单纯以高效、安全为唯一标准的机械化而非人性化的解决方式，人的体验与主动参与性在这种模式下被彻底绑架。

创造一种将"户外登山"与高效功能有机结合的都市摩天楼的新模式是校友大厦的设计目标，完全开放的户外"中庭"是这一目标的副产品。形式的独特完全是自然产生的结果。

深圳外国语学校国际部

建筑面积：35 000平方米
项目地点：广东 深圳
项目状态：概念设计

长期以来，国内中小学校的校园设计往往只是满足于遵守严格的日照以及相关规范的要求。使校园的建筑设计仅仅停留在为教育提供一个物质平台的高度上，而忽略了建筑空间的设计与场所的塑造具备主动参与教育的潜力。例如南向单面走廊，教室线性排列的常规设计无疑可以有效地创造教室采光通风的均好性，但是狭长的功能化的外走廊却不能充分促进不同班级学生之间的自发交流。而院落式的点式空间更有利于促进自由的多向交流。精心而科学设计的建筑空间可以有效地激发少年儿童的创造力和想象力；自由而尺度亲切的场所可以促进学生间的交流；轻快活泼的建筑形式还可以强化孩子们的参与意识与体验。

深圳远致创业园

建筑面积：638 000平方米
项目地点：广东 深圳

项目基地位于莲花山、笔架山之间，空间距离很近。但彩田立交桥阻断了两山之间的人行流线，所以在莲花山和笔架山之间建立步行连接，通过重新整合两个公园的出入口，使它们成为完整的带状公园。在基地中心位置设置缆车站，游客可乘坐缆车快速直达莲花山、笔架山山顶。

现在的基地周围的城市片区被道路分割成孤岛。建立基地内部的带状公园与周边城市孤岛的步行联系，在连接城市孤立街区的步行交叉点形成公共活动空间。

基地内部与周边存在大量的机动车通行要求，容易与穿越基地的人行流线产生冲突。通过步行系统架空实现与地面机动车行交通的人车分流。

清華大學 曾昭奋 建筑研究所

清华大学曾昭奋建筑研究所是中国加入WTO之后，面对国际及国内日渐开放的建筑市场，为了适应国际竞争及促进中国的建筑创作而设立的建筑创作团体。旨在给蓬勃发展的建筑市场提供具有市场价值及文化品位的原创性建筑作品。

研究所由中国建筑理论家、前《世界建筑》主编曾昭奋教授主持；中国建筑设计泰斗中国工程院院士莫伯治教授曾任技术总顾问。

研究所还拥有一批硕士、博士学位的新锐设计师，并有多位清华大学、华南理工大学的著名学者、教授提供技术支持。

研究所与国内外多个著名设计机构发展多元化的合作，如香港巴马丹拿、贝尔高林、美国SBA国际设计集团、广州美术学院集美设计公司等。

研究所业务以工程设计咨询及提供建筑设计方案为主。其工作范围，包括城市规划、项目策划管理、总体规划、各类大中型建筑设计、园林与室内设计、建筑教育与培训等。

地址：广东省广州市天河区燕岭路89号1901
邮编：510007
电话：+86–13825025148
传真：+86–20–62818605
邮箱：qhk555@163.com
Add: Room 1901, No. 89, Yanling Road
Tianhe District, Guangzhou, Guangdong
P.C.: 510007
Tel: +86–13825025148
Fax: +86–20–62818605
E-mail: qhk555@163.com
联系人：柯维扬、黄珍祥

Tsinghua University ZZF Architect was founded by Prof. Zeng Zhaofen, one of China's foremost architects and the former chief editor of *World Architecture* magazine. ZZF Architect aims to create architectural designs that are both commercially successful and culturally creative, while embracing international competition and exchange.

The firm is headed by Prof. Zeng, and the chief technical adviser is Prof. Mo Bozhi Who is an authority in China's architecture field and a member of Chinese Academy of Engineering. The firm is consisted of lots of architects with master's or doctor's degrees; and is supported by leading scholars and professors from Tsinghua University and South China University of Technology.

ZZF Architect has developed various cooperation with some of the most famous architectural design firms, including P&T GROUP, Hong Kong; Belt Collins; U.S.A. SBA. Guangzhou JIMEI GROUP,etc. The firm specializes in providing architectural design and engineering design consulting. It also serves clients in the following areas: urban planning, project management, integrated planning, architectural design, garden and interior design architectural education and training etc.

清華大學 曾昭奋建筑研究所

酒店建筑

“双贝含珠”日景

“双贝含珠”度假酒店

建设地点：广东 阳江
项目规模：400间，6万平方米，投资6亿元，
　　　　　白金五星级
设计时间：2010-2011年
项目状态：在建
设 计 人：柯维扬、苏安娜、柯剑敏、蒋 敏
设计构思：

贝壳、珍珠自古是财富的美丽象征。该设计采用“象征主义”手法，诠释设计的美好与吉祥，并提示作品的地域环境特征。它如贝如帆，如珠如钻，亭亭玉立于祖国南海之滨。是设计者研究了亚洲顶级酒店之后满怀激情的创新之作！

“含珠”咖啡厅

鸟瞰图

清华大学 曾昭奋 建筑研究所 酒店建筑

东莞塘厦三正半山酒店

东莞塘厦三正半山酒店

建设地点：广东 东莞
项目规模：8万平方米，投资8亿元，白金五星级，客房500间
设计时间：2001—2002年
项目状态：2005年建成
设 计 人：柯维扬
合作设计人：蒙震旦
获　　奖：
2006年获亚洲十大精品酒店奖（中国大陆仅上海"香格里拉酒店"与本案）
设计构思：
与自然对话、与历史对话是本案的中心设计理念。本案坐落于山林湖泊之间，设计将自然山水的主题作了生动的诠释，让现代人的心灵有了皈依，建成后深受市场追捧！

二号总统别墅

清華大學 曾昭奋建筑研究所

高层建筑

广州本田汽车大厦（圣丰广场）

建设地点：广东 广州

项目规模：20万平方米，投资20亿元，

准五星级及高层办公楼

设计时间：2005–2006年

项目状态：2011年建成

顾问设计师：柯维扬

合作设计人：美国SBA设计公司、广东省建筑设计研究院

设计构思：本案主要表达作品的工业产品设计内涵及作品的时尚以及亮丽外观！

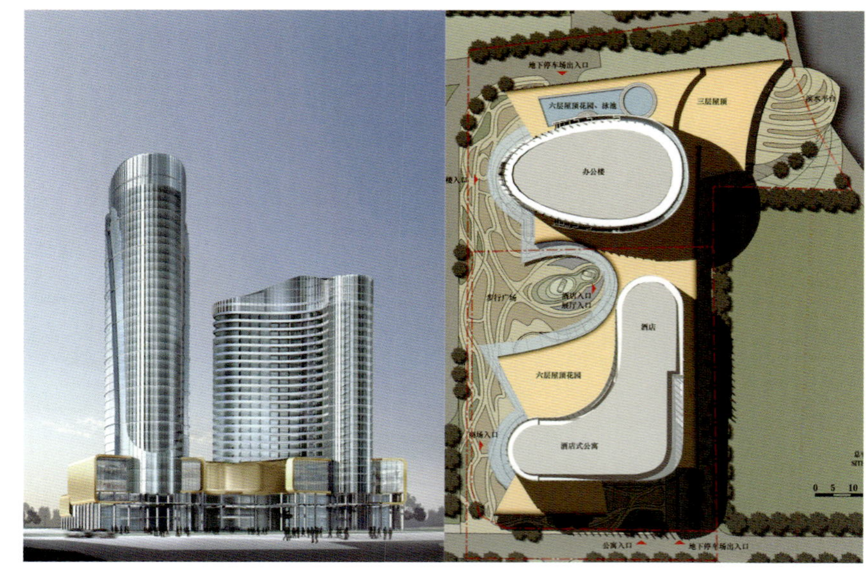

主入口透视图　　平面图

清華大學 曾昭奋 建筑研究所

旅游与博览建筑

阳江闸坡大角湾景区酒吧街

阳江闸坡大角湾景区改造工程

建设地点：广东 广州

项目规模：3万平方米，投资8 000万

项目状态：在建

设 计 人：柯维扬、陈秋宇、蒋 敏

设计构思：

海浪、沙滩、渔火、仙人掌……海洋的美丽元素引入本案设计，隐喻本案的海滩环境特征。如酒吧街设计犹如卷起的蓝色浪花。还将引入高科技元素，以4D影院、游客互动广场等表达旅游活动的当代发展！

闸坡大角湾景区大门

东南亚风情园

关山月艺术馆

建设地点：广东 广州

项目规模：1万平方米，用地1万平方米

项目状态：方案完成

设 计 人：柯维扬、蔡文齐

阳江关山月艺术馆

曾昭奋 建筑研究所

探索性别墅

探索性别墅1号至4号

富裕起来的中国城乡正复制着千篇一律的别墅，看着令人生厌！在这里设计者做一点另类的探索，给人们一个美丽的提醒：原来家可以这样设计……

项目地点：广东 肇庆
设计时间：1992–1994年
项目状态：方案完成
设 计 人：柯维扬、缪 军、蔡文齐
艺术顾问：廖冰兄

（本案别墅共100栋，每栋均由中国著名画家如刘海粟、关山月等题名）

1号别墅

以型钢、玻璃、铝材等现代建筑材料诠释中国西南干栏式民居的现代发展。

2号别墅

屋顶与露台以优美的弧线造型力求与山野环境对话，用庭院、灰色中国瓦与白墙诉说现代别墅的民族内涵……

3号别墅

格栅屋顶有通风及遮阳功能，探索现代别墅与亚热带气候的适应性。屋顶如白云，又如贝壳，适合山野及海岸环境。

4号别墅

树叶造型屋顶尝试用地方材料表达别墅的乡居特色。本设计更适合于山间会所。（商务别墅）

清華大學 曾昭奋建筑研究所

居住建筑

二期主入口

阳江伴湖花园

建设地点：广东 广州

项目规模：15万平方米，用地5万平方米

设计时间：2011年

设 计 人：柯维扬、陈秋宇、蒋 敏、黄珍祥

设计构思：本案主要表达现代居民的生活方式及对环境的诉求，同时注重当地的风水文化及居民对色彩的心理要求。

一期次入口

鸟瞰图

清華大學 曾昭奋 建筑研究所

室内设计

大亚湾核电站 总统别墅接待厅

大亚湾核电站总统别墅及室内设计

建设地点：广东 深圳

项目规模：建筑面积7 000平方米

投资8 000万元

项目状态：已建成

设计时间：2005年

设 计 人：柯维扬、刘 英

设计构思：

以中国特有的“人字形图案”表达现代室内设计的民族内涵。

商务套房

高级餐厅

贵州省建筑设计研究院建于1952年，是国内最早成立的综合甲级设计单位之一，在不断的竞争和超越中走过了半个多世纪。

贵州省建筑设计研究院具有国家建设部及国家计委颁发的工程设计、工程勘察、工程咨询、城镇规划、市政设计、工程监理、造价等多项资质。

贵州省建筑设计研究院人才荟萃，现有职工500余名，其中有各类高、中级工程技术人员300余人，享受国务院及贵州省政府特殊津贴的专家有14名，贵州省管专家1人，贵州省核心专家1人、教授级专家12名，具有国家注册资格的从业人员160名以上。

设计院除管理和服务部门外，下设多个综合设计所，包括山地建筑设计研究所、规划设计所、城市设计所、造价所、市政设计所、装饰设计所、监理所，以及勘察分院（贵州工程勘察院）、建筑公司、项目管理公司、铭仁咨询公司等多个具有独立法人资格的公司。

自20世纪80年代以来，设计院完成工程10 000余项，超过500个工程项目获得国家、部、省、地厅级优秀设计、优秀勘察、优秀标准设计、科技成果和科技进步奖。设计院还参与编制了国家和地方的技术标准、规范40余项。

设计院为全国优秀勘察设计单位，荣获建设部授予的“全国科技创新先进单位”、“全国建设系统精神文明先进单位”、“全国工程勘察先进单位”和中共贵州省委授予的“贵州省思想政治工作先进单位”以及中华全国总工会颁发的“模范职工之家”等荣誉称号。

设计院将一如既往贯彻“坚持社会责任至上、客户价值至上、员工发展至上，树立行业风范，追求卓越品质，打造企业品牌”的宗旨，参与竞争，努力实践，以创贵州品牌企业为目标，竭诚为社会各界和广大客户提供最优质的服务。

Guizhou Provincial Architectural Design & Research Institute, one of the national comprehensive design units with the Grade A qualifications, established in 1952, has survived over half century in market competitions and outdoing ourselves.
The institute has gained the certificates issued by the former Ministry of Construction and State Planning Commission, including engineering design, engineering survey, engineering consulting, urban planning, municipal engineering design, project supervision and construction cost.
The institute is now employing over 500 staffs, among which there are over 300 senior or medium technicians, 14 experts enjoying the State Council or Guizhou Provincial Special Allowance, a provincial expert, a key expert of Guizhou, 12 professor grade experts, and over 160 employees with the state registered certificates.
Besides the management and service, the institute has several comprehensive design departments, including mountainous architectural design, planning and design, city design, construction cost, municipal engineering design, decoration design, and project supervision. And it also has several independent agencies such as Guizhou Engineering Survey Institute, Mingren Engineering Consulting Company and construction, project management companies.
Since the 1980's, the institute has accomplished over 10,000 projects, among which over 500 won awards as "Excellent Design", "Excellent Survey", "Excellent Standard Design", "Scientific Achievement Award" and "Scientific Progress Prize" awarded by the national or local agencies. It also participated in the drafts of over 40 national or local technical standards and specifications.
As one of national excellent survey & design units, not only the institute was rewarded by Ministry of Construction as "Advanced Unit for National Technical Innovation", "Advanced Unit for National Spiritual Civilization in National Construction Field" and "Advanced Unit of National Engineering Survey", it was also rewarded "Advanced Unit in Ideological and Political Construction" by CPC Guizhou Provincial Committee, and the title of "Home of Exemplary Workers" by All China Federation of Trade Unions.
With the purpose of putting social responsibility first, customer value first and employee development first, setting refined manners in the industry, pursuing better quality and creating the brand of the company, the institute tries to provide best service for the society and clients in the market competition and practice.

地址：贵州省贵阳市遵义路48号 **电话**：+86-851-5573792 **传真**：+86-851-5572433
邮箱：gzsjy@vip.sina.com **网址**：www.gadri.cn

贵阳龙洞堡国际机场扩建（与北京民航院联合设计）

按旅客吞吐量1 550万、货邮22万吨、起降14.63万架次进行规划设计。主要工程内容为新建航站楼11万平方米，扩建站坪26万平方米，新建停机位25个，停车场（楼）10.5万平方米及其他配套设施。

设计理念

整体性、综合性——航空港设计新理念

地域性——传递“多彩贵州”和“林城贵阳”的可识别性

时代性——高技派绿色建筑

生长性——单元线形生长模式

Expansion of Guiyang Longdongbao International Airport (co-design with Beijing Civil Aviation Institute)

Planning and design according to the throughput of 15,500,000 passengers, cargo volume of 220,000 metric tons, arrival or departure of 146,300 flights. The project mainly consists of a new airport terminal with the area of 110,000 m^2, a station site expanded by 260,000 m^2, 25 new aircraft parking stands, a parking lot of 105,000 m^2 and other supporting facilities.

Design Concept

Integrity, Comprehensiveness— New Design Concept for Airport

Local Features—Identification Marks for “Colorful Guizhou” and “Forest City of Guiyang”

Modernity—High-tech Green Architecture

Growth—Unit Linear Growth Mode

花溪迎宾馆工程

设计理念

一、环境生态特色：体现尊重环境的设计理念；

二、依山就势特色：体现山地建筑的个性特色；

三、地域文化特色：体现地域文化和时代精神兼容的建筑风格；

四、绿色建筑技术特色：设计采用绿色建筑技术和材料；

五、营造“花”与“溪”的景观意境；设计借景于自然，融入于环境；

六、空调系统耗电量指标27.6W/平方米，在目前建筑设计领域内属领先水平。

获奖情况

国家铜质奖、全国建筑创作大奖、省优秀设计一等奖。

Huaxi Yingbin Hotel Project

Design Concept

1. Environmental and ecological features: give expression to the concept of protecting the environment;
2. Make use of the natural conditions: give expression to the individual characteristics of mountain buildings;
3. Regional culture features: architectural style of combining the regional culture and the spirit of the times;
4. Green architectural techniques: adoption of green architectural technology and materials;
5. Create the poetic image of “flowers” and “brooks”, borrowing scenery from the nature to integrate with the environment.
6. The power consumption index for air-conditioning system is 27.6 W/m^2, being in the lead in current architectural design field.

Awards

National Bronze Medal; Award for National Architecture Creation; First Prize for Excellent Design of Guizhou Province

贵州省委办公业务大楼

建设地点：贵州 贵阳
设计时间：2006—2007年
竣工时间：2009年
验收时间：2009年
总用地面积：26 681.8平方米
总建筑面积：55 765.66平方米
地上面积：45 105.66平方米
地下面积：10 660平方米
容 积 率：1.69

项目概况

贵州省委办公业务大楼坐落在风景如画的贵阳市南明堂内，北有一山——观风山，南有一水——南明河，正所谓山南水北。南明河水缓缓地将省委大院沿西、南、东三个方向环抱。从总体来看，整个大院最高点在观风山顶，地势由北至南逐渐降低，坐北观南。

设计理念

1. 设计遵循“着眼长远、适度超前、统一规划、分步实施”的原则；
2. 办公大楼的体量采用南面多层、北面高层结合内庭的梯级过渡方式；
3. 充分利用地形高差，尽量做到房间自然通风、采光；
4. 采用双子廊和单廊、内廊相结合的方式，提高平面利用系数，打破传统行政办公建筑单调的平面布局，提高办公的舒适性；
5. 平面各功能区的组织力争联系顺畅、注重私密、节能高效、疏密结合；
6. 建筑形体力求雅致、简洁、庄重、大方；建筑色彩追求稳重、明快、和谐，从而形成具有鲜明特色的大楼形象。

关岭游客服务中心

贵州省利用世界银行贷款实施文化与自然遗产保护和发展项目之一。

设计理念

立足地域精神和文化多样性，并延伸到社区发展原住民生计在内的主体性规划与设计。

项目涉及24个重要民族村寨和古镇、6个国家级风景名胜区与地质公园的前期设计和系列子项目的全部设计，项目投资6亿元。

贵州省民主党派和政协委员会活动中心

建筑面积：38 927平方米
建筑高度：99.9米
该项目为一栋集办公、会议及活动用房为一体的高层办公楼。

项目主要特点

1. 难度大：一幢建筑分三期实施，34米跨井字楼盖分两次浇筑；
2. 布局巧：塔楼中各种会议室达43间，在高层建筑中尚不多见；
3. 功能细：辅助用房有大量功能细化设计，使建筑更具人性化；
4. 创意新：以“纳谏之门”的造型寓意政协广纳民意的内在含义，117.3米高度设31.2米跨钢网架，外墙石材改为仿石铝塑板，外观浑然一体；
5. 用料新：采用磷石膏墙板、氟碳复合铝板、张拉膜等新材料。

获奖情况

贵州省优秀设计一等奖
中国建筑鲁班奖

贵阳市科学技术馆

建设地点：贵州 贵阳
建筑面积：21 000平方米

设计理念

构思来源于“魔方”——形体简单，却是变化无穷。以此概念衍生出对该项目的理解和诠释——强调科技馆的可变性和趣味性，给人以新奇感，体现简单与复杂的和谐统一，也阐述了一种对深奥的科技知识进行通俗易懂的表达的含义。以此为源引入“魔方”的各种变形，在建筑体系中加以提炼和组合，最终得出充满神奇“魔力”的建筑。

这个充满神奇“魔力”的建筑，本身就是一个高科技的产品。它所展示的，除了科技的外观和内涵外，在更多意义上，由于设计中公众性和社会性理念的充分引入，它更成了日新月异的科技发展和社会人文发展的风向标。

设计中，对于这种“魔力”的诠释，其一体现在建设平面的多种变化中，其二体现在建筑体量的穿插变幻上。这个建筑，在城市肌理上是一个焦点，在公众视野里是一个科技发展的载体，在社会市场化发展中是一个新的模式。

该方案体现的是一个集“科技性、艺术性、群众性”为一体的标志性建筑。

开磷大厦（开磷集团总部）

建筑面积：38 000平方米　高24层　高层生态型办公建筑

黄果树演艺中心

用地面积：8 487平方米
建筑面积：4 050平方米
占地面积：1 903平方米
容 积 率：0.48
绿 地 率：33%
密　　度：0.22
建筑高度：19.95米

设计理念

走现代建筑地域化道路，融入贵州安顺地区屯堡文化和当地布依族服饰文化，用现代建筑科技予以充分诠释。

造型如布依族头饰，与山体镶嵌；又似当地屯堡古堡，充分展示建筑地域性特征。用现代的索膜结构、木质龙骨玻璃幕墙结构与当地的白棉石地材干挂外墙相结合，体现传统与现代的冲撞和有机相融。使建筑极具标志性和地域性，不失为一条西部建筑师创作的探索之路。

花溪新区行政区行政中心

设计理念

（1）布局：

在行政区地界12公顷范围内，总图行政楼定位靠南部规划大道，坐北朝南，以水为中轴线，对称布局。总图以“天下为公”的“公”字为母题，对称相连成蝴蝶状平面，取“蝶舞花溪”之意。以现代园林手法围合三个中庭，一层架空，小桥流水贯穿其中，融入现代行政办公理念“开放、透明、法制、参与”。让园林化、文脉化、生态化、现代化、网络化原则得以实现。

平面构思体现现代办公理念，内外廊结合，小办公、大办公、景观办公结合，建筑与庭园互为背景，全天候中庭与自然中庭结合，创造丰富使用空间和景观空间。

（2）立面造型构思：

在限高35米的情况下，立面力求在舒展方向做文章，力求体现“清新亲和、文化味、地方特色结合现代感”的造型特色，而不追求豪华、雄伟的普通办公楼设计手法。

在面宽达160米宽的主体立面上，立面采用双梁双柱架构，让传统的干栏式梁柱线条体现在凹缝空间上；立面六层后退，新颖的坡顶作为收头，再现传统民居的第五立面，新的金属坡顶语言体现了丰富的文脉特色。在两翼采用横向分格，以细腻的点窗、隐框幕墙加电动百叶手法，配合立面干栏式梁柱。在传统花溪建筑风格基础上力求用新的建筑语汇、高新的材质，去诠释花溪文化，让人耳目一新。

贵州大学新图书馆

建设地点：贵州 贵阳
建筑面积：30 000平方米

设计理念

现代大学精神与中国传统书院灵魂相结合。

花溪新区行政区公检法大楼

建设地点：贵州 贵阳
用地面积：39 886平方米
建筑面积：20 180平方米
占地面积：4 620平方米
容 积 率：0.42
绿 地 率：45%
密　　度：0.12
楼　　高：七层
建筑高度：30.5米

设计理念

作为花溪新区行政中心另一个重点项目，强调行政区域中轴线对位关系；总图布局为弧形，与行政中心形成合抱之势，三权机构各自独立，又相互联系，造型设计采用与行政中心不同的风格，贯彻地域建筑现代化的《北京宪章》精神，造型立意用三个独立正义之门穿套在国门之内，用朴素端庄的高直语言和柔美造型，既强调三权机构的庄严、正直，又隐喻二十一世纪公检法机构的人文主义法制理念。

贵州省珏石矿业投资开发有限公司

贵州省珏石矿业投资开发有限公司是一家集开采、加工、销售、施工、装饰装修为一体的综合性企业，主要致力于为资源节约型和环境友好型建设经济找到新的途径和创新方式。

低碳产业示范项目——珏石屋

宗旨："科学综合利用资源，精心营造绿色建筑，做好低碳产业示范，实现循环经济发展"为加快建设资源节约型和环境友好型社会开创新途径作出新贡献；

目标：用自有的专利技术、企业标准、商标、科研项目等知识产权，营造节能、减排、低碳、环保、绿色安全的建筑物，设计生产建造出全寿命周期最长，功能性价比最优，施工建造工期最短，安全性、舒适性俱佳，天下居士都买得起的房子。

理念："以石造房"，"以矿为厂"，"以功能兼备为主"，按五个"一"：建筑、材料、结构、抗震、装修、施工一起设计；结构、材料、装饰一体实现；建造用砌块、构件、石材的开采、加工、制作一步成型；施工安装、水、电、装饰、装修一集完成；建筑设计、结构设计、材料设计、装饰设计、施工成型一气呵成。简化传统产业链，集合多行业利润，形成高效利用资源、高收益回报的新兴产业。

用材：天然石头，无机胶，无毒害材料。钢、木、竹、天然纤维等、

营造模式：集成设计，一图成屋，石石复合、石混凝土复合、石钢复合、石木复合共存，小板、中板、大板、板柱混合结构、抗震体系并举；标准化、模块化、工厂化、个性化、艺术化、集成化、用传统材料加新技术，开创新建筑领域；

市场定位：危房改造，新农村、新城镇、旅游景点、安居工程、廉租房建设，以农村包围城市，从国内市场走向国外市场；做房地产的上游，把产品卖给房地产开发商、卖给买不起商品房的人、卖给政府、卖给慈善机构、用于抢险救灾、安民，卖给有"家业传世代，祖居永流芳"之高雅情操的人。

发展方向：做建材、建筑、建工、装饰行业的综合生产、供应、服务商，产品从低层到多层，再到高层，从普通民房住宅到中高档山庄别墅、宾馆酒店。改变现状"盖"房子为"造"房子，消除粉尘、污水、噪音、建筑垃圾污染环境、伤害人类的顽疾，聚建筑艺术、文化展示、民族特色、旅游景观、文物传承于一体的传世建筑物；

社会效益：全寿命周期长150年以上，节能90%，减排95%，废水量重复利用达95%，节土100%，节木60%，缩短施工期70%，可以让小水泥厂、采石厂、砖瓦厂、砂石场、预制厂等"五小"行业自动淘汰关停，其简约、清洁的生产、施工过程可以节能减排数亿吨，将成为建材、建筑施工、装饰装修行业中最大的节能减排项目。

经济效益：与传统建筑相比，降低造价20%，提高成材率150%，所耗同量资源产出的产值和经济效益是水泥、砂石、石灰、砖瓦的三倍以上；整个产业链的利税是传统产品产业的双倍以上。

珏石屋的问世：

（1）是对传统建筑材料功能的最佳优化应用；

（2）是将建筑设计转化成建筑走工业化集成的高效简化方式；

（3）是现有盖房子装修工程做到清洁施工的有效净化方法；

（4）是让建筑物展示天然、典雅、恒久又具建筑文化艺术的美化视觉传承之典范。

地址：贵州省贵阳市中山东路42号银座大厦10层3-4号
电话：+86-851-5833476
传真：+86-851-5833472
邮箱：guizhoujueshi@yeah.net

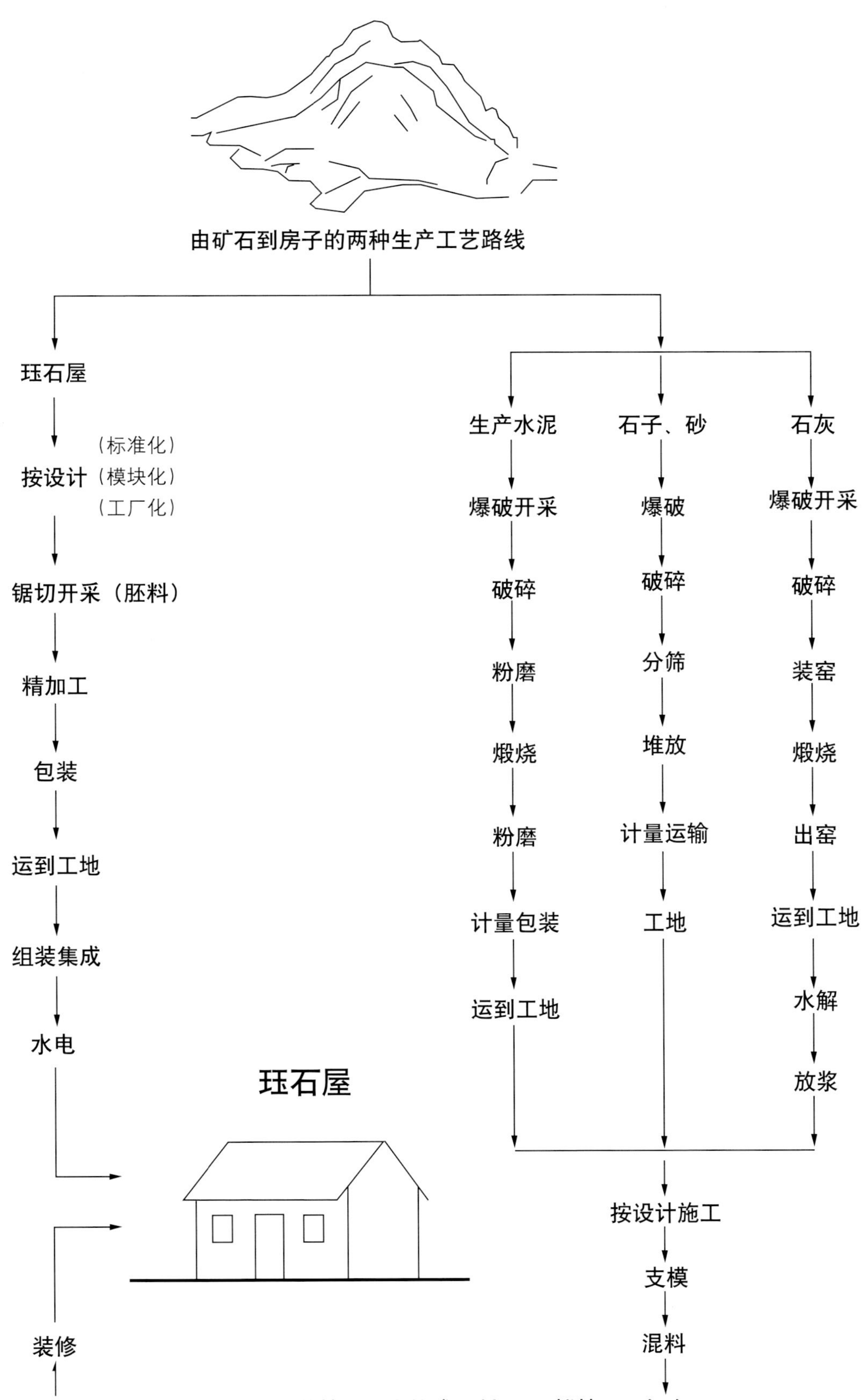

从以上两种不同的生产方式及工艺路线中看出，珏石屋计划实施的生产过程是在走一条简约、创新、节能减排、绿色环保、提高建筑使用寿命，组合材料、建筑施工、装饰等部品集成化、清洁生产化、高效工业化的发展道路；建成完整的产业链，快速形成产业集群；使传统的建筑材料生产、建筑设计施工、装饰装修行业“千锤百磨”，多头、分散、脱节现象得以改变。让最古老、传统、普通、价廉的石头“原质修身”一步到位，还其统领全世界古老、经典、优美建筑物的至尊地位。在21世纪中闪现出令人瞩目的石文化及建筑艺术之光！

据初步估算，如按“全寿命周期”的分析方法计算珏石屋寿命（150年计），从建筑材料生产、建筑设计施工、建筑运行应用、拆除再建等方面综合考虑，与传统建筑材料（如红砖、混凝土砌块）建筑比，可节能90%，减排95%，节土100%，节水90%，节木60%，缩短生产施工时间70%，降低造价20%以上，消耗同量资源带来的经济效益是水泥、石灰、石子、砂的三倍以上，其社会效益更加显著。

如果全面开发应用，将岩石直接加工成石砖、板、砌块、构件、柱、梁、门框、窗、楼梯、屋面等部品，使结构功能和装修饰面效果整体化，与其他多种功能材料复合，通过集合，结构胶粘接，螺栓、钢构连接及增强抗震材料的科学组合，集成组装成建筑主体。其“原质真身”完全能更加优化，代替水泥、石子、砂、石灰、混凝土制品、年、黏土砖等建筑材料。其简约、清洁的生产过程可以节能减排数亿吨，将成为建材、建筑施工、装修行业中最大的节能减排项目。是推进产业结构调整、加快经济发展的重要途径；更能体现出节土、节水、节木、节约资源的优势；对实现循环经济、全面协调可持续发展具有深远意义；并为加速建设资源节约型和环境友好型社会作出重大贡献。

珏石屋效果图

“绿色生态建筑体系”项目模式

节能、减排、绿色、生态建筑形式一直是建筑行业创新的一个焦点。人类不断地向环境排放污染物质。但由于大气、水、土壤等的扩散、稀释、氧化还原、生物降解等的作用，污染物质的浓度和毒性会自然降低，这种现象叫做环境自净。如果排放的物质超过了环境的自净能力，环境质量就会发生不良变化，危害人类健康和生存，这就发生了环境污染。针对以上一系列的问题，作为创新型企业，珏石公司正努力构建一个“节能、绿色、环保、低碳”的循环经济建筑体系：“绿色生态建筑体系”。

“绿色生态建筑体系”主要是在建筑物外墙及楼面以下周边种植绿色植物，以实现土地的回收利用，增加绿化面积，净化空气自循环系统、水自循环系统，达到“自给自用、节能减排”的低碳建筑经济体系。以下为本项目的几大特点：

一、空气自循环系统：主要通过增加外墙及楼面的绿色种植面积，通过大量的植物排放出足量的氧气，并吸收人类活动的各种废气（二氧化碳），实现空气的自循环。标准的绿化率，一般达到40%以上就可以实现空气净化的良性循环，而“绿色生态建筑体系”项目在通过图纸和样品房的实际论证后最高绿化率可达到200%，例如修建的一套占地面积120平方米、三层楼的样品房，通过“绿色生态建筑体系”的建筑方式，回收的绿化面积可达到240平方米左右。所以本项目的绿化率远远高于标准，为空气的良性循环提供了有力的保障。

二、水自循环系统：水循环系统主要是实现生活用水、生产用水的再生循环利用，有效节约水资源。建筑物周边及楼面建立排水、净化水、储水等设施，具体为：

1. 楼面建造露天储水池，可以用于养殖各种水生动物，水来源主要为天然积水，多余的水可以通过排水管排到楼下的净水系统中。

2. 建筑物外围建造两个净水池，主要用于净化天然积水和生活用水。净化方式：自然净化。第一个净化池净化后的水排到第二个净化池里，通过第二个净化池二次净化后的清水排到储水池里面，而净化池里面沉淀后的残余物排到沼气池或化粪池。水来源主要为：天然积水和生活、生产用水外。

3. 净水池清水排放口处建造储水池，主要用于存储净化后的清水，通过抽水设备抽到生活用水池里，实现循环利用。水来源主要为：净化后的清水。

三、保温系统：通过大量的绿色植物、楼面储水及墙面排水以及外墙种植物遮挡阳光等系列措施可以有效地控制室内温度。

四、节能系统：传统的生活能源主要以电力能源为主，而“绿色生态建筑体系”项目的生活能源主要以太阳能、沼气、电能为主，具体方式为：

1. 在楼面设置太阳能接收器，将太阳能转化为电能，主要用于加热生活用水和通过电能存储设备进行存储后备用。

2. 建筑物外建造沼气池，将各种植物和生活垃圾的腐烂、发酵产生的沼气，利用管道

珏石屋效果图

输送到专业设施上进行生活照明和烧水、煮饭等生活活动。

3．电能：主要是在沼气和太阳能能源不足的基础上进行补充利用，例如冬天阳光不足的情况下，就可以利用电能进行补充利用。

三种能源以太阳能、沼气为主，电能为辅，达到充分节约能源的目的。

五、绿化率：标准的绿化率为40%，可以达到空气、保温等良性循环，“绿色生态建筑体系”项目的绿化率在图纸和样品房的论证中最低可达到110%，最高可达到140%，有效地利用外墙面积和楼顶面积，回收了一定比例的种植面积，从外观上也有力地增强了建筑物的视觉美化程度。

以上为“绿色生态建筑体系”项目的特点，但本项目在实施过程中也存在各种各样的问题，主要体现在以下几个方面：

1．后期维护：本项目主要针对小高层建筑，层高以3米计算，三层建筑就有9米高，对外墙各种种植物的维护，就存在高空作业的危险性，所以在后期维护上就存在以下两个方面的问题：

（1）维护的机械设施问题：可选用高层楼梯、移动升降式楼梯，或者为本项目设计一种专用的维护设施，选择哪种设施安全系数较高是必须面对的问题（一般人员是否能操作）。

（2）维护的安全性问题：涉及高空作业时，就必须考虑一个安全系数问题，高空作业严格规定必须要专业的操作人员才能进行，但本项目的建筑形式均已居住建筑为主，日常维护只能靠房屋居住主人或家属自己维护，这样就存在一定的危险性，因为他们都不是专业的维护人员，即使进行简单的培训，也不能达到专业操作人员的要求，所以还得请专业的后期维护人员（是否现实）。

2．造价成本增加：成本的增加主要表现在以下几个方面：

（1）建筑物墙体外墙的支撑结构：因为扩大了外墙植物的种植面积，就必须增加外墙支撑结构，且还必须考虑到安全性，对各种结构要有目的性的加固，以达到承受重量和保证使用年限的效果。

（2）防水、防腐、防潮措施：因外墙和楼面长期处于植物和水的侵蚀下，所以必须在修建期间进行更大强度的防水、防腐、防潮处理。

（3）后期维护：必须使用专业的维护设施和外聘专业的维护人员，这都需要增加一定的使用成本。

“绿色生态建筑体系”项目是一种新型的建筑模式，在创新本项目期间还有很多有待验证的问题。希望在以后建筑实施过程中得到更大的改进，使得本项目更加完善，成为建筑行业“绿色生态”建筑的先驱。

珏石屋效果图

香港华杰小区效果图

香港华杰小区效果图

香港华杰小区效果图

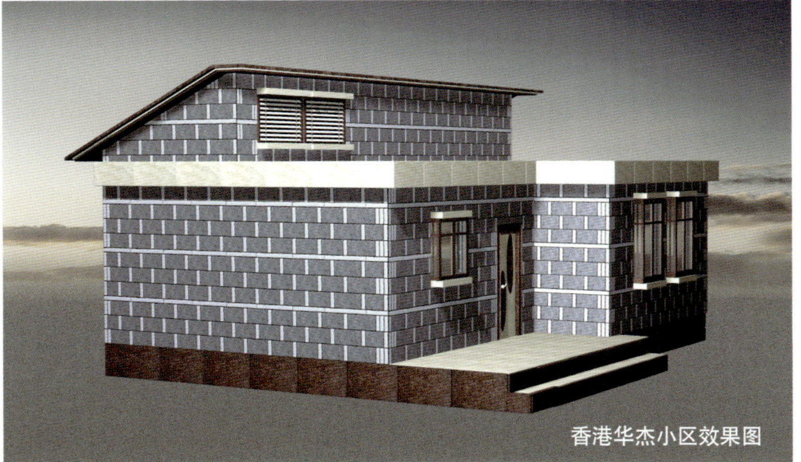

香港华杰小区效果图

江西全南韬略运动器材厂管理层宿舍效果图

保定市建筑设计院

Baoding Institute of Architecture Design

保定市建筑设计院始建于1952年，是河北省成立最早的综合甲级建筑设计院之一。建院50多年来，完成了近万项工程项目的勘察设计、监理任务，项目遍及我国各省、市及自治区，连续多年被评为“重合同守信用单位”、“建设系统信誉、信用AAA单位”、“优秀勘察设计院”、“河北省诚信示范单位”等荣誉称号。

目前，我院现有各类专业技术人员194名，具有中高级职称人员126名，其中教授级高级工程师9名、高级工程师53名、工程师67名；国家各类注册师68名，其中一级注册建筑师8名、一级注册结构工程师20名、注册监理工程师17名、注册岩土工程师7名、注册咨询师6名、注册造价工程师2名、注册设备师8名。专业范围包括建筑设计及规划、结构、给排水、电气照明、弱电通信、暖通、电子计算、技术咨询等。

作为河北省最好的地市级建筑设计单位之一，我院以“精心勘察设计、公正科学监理、优质诚信服务、实现顾客满意”为己任，坚持以繁荣建筑创作为宗旨，不断完善创新设计理念，力创建筑设计精品，在工程设计和科研方面屡获国家级、部级和省级优秀奖，在文化建筑、体育建筑、医疗建筑、教育建筑、居住建筑以及空间结构等设计领域具有独特的设计优势，而严格、规范的ISO：9001质量体系认证管理更使我院的设计质量为业界广泛认同。

我院具有住房和城乡建设部颁发的建筑行业（建筑工程）甲级资质证书、工程勘察专业类岩土工程（勘察、设计、咨询、监理）甲级资质证书、房屋建筑工程监理甲级资质证书、施工图审查一级资质证书和建筑行业人防设计乙级、市政公用行业（道路）设计乙级、城市规划乙级，以及咨询、市政公用行业（热力、风景园区）丙级资质证书等。可承接各类大中型民用及工业建筑勘察、设计以及工程建设可行性研究，人防工程设计及相应的咨询与技术服务，市政公用行业给水、排水、热力、风景园林等工程设计及相应的咨询与技术服务，工程总承包、工程建设监理、施工图审查等业务。

我院注重推行和运用现代化设计手段，配备有电脑网络管理系统、协同设计系统、办公自动化系统，并建立了一套完整严密的管理体系，形成了团结求实、严谨高效、进取奉献、严守职业道德的院风。

我院将不断开拓国内市场并争取开拓国外市场，以造就一流人才、创造一流设计、提供一流服务为目标，竭诚为各界服务。

保定市高新区小学

高新区小学位于朝阳路以东、鲁岗路以北、沈庄以东，北临二环路防洪堤。规划总用地25 722平方米，新建二期建筑20 386平方米，原有一期建筑12 500平方米。

因用地紧张，设计采用“日”字形布局，形成两个院落，房屋朝向为东西向，这样既充分利用了土地，又为学生提供了多重室外空间。整个用地设置两个出入口，东南角为主要出入口，西边为次要出入口，两出入口相连，交通便捷。一期为“L”形布局，二期为“日”字形布局，中间为200米的运动场及风雨操场，方便学生的室外活动。一期建设的于东边设置文化景观廊与二期拟建建筑前的广场相连，以形成学校的文化景观空间，在学生上下学时有良好的文化氛围。

二期拟建建筑，呈“日”字形、中轴线设计，东南角为主要出入口，中间为学生主要出入口。西边设置两院落的消防口。普通教室采用单廊南北向布置，专业教室及教学辅助用房布置在东西向，靠近运动场布置行政办公用房及展室、图书室等功能室，在东南角设置入口大厅。

造型结合太阳能技术进行设计，太阳能技术的运用采用屋顶及立面两种形式，让太阳能板成为造型的一部分，变被动为主动，既美观又具有节能的功效。

高阳县医院门诊病房综合楼

保定市教育局直属九年一贯制学校

高新区写字楼

该项目位于高新区朝阳北大街与三号路的交口处，地理位置优越，交通方便。总建筑面积61 674.86平方米，建筑用地面积29 953平方米。地上19层，地下1层。

大学科技园研发中心

建设地点：

本案位于保定市北二环路以北、朝阳北大街以西、向阳北大街以东。基地东面毗邻新华政法学院，南面是50米的城市绿化带，北面是花庄变电所和保定吉达电力公司，西边界邻近一条高压走廊。东南面具有良好的景观界面。

结构类型：框架结构。

建筑工程设计等级：一级。

项目介绍：

该项目地下一层为车库和各种设备用房，地下可停车163辆。首层为园区服务、展示、后勤用房，2层到25层为研发用房。裙房2层，为长方形布局。主楼25层，配楼6层，后退为“L”形布局，在入口处形成环抱和欢迎的姿态。建筑造型考虑与周边环境的和谐性，立面以落地竖条玻璃窗为模块，强调建筑竖向肌理，形象简洁、鲜明有力，造型挺拔。运用折板的形式将主楼与裙房连为一体，整体建筑庄重大方，通过环境与建筑的有机穿插与结合，达到和谐共存的目的。

Tangshan Haoyu Architecture Design Co., Ltd.

唐山昊宇建筑设计有限公司

唐山昊宇建筑设计有限公司成立于2008年，具有河北省住房和城乡建设厅核定的建筑行业（建筑工程、人防工程）乙级设计资质，可承担建筑装饰工程设计、建筑幕墙工程设计、轻型钢结构工程设计、建筑智能化系统设计、照明工程设计和消防设施工程设计相应范围的乙级专项工程设计业务，可从事资质证书许可范围内相应的建设工程总承包业务及项目管理和相关的技术与管理服务。公司的前身为北京筑福建筑设计有限责任公司唐山办事处。现有员工65人，其中高级工程师6名，工程师16名，全国注册工程师16名。公司主要从事民用建筑工程的策划与设计，尤其对房地产住宅开发项目及校舍鉴定、酒店、办公楼、商场、学校的规划与设计有着深入细致的研究。

自公司成立以来，以积极的姿态活跃于设计领域，已完成多项不同类型的建筑设计任务。从整体设计的原则出发，在对社会、城市、技术及公众需求的综合考虑上，以及对地方性大众文化的充分了解方面，公司力求引导与提升市场对建筑艺术的理解，创作了不少优秀作品，得到了社会各界的一致好评。公司内部不断完善管理机制，并进行人性化管理。“人情感动人，利益驱动人，事业成就人”是公司的管理理念，从而吸引了大批技术精英的加入。公司为员工提供了一个健康发展，能够充分展现自我能力的良好环境。

唐山昊宇建筑设计有限公司立足于中国东方文化，融汇了国际先进的设计理念，在设计项目管理、专业技术方面打造了一支特别能吃苦、特别能战斗的高效率团队，创造了显著的社会效益和经济效益。公司实力不断壮大，业务量不断提高，以质量求生存、工期作保证、服务创一流的经营方式敏锐感知客户的需求，协助客户追求卓越的市场业绩，致力于专业领域的探索，力求让综合实力和先进技术产生最现实的社会效益，鼓励创新，面对复杂问题寻找多种解决方案。公司提供从策划、方案、初设到施工图的整体建筑设计服务，一直在为创造超越客户期望的产品和服务而努力，力求成为建筑设计行业的先行者。

Tangshan Haoyu Architecture Design Co., Ltd. which was established in 2008, has Grade B design qualification in construction field (include building engineering, civil air defense engineering) approved by the Department of Housing and Urban Rural Development, Hebei Province, can take architecture engineering, architecture curtain wall engineering design, light steel structure engineering, building intelligent systems design, lighting, fire safety engineering design and engineering of the range of Grade B special engineering design services, can engage in corresponding construction general contracting business and project management and related technical and management services in the permitted extent in the qualification certificate. The building formerly known as Beijing Zhufu Co., Ltd. Tangshan Architectural Design Offices and currently there are 65 staff, including 6 senior engineers, 16 engineers, 16 registered engineers. The company is mainly engaged in civil engineering planning and designs; particularly for the planning and design of residential real estate development projects and school identification, hotels, office buildings, shopping malls, schools, and has an indepth and meticulous research.

Since its establishment, active in design field with a positive attitude, the company has completed a number of different types of architecture design tasks. From the overall design principle, taking the society, city, technology and public demand into account, fully understanding the popular culture of the local, the company sought to guide and enhance the understanding of the market for architecture art, created a lot of good works, has gained the praise from the society. The company continuously improves the internal management mechanism for a user-friendly management. "Human emotion moves people, profit drives people, career makes people successful" is the company's management philosophy, which attracted a large number of technical elites. The company also provides employees with a good environment of a healthy development, showing their ability.

Tangshan Haoyu Architecture Design Co., Ltd. based on China oriental culture, a combination of advanced design concepts, has built up with a special ability to endure hardship, to fight a high efficient team in design project management, professional and technical aspects, creating a significant social and economic benefits. Company has growing strength, increasing business volume, survive by quality, time insurance and first-class service mode of operation, keen perception of customer needs, to help our customers achieve market performance excellence. We are committed to exploring areas of expertise, so as to provide overall strength and advanced technology to produce the most realistic social benefits. We encourage innovation, find multiple solutions to complex problems. We offer the overall architecture design services from the planning, programming, preliminary design to construction drawings, making effort to create products that exceed customer expectations and services, striving to become a pioneer in the architecture design industry.

地址：河北省唐山市高新技术开发区大陆阳光103楼7层
电话：+86-315-8308336
网址：www.haoyusheji.cn
邮箱：sheji2008sheji@163.com

Add: The 7th Floor, Continent Sun Building 103, High-tech Development Zone, Tangshan City, Hebei Province
Tel: +86-315-8308336
Http://www.haoyusheji.cn
E-mail: sheji2008sheji@163.com

唐山君德城上城规划设计方案
Tangshan Junde City on City Planning and Design Program

建设地点：河北 唐山
用地面积：215 700平方米
项目类型：集住宅、商业、教育为一体的高档居住小区

Location: Tangshan, Hebei
Land Area: 215,700 m^2
Project Type: High-grade residential area including residence, commerce, education

唐山迁安市中部南关
Tangshan Qian'an City Central South Gate

建设地点：河北 迁安
Location: Qian'an, Hebei

许各寨城中村改造
Xu Ge Zhai Village in the City Reconstruction

建设地点：河北 唐山
用地面积：55公顷
建筑面积：1 350 000平方米
项目类型：集住宅、商业、教育为一体的高档居住小区

Location: Tangshan, Hebei
Land Area: 55 ha
Building Area: 1,350,000 m^2
Project Type: High-grade residential area including residence, commerce, education

唐山如意湖畔馨苑
Tangshan Ruyi Lakeside Xinyuan

建设地点：河北 唐山
总建设用地面积：28 258.66平方米
地上总建筑面积：84 351.35平方米
容 积 率：2.99

Location: Tangshan, Hebei
Construction Area: 28,258.66 m^2
Floor Area on the ground: 84,351.35 m^2
Floor Area Ratio: 2.99

韩城中心商业地块华北铁合金物流交易中心设计方案
Hancheng Center Commercial Block, North China Ferroalloy Logistics and Trading Center Design Program

建设地点：河北 唐山
建设用地面积：9.33公顷
建筑面积：354 800平方米
容 积 率：3.8

Location: Tangshan, Hebei
Construction Area: 9.33 ha
Floor Area: 354,800 m^2
Floor Area Ratio: 3.8

Hainan Hualei Architectural Design Consulting Co., Ltd.

海南华磊建筑设计咨询有限公司

海南华磊建筑设计咨询有限公司成立于1988年，是海南建省拥有建设部甲级设计资质证书最早的民营企业，具有独立的企业法人资格，法定代表人于瑞为国家一级注册建筑师、国家注册公用设备工程师。现公司持有住房和城乡建设部颁发的甲级建筑勘察设计资质证书、乙级市政设计资质证书。主要承担建筑工程、市政工程、室内外装饰工程、小区规划设计，以及项目前期策划、工程咨询、项目可行性研究等业务。公司现有技术人员270余名，其中国家注册工程师40余人，人员技术素质优秀，专业配套齐全，是一支技术过硬、专业组成合理、服务一流的设计团队。

公司已于2003年通过国际ISO9001：2000质量体系认证。公司组成包括总公司外，海南省内还下设有13个分公司及武汉、杭州、重庆、天津、兰州、福州、深圳、广西等8个省外分公司；总公司设有业务部、质检部、人力资源部、综合部等机构，并引进了先进的管理软件，对公司经营实施全方面信息化管理。内部执行严格的质量管理制度，对每个项目的设计质量层层把关，逐级审核，确保高质量地完成各项设计任务。同时，为保证公司的可持续发展，公司注重对人才的发现、培养和教育，对专业技术人员进行不定期的专业继续再教育，积极参加各类新技术的推广与应用。

强调设计的后期服务，以达到客户最高满意度。公司以"精心设计，诚信服务，人才为本，持续改进"为十六字方针，十几年来，已完成各种设计项目近千例，海南市场的占有率名列前茅。同时，近几年来，公司领导积极拓展内地市场，内地工程量也不断扩大。其中包括不少获奖项目、大型公建、有影响的地方标志性建筑及有一定水准和一定规模的住宅小区，为公司赢得了良好的社会效益与经济效益。

公司拥有数位知名资深专家，承担着中南地区标准图集、海南省施工图标准图集、海南省地方节能标准的编制工作，也担负着政府部门的顾问专家、项目评审专家、审图专家及注册建筑师再教育培训的工作，为社会及本行业发展作出很大的贡献。

凭借着公司的业绩和整体实力，公司已跻身为中国建筑协会成员单位及中南建筑协会会员单位。今后公司会继续努力创造企业品牌，为城市发展创作更加优秀的建筑作品。

Hainan Hualei Architectural Design Consulting Co., Ltd. founded in 1988, is the first private enterprise that got the Grade A design qualification certificate approved by the Ministry of Construction (now is Ministry of Housing and Urban-Rural Development) since the foundation of Hainan province, has independent legal personality. Its legal representative is Yu Rui, who is the First Class registered architect of China, registered public facility engineer of China. At present, the company has Grade A building exploration and design and Grade B civil engineering design qualification certificate issued by the Ministry of Housing and Urban - Rural Development, mainly engaging in the construction, municipal engineering, interior decoration project, community planning and design and project planning, engineering, consulting, project feasibility studies and other services. The company which has more than 270 technical staff, including more than 40 registered engineers of China, technical staff with outstanding quality, is a design team with high technical ability, skilled, professional support and reasonable structure, and satisfactory service.

The company has passed international ISO9001: 2000 quality system certification in 2003. It is composed of headquarter, 13 branches in Hainan province, and 8 branches outside Hainan including Wuhan, Hangzhou, Chongqing, Tianjin, Lanzhou, Fuzhou, Shenzhen, Guangxi. There are operations, QA, Human Resources, General Department and other agencies in headquarter, and it introduces advanced management software, implements information management of the company's business in all aspects. Strict internal quality management system, quality checks at each level of each project design, step by step review are to ensure high quality completion of the design task. Meanwhile, to ensure the sustainable development of the company, it focuses on the discovery, training and education of talent, re-educate professional and technical personnel from time to time, and actively participates in all kinds of new technologies and applications.

With emphasis on the services of post-design, it aims to achieve the highest customer satisfaction. With the principle of well-designed, integrity, talent-oriented, continuous improvement, the company has completed a variety of design projects nearly one thousand for decade, taken the top market share in Hainan. In recent years, the company leaders have developed the market actively, the quantities are also expanding. Including many awarded projects, large-scale public buildings, influential local landmark buildings and residential communities with a certain level and scale, they have gained a good social and economic benefits for company.

There are several well-known senior experts in the company, who are responsible for the standard atlas of south-central region in Hainan Province, the standard atlas construction plans in Hainan Province, the preparation of local energy efficiency standards, also charged as advisers to government departments, project evaluation experts, trial plan experts and the work of the registered architects' re-education training, make a contribution to he society and the industry.

With its performance and overall strength, the company has ranked as a member unit of China Construction Association and Middle South Building Association. In the future, the company will make efforts to create corporate brand and excellent buildings for city development.

地址：海南省海口市海甸岛三东路金谷大厦2楼
电话：+86-898-66270303
传真：+86-898-36365414
邮箱：hnhualei@sina.com
网址：www.hn-hualei.com

Add: Second Floor, Jingu Building, Sandong Road, Haidian Island, Haikou City, Hainan Province
Tel: +86-898-66270303
Fax: +86-898-36365414
E-mail: hnhualei@sina.com
Http://www.hn-hualei.com

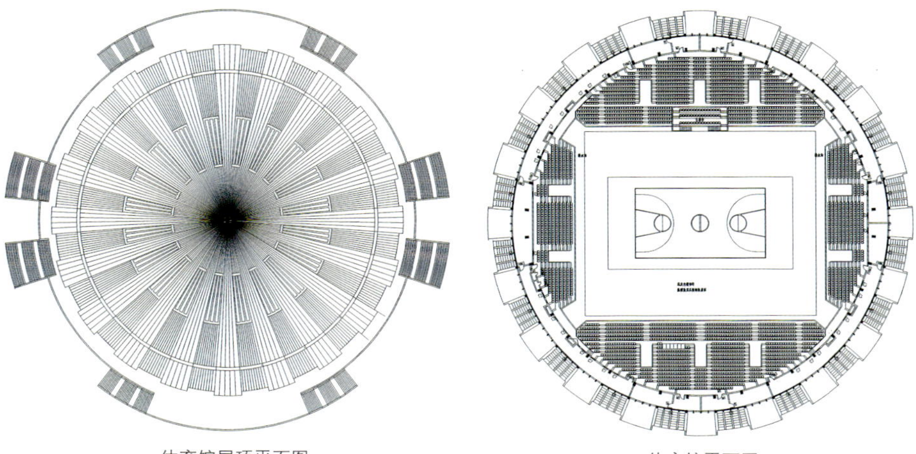
体育馆屋顶平面图　　体育馆平面图

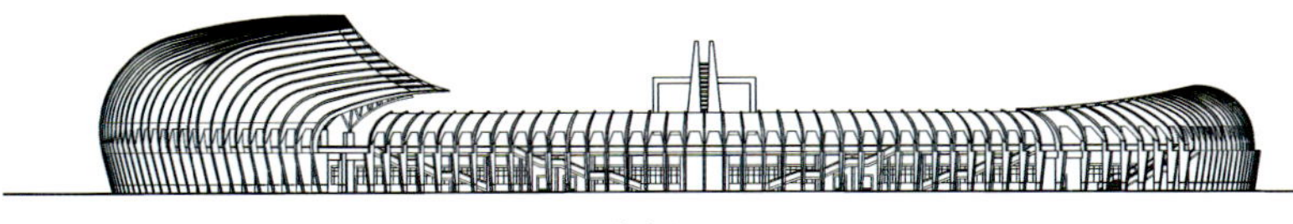
体育场立面图

游泳馆立面图

体育馆立面图

三亚体育中心工程

三亚市体育场：建设地点位于三亚市荔枝沟，建筑面积16 518平方米，容纳人数16 000人，建造时间为2010年，设计单位为海南华磊建筑设计咨询有限公司，施工单位为海南省第五建筑工程公司。

三亚市体育馆：建设地点位于三亚市荔枝沟，建筑面积12 764.8平方米，容纳人数3 000人，建造时间为2010年，设计单位为海南华磊建筑设计咨询有限公司，施工单位为二十三冶建设集团有限公司。

三亚市游泳馆：建设地点位于三亚市荔枝沟，建筑面积4 621.3平方米，容纳人数326人，建造时间为2010年，设计单位为海南华磊建筑设计咨询有限公司，施工单位为二十三冶建设集团有限公司。

Sanya Sports Center Project

Sanya City Stadium: Construction site is Lizhigou Sanya City, building area is 16,518 square meters, capacity of 16,000 people, built in 2010, designed by Hainan Hualei Architectural Design Consulting Co., Ltd., Hainan Province, constructed by Hainan Province No.5 Construction Engineering Company.

Sanya City Gymnasium: Construction site is Lizhigou Sanya City, building area is 12,764.8 square meters, capacity of 3,000 people, built in 2010, designed by Hainan Hualei Architectural Design Consulting Co., Ltd., Hainan Province, constructed by Ershisanye Construction Group Co., Ltd.

Sanya Natatorium: Construction site is Lizhigou Sanya City, building area of 4,621.3 square meters, capacity of 326 people, built in 2010, designed by Hainan Hualei Architectural Design Consulting Co., Ltd., Hainan Province, constructed by Ershisanye Construction Group Co., Ltd.

2010—2011
主要建筑设计作品
Main Architectural Design Works
Annual Review of Chinese Architectural Design Works

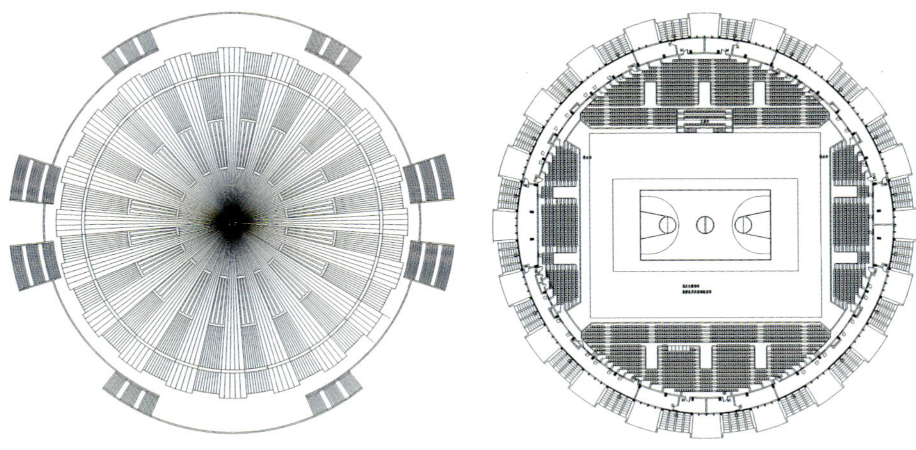

体育馆屋顶平面图　　体育馆平面图

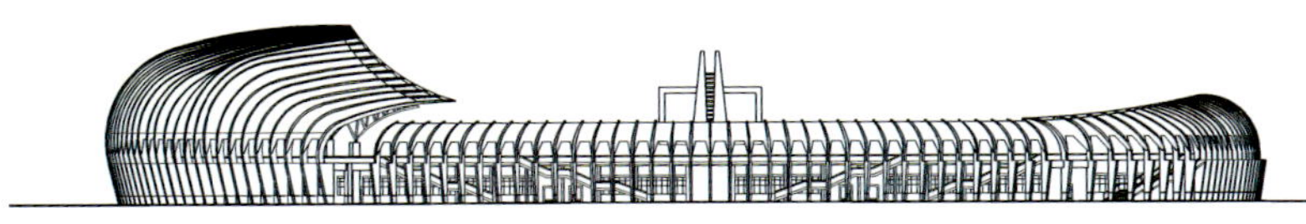

体育场立面图

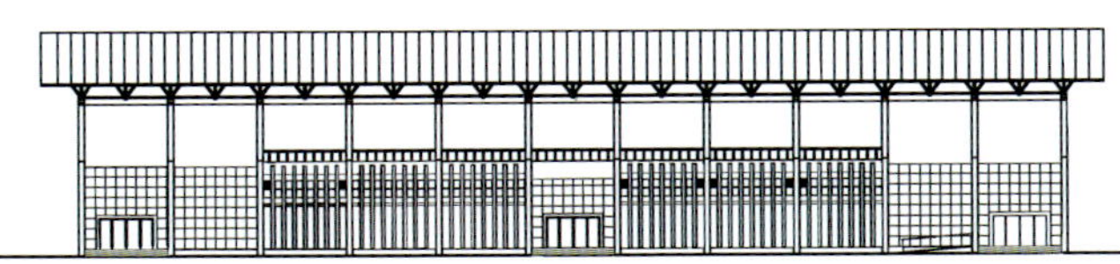

游泳馆立面图

体育馆立面图

三亚体育中心工程

三亚市体育场：建设地点位于三亚市荔枝沟，建筑面积16 518平方米，容纳人数16 000人，建造时间为2010年，设计单位为海南华磊建筑设计咨询有限公司，施工单位为海南省第五建筑工程公司。

三亚市体育馆：建设地点位于三亚市荔枝沟，建筑面积12 764.8平方米，容纳人数3 000人，建造时间为2010年，设计单位为海南华磊建筑设计咨询有限公司，施工单位为二十三冶建设集团有限公司。

三亚市游泳馆：建设地点位于三亚市荔枝沟，建筑面积4 621.3平方米，容纳人数326人，建造时间为2010年，设计单位为海南华磊建筑设计咨询有限公司，施工单位为二十三冶建设集团有限公司。

Sanya Sports Center Project

Sanya City Stadium: Construction site is Lizhigou Sanya City, building area is 16,518 square meters, capacity of 16,000 people, built in 2010, designed by Hainan Hualei Architectural Design Consulting Co., Ltd., Hainan Province, constructed by Hainan Province No.5 Construction Engineering Company.

Sanya City Gymnasium: Construction site is Lizhigou Sanya City, building area is 12,764.8 square meters, capacity of 3,000 people, built in 2010, designed by Hainan Hualei Architectural Design Consulting Co., Ltd., Hainan Province, constructed by Ershisanye Construction Group Co., Ltd.

Sanya Natatorium: Construction site is Lizhigou Sanya City, building area of 4,621.3 square meters, capacity of 326 people, built in 2010, designed by Hainan Hualei Architectural Design Consulting Co., Ltd., Hainan Province, constructed by Ershisanye Construction Group Co., Ltd.

2010—2011
主要建筑设计作品
Main Architectural Design Works
Annual Review of Chinese Architectural Design Works

China Railway Siyuan Survey and Design Group Co., Ltd.

中国铁建 中铁第四勘察设计院集团有限公司

中铁第四勘察设计院集团有限公司（简称铁四院）成立于1953年，是国家大型综合性勘察设计单位，拥有工程技术人员3 272人，其中教授级高工189人、各类执业注册工程师700余人。是首批获得国家工程设计综合资质、甲级测绘资质的国家高新技术企业，是国家铁路建设投资评估单位。

半个多世纪以来，完成全国铁路勘察设计任务的1/3。勘察设计了武广、京沪等高速铁路，设计建成的高速铁路占全国已投入运营的高速铁路总里程的60%以上。

铁四院创立了路网规划、铁路枢纽、高标准铁路、复杂山区铁路、铁路站房、水底隧道、轻轨、地铁、桥梁、软基处理和环境评估等设计品牌。主持了数十项国家、行业规范与标准的编写，建院以来获国家级大奖67项，获得省部级奖302项，其中143项工程设计入选中国企业新纪录。被评为建国60周年全国勘察设计行业“十佳自主创新企业”，获“全国最佳诚信单位”、“全国五一劳动奖状”称号。

China Railway Siyuan Survey and Design Group Co., Ltd. (Railway Siyuan for short), founded in 1953, is a large comprehensive survey and design unit, with 3,272 engineers and technicians, including 189 senior engineers, more than 700 all kinds of registration engineers. It is a high-tech enterprise with National Engineering Design Comprehensive Qualification and Grade-A Qualification of mapping, and is also the assessment unit of state investment in railway construction.

Over half a century, it has completed one third of the task for the national railway survey and design. Survey and design of the Wuhan-Guangzhou, Beijing-Shanghai high-speed railway, and high-speed railway completed account for over 60% of the total mileage of the national high-speed railways which have been in operation.

Railway Siyuan creates road network planning, railway hub, high standard railway, complex mountain railway, the railway station house, underwater tunnels, light rail, subway, bridge, soft ground treatment and environmental assessment, and so on. It presides over dozens of national, industry codes and standards writing, since its establishment, has 67 state-level awards, and 302 provincial and ministerial awards, 143 engineering design selected in New Record of Chinese Enterprises. It was entitled Ten Excellence Independent Innovation Enterprises in the 60th Anniversary of the National Survey and Design Industry, and won the title of The Best Integrity Unit and National Labor Award.

邮编：430063　P.C.: 430063
传真：+86-27-86811444　Fax: +86-27-86811444
电话：+86-27-86812844　Tel: +86-27-86812844
网址：www.crfsdi.com　Http://www.crfsdi.com

南京南站

总建筑面积：38.7万平方米
亚洲最大的铁路枢纽站之一

南京南站的现代性，体现在其"桥建合一"的结构体系上。地上3层，地下2层，加上周边的城市配套设施，既满足了旅客乘坐火车的需要，又使长途汽车、公交车、出租车、地铁等现代交通工具，都能在此实现"无缝"换乘。

南京南站借鉴了中国的传统建筑意向和木构特点，以及南京城墙的肌理特征，以南京中华门（明代称聚宝门）的三重空间序列、宫廷建筑的重檐木构、雨花石的色彩斑斓为元素组织室内外空间。中轴序列，延续六朝古都的瑞气；重檐木构，彰显文化名城的神韵；雨花璀璨，蕴涵江南山水的灵秀。"古都新站"的设计理念很好地体现了铁路客站"文化性"的设计原则，代表了中国铁路客站设计的方向和趋势。

武昌站

铁四院设计的武昌站采用“上进下出”的流线，按线侧下式设计，总建筑面积110 258平方米，其中站房46 733平方米、无站台柱雨棚63 525平方米。

设计从楚城和楚台入手，结合现代铁路站房“以流为主”的空间组织特点，将站房设计成叠台形，运用现代的材料和当地出产的石材干挂相结合，塑造出一个稳重而又有气势的空间形象。外窗借用编钟的形式，与墙体一起，形成连续的韵律。同时借用“廊”的概念，将雨廊与站房结合起来，形成一个灰空间，较好地解决了西晒和交通流线问题，同时也丰富了建筑的室内外空间，完善了建筑造型。建筑独特的屋顶与吊顶连接成一个整体，与下部支柱“脱开”，无论白天还是晚上，仿佛悬浮于半空中，展现了空灵之美。

武汉站

武汉站是我国六大铁路客运中心之一，客站建筑面积33.2万平方米，站房建筑面积11.2万平方米，雨棚建筑面积13.45万平方米，建筑中央屋顶最高处为59.3米，雨棚最高点33米。站房地下3层，地下1层，局部设有夹层。

武汉站建筑造型结合武汉独具的地域文化特色，富有“千年鹤归”、“中部崛起”、“九省通衢”三层寓意。对车站流线进行了大胆创新，运用立体布局、视觉引导、绿色通道等先进理念，为实现“零时间候车、零距离换乘”创造了条件。首度在铁路站房中采用116米的超大跨度结构体系和“桥建合一”的综合结构体系。采用了地源热泵系统、屋顶自然采光、太阳能光伏发电等多项环保节能措施。

广州南站

广州南站是我国目前规模最大、功能最复杂的全高架站桥合一的铁路客运站。车站衔接武广、广深港、广珠城际、贵广各线路，并开行至香港直通旅客列车，设15座站台、28条到发线。最高旅客聚集人数7 000人，高峰小时旅客发送量28 400人。客运用房21.5万平方米，雨棚覆盖面积20.7万平方米，站台下停车场库14.4万平方米。

车站建筑外墙南北向宽448米，东西向进深398米。无站台柱雨棚覆盖范围南北长580米，东西宽475米，建筑最高点距地面52米。

站房建筑造型为体现车站所在地域的特色，建筑以具有南国特色的芭蕉叶作为屋面造型题材，结合站场排布和内部功能布局，分为相同的几个单元，由东向西逐渐升高，建筑与站场融为一体。

站房设计采用高架候车与地面站厅相结合的布局，分为高架、站台、地面三个层次，各类客流根据不同的旅行特点分层、分方向进出站。武广客运专线及直通车在高架层设置候车室，在地面出站"上进下出"；城际铁路在地面出站层设置站厅，"下进下出"，形成便捷、快速的进出站通路，并与地铁结合紧密。

湖南大学设计研究院有限公司

DESIGN AND RESEARCH INSTITUTE OF HUNAN UNIVERSITY CO.,LTD

湖南大学设计研究院有限公司(改制前为湖南大学设计研究院)是国家批准的甲级设计研究单位。拥有建筑设计甲级、市政道桥设计甲级、咨询甲级资质，同时具有施工图审查一类，城市规划甲级，旅游区规划甲级，风景园林、给排水、环境工程公路设计乙级以及工程勘察乙级资质，并于2000年通过了ISO9001质量体系认证。

本院现有专职技术人员200余人，其中国家一级注册建筑师、一级注册结构工程师、注册规划师、注册设备工程师、高级工程师等共约120人。技术力量雄厚，专业配置完善，设备先进。对外承接各类高层及大中型民用建筑设计、市政道路、桥梁设计、工业建筑、城镇规划、旅游区规划设计、室内设计、施工图审查、工程咨询、项目代建管理和工程总承包等业务。

本院依靠自身实力和高校人才技术优势，注重设计实践与理论研究相结合、工程技术与建筑艺术创新，注重业主利益和社会效益，以高效率、高质量的设计赢得了社会各界的赞誉与好评。近年来在省内外承担了大量有影响力的项目。

综合办公建筑，包括长沙亚大数码港、恒隆国际、交警大楼、湖南广播电台、湖南省人民银行综合楼、湖南电力院大楼、湖南商务会馆、衡阳广播电视中心、岳阳市公安局指挥大楼，广东省湛江市市委市政府大楼、广西公安厅技术大楼、湖南公安信息大楼等。

居住建筑，包括长沙北辰新河三角洲超高层住宅、长沙中信城市花园（二期）、梦洁金色屋顶、郡源广场、永琪西京、长沙阳光100（3号地块北区）、长沙卓越蔚蓝海岸（二期）、长沙双盈卧龙湾（二期）、长沙嘉华城等高档社区的规划设计，以及长沙卓越麓山别墅、香格里麓山别墅（二期）、橘郡花园别墅、株洲惠天然、衡阳凯星名城、岳阳南湖“月影湾”、郴州市人民电影院综合楼、河南信阳公务员小区等规划与单体方案设计。

商业建筑，包括长沙王府井商业广场、阿波罗商业广场、友谊商店、黄兴南路商业步行街（西厢南段）、长沙奥特莱斯商业街、株洲中国城、株洲中央商城、株洲世贸、益阳福中福国际商贸城等。

酒店建筑，包括长沙皇冠假日大酒店（五星）、湖南影视会展山庄酒店（五星）、华雅华天大酒店（方案，五星）、长沙立达人酒店（五星）、长沙开源新城国际大酒店（五星）、益阳大酒店（五星）以及索溪峪大酒店（四星）等。

教育建筑如：湖南大学校园规划及单体设计，湖南师范大学图书馆及音乐培训楼、国防科大计算机学院银河楼、湘潭大学校园规划及体育馆设计，湖南城市学院校园规划及单体设计、湖南工业大学音乐楼、南华大学新校区规划及单体设计，湖北咸宁职业技术学院主校区规划设计，江西理工大学体育馆、河南信阳师范学院体育馆等设计。

此外，还完成了长沙高新技术开发区、长沙经济开发区等十多项大型规划和单体建筑设计。

在市政道桥、公路设计方面完成了很多有影响力的项目，如长沙机场高速路、常德机场高速路、常德大道，长沙二环龙王港立交桥、长宁路立交桥、浏阳河洪山庙大桥等。在景观工程设计方面完成了长沙芙蓉路、浏阳河路、岳阳楼景区、长沙高新技术开发区城市设计和湘江大道南段滨江风光带等多项大型景观与环境项目。

此外，我院每年都有一大批项目获国家、部、省级优秀工程设计奖或竞赛投标奖。

地址：湖南省长沙市岳麓山湖南大学一舍
邮编：410082
电话：+86-731-88821068
传真：+86-731-88824092
网址：www.hdsjy.cn
邮箱：hdsjy@vip.sina.cn

总 经 理：唐国安（兼总建筑师）
副总经理：刘子毅（兼副总建筑师）
郦世平（兼总工程师）
王新夏
池 峰
项丹强

1–2 湛江市行政中心办公楼
3–4 衡阳市广电中心
5 岳阳市公安局指挥中心大楼
6 湖南省电力院

1

3

2

5

4

6

曾益海

国家一级注册建筑师
国家注册规划师
教授级高级工程师
中国建筑学会会员
香港建筑师学会会员
湖南省建筑师学会副理事长
中铝国际长沙有色冶金设计研究院有限公司总建筑师

Class 1 Registered Architect (PRC)
State Registered Urban Planner
Senior Engineer (Professor Level)
Member of Architectural Society of China (ASC)
Member of the Hong Kong Institute of Architects (HKIA)
Deputy Director of Hunan Province Architect Association
General Engineer of Changsha Engineering and Research Institute Ltd. of Nonferrous Metallurgy OF CHINALCO

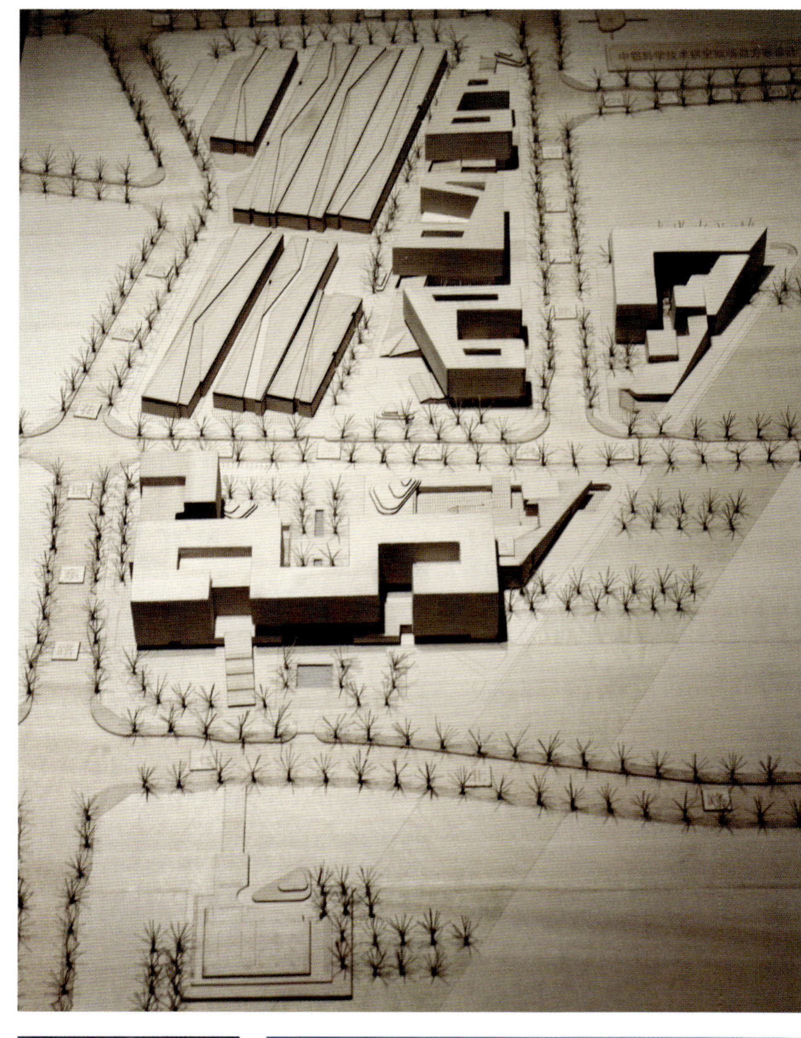

■ 中铝集团科学技术研究院
Science and Technology Institute of CHINALCO
项目位置: 北京
项目规模: 30万平方米
设计时间: 2010年

曾益海工作室 *ZENG YIHAI STUDIO*

Tel: +86-731-84397140
Fax: +86-731-84458515
E-mail: zengyihai@yahoo.cn
Http:// www.cinf.com.cn

中铝国际
长沙有色冶金设计研究院有限公司
Changsha Engineering and Research Institute Ltd. of Nonferrous Metallurgy of CHINALCO

■ 花瑶风景名胜区游客接待中心
Huayao Scenic Areas Reception Center
项目位置：湖南 隆回
设计时间：2010年

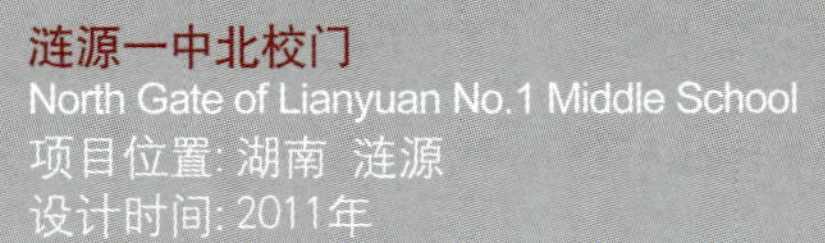

■ 涟源一中北校门
North Gate of Lianyuan No.1 Middle School
项目位置：湖南 涟源
设计时间：2011年

乌鲁木齐环保局办公楼
Urumchi Pro-environment Bureau's Office
项目位置：新疆 乌鲁木齐
项目规模：4.6万平方米
设计时间：2011年

中国黄金集团拉萨基地
China National Gold Group Corporation, Lhasa Base
项目位置：西藏 拉萨
项目规模：7.4万平方米
设计时间：2011年

中铝科技大厦

Building of China Aluminum Tech.

项目位置: 湖南 长沙
项目规模: 13万平方米
设计时间: 2011年

湖南省总工会项目
Trade Unions of Hu'nan Province
项目位置：湖南 长沙
设计时间：2010年

内蒙古鄂尔多斯市准格尔旗包子塔景区

Baozita Scenic Spots in Zhungeer, Ordos, Inner Mongolia

项目位置：内蒙古 鄂尔多斯

设计时间：2011年

中机国际工程设计研究院有限责任公司

China Machinery International Engineering Design & Research Institute Co., Ltd.

中机国际工程设计研究院有限责任公司（原机械工业部第八设计研究院），创建于1951年5月5日，是集工程咨询、工程设计、工程总承包、工程监理和项目管理于一体的大型甲级综合设计研究院，坐落在风景秀丽的历史文化名城——湖南长沙。

经过60年的风雨磨砺，中机院已发展成为工业工程、民用建筑、市政环保等领域极具竞争实力和良好声誉的高新科技企业。具有机械、军工、冶金、建筑、轻纺、市政、环境保护、风景园林等行业工程设计咨询甲级资质；电力、化工石化、医药、农林、商物粮、石油天然气等行业工程设计咨询乙级资质；房屋建筑、市政公用工程监理甲级、机电安装工程监理乙级资质；工程造价甲级资质；城市规划、建设项目环境影响评价资质；压力管道、压力容器设计许可证；具有对外承包工程经营资格、进出口企业资格证书和独立的进出口经营贸易权；质量管理体系认证证书和AAA企业信用等级证书。

作为科技型企业，中机院拥有高素质的科研、设计人才700余人，涵盖30多个专业门类。建院60年来，已累计完成国内外工程咨询、设计、监理、项目管理或总承包项目13 000余项，取得科学研究成果近500余项。科研设计成果遍及全国各地和世界十多个国家和地区；获得国家、部、省级优秀科技成果、优秀设计奖300余项，拥有国家专利技术和专有技术50余项，是国家多项技术标准、规范的主编及参编单位。

中机院秉承中华文化的深厚底蕴，“共创、共建、共赢、共享”是中机院企业文化的精髓，它激励全体员工发扬企业精神，求真务实，继往开来，以高度的责任感服务于社会。

China Machinery International Engineering Design & Research Institute Co., Ltd.(originally the Eighth Design & Research Institute of the Ministry of Machine-building Industry) is established in May 5th, 1951. It is a large-scale Grade-A comprehensive design and research institute at the State-level with business scope covering fields such as engineering consultation, engineering design, general contracting, engineering supervision and project management. The Institute is situated at the picturesque historic city of Changsha.

After development of almost 60 years, our Institute is now a sci-tech oriented enterprise that is highly developed with competition strength and reputation both at home and abroad in such fields as industrial engineering, civil architecture engineering, municipal and environmental protection engineering, and so on.

Our Institute is authorized with Grade-A engineering design and consultation certificate in such areas as machine-building, military projects, metallurgy, architecture, textile, municipal services, environmental protection, landscape; Grade-B engineering design and consultation certificate for industries as power generation, chemical and petrochemical engineering, pharmaceutical industry, agriculture and forest, commercialized food circulation and storage, petroleum and natural gas; Grade-A engineering supervision certificate in building architecture, municipal public project; Grade-B engineering supervision certificate in electromechanical installation; Grade-A certificate in engineering cost; evaluation certificate of urban plan, environmental effect of construction project; design permission certificate in pressurized vessels and pipelines; It is also empowered with business certificate of foreign contracting projects and independent trading rights for import and export by the state. The quality control system of the Institute has been successively confirmed with ISO9000 (Version 94 and Version 2000) and are now undergoing further perfections.

As a sci-tech oriented enterprise, our Institute is in possession of 700 highly-qualified research and design personnel, covering more than 30 faculties and specialties. We have successfully accomplished more than 13,000 projects of engineering consultation, design, supervision, project management or general contracting home and abroad, together with 500-odd sci-tech research results. Our sci-tech research and engineering designing projects now spread all over the country and reach more than 10 other countries and regions with conferment of more than 200 prizes at the State, ministerial and provincial levels. We now possess 50-odd State patents and patented technology, and are the editing or co-editing unit of many state technical standards and codes.

By inheriting the profound heritage of Chinese culture, we have established our corporate culture of “Creation, Construction, Benefit and Sharing ”. With constant stimulation on creativeness and service spirit of our staff and strong sense of social responsible, we provide our customers with first-class consulting services in investment decision-making, engineering design, general contracting, engineering supervision, technical consultations etc.

中国（长沙）工程机械交易展示中心

项目地点：湖南 长沙
项目状态：方案设计
设计时间：2009年
总建筑面积：951 410平方米

南湖一号综合体

项目地点：湖南 长沙
项目状态：建设中
设计时间：2010年
总建筑面积：111 223.63平方米（其中地下20 306.16平方米）

蒙古国体育中心

项目地点：蒙古 乌兰巴托
项目状态：待建
设计时间：2010年
总建筑面积：110 000平方米
设计理念：该设计以蒙古包的形态为出发点，经过折叠倒转等手法，通过现代的材料、时尚的元素、干净的色彩来表现建筑传统中内在的韵味。

湖南理工学院综合实验楼

项目地点：湖南 岳阳
项目状态：2010年竣工（2011年湖南省优秀设计二等奖）
设计时间：2008年
总建筑面积：24 000平方米

托斯卡纳

项目地点：湖南 长沙
项目状态：已建成
设计时间：2006年
建筑面积：20万平方米

中城·丽景香山

项目地点：湖南 长沙
项目状态：已建成
设计时间：2005–2007年
建筑面积：45万平方米

青竹湖畔住宅小区南区

项目地点：湖南 长沙
项目状态：一、二期已建成，三、四期正在设计施工
设计时间：2008–2011年
建筑面积：45万平方米

长沙铜官窑遗址博物馆

项目地点：湖南 长沙
项目状态：建设中
设计时间：2011年
总建筑面积：13 000平方米
设计理念：铜官窑遗址博物馆建筑顺应山势，让人在建筑中的行为过程仿佛“龙窑”生产的过程，预示一件陶瓷器物从泥土中来，又回到泥土中去的命运轮回，而建筑的过程仿佛也是瓷器的一个生命历程。

军魂楼

项目地点：湖南 湘潭
项目状态：建设中
设计时间：2011年
总建筑面积：13 000平方米

长沙市总工会帮扶中心大楼

项目地点：湖南 长沙
项目状态：建设中
设计时间：2010年
总建筑面积：15 000平方米
设计理念：建筑主体7层，局部8层，地下1层。办公大楼平面呈方形布局，现代风格的立面设计简洁大方，在建筑的入口挑高处设张拉膜雕塑，利用现代材料的处理来赋予办公建筑新的特色。

桃江大汉龙城

项目地点：湖南 益阳
项目状态：待建
设计时间：2011年
总建筑面积：200万平方米

湖南移动枢纽楼环境景观设计

项目地点：湖南 长沙
项目状态：已建成
设计时间：2011年
景观面积：1.3万平方米
设计理念：提炼蒙德里安简朴、几何的抽象主义创作艺术，融合现有住宅与办公楼环境景观，营造品质卓越的和谐办公环境。

邵东县妇幼保健院（邵东县妇女儿童医院）

项目地点：湖南 邵阳
项目状态：建设中
设计时间：2011年
建筑面积：7.2万平方米

中机国际工程设计研究院技术研发中心

项目地点：湖南 长沙
项目状态：待建
设计时间：2011年
总建筑面积：10万平方米

怀化市妇幼保健院（怀化市妇女儿童医院）新院建设项目

项目地点：湖南 怀化
项目状态：方案设计
设计时间 2011年
建筑面积：8.6万平方米

哈尔滨工业大学建筑设计研究院

THE ARCHITECTURAL DESIGN AND RESEARCH INSTITUTE OF HIT

哈尔滨工业大学建筑设计研究院成立于1958年，是持有多项国家甲级资质的国有大型勘察设计机构，综合实力跻身国内同行前列，位列全国民用建筑设计市场排行榜前十名，并荣膺“中国十大建筑设计公司”、“中国勘察设计协会优秀设计院”等荣誉称号。

单位现有员工500余人，技术人员占全院总人数的90%，其中50%拥有硕士及以上学历，注册人员占技术人员总数的35%，拥有中国工程院院士、国家设计大师等一批资深专业人士以及体育、博览、医疗、教育建筑，钢结构，水处理等设计技术领域的权威专家。高端人才与先进技术的融汇集成，把该院铸就为敢打敢拼、善战能胜的精英团队。

单位业务涵盖建筑设计、城市规划、室内外环境景观与市政工程设计、交通工程设计，同时延伸到投资策划、工程咨询、岩土勘察、项目代建等相关产业。在复杂的公共建筑工程和大型城市住区的规划设计方面成就突出，钢结构、环保、节能、智能等新技术开发和应用，也处于国内先进水平。

单位设计作品遍布祖国各地并远及俄罗斯等国家和地区，已完成哈尔滨工业大学主楼、中国抗美援朝纪念馆、北京石景山体育馆等经典代表作品和哈尔滨国际会展体育中心、北京四季滑雪馆等高技术难度的标志性工程以及圣彼得堡“波罗的海明珠”等深具国际影响力的大型项目。

单位严格执行ISO9001:2000质量管理体系，积极投保工程设计责任保险，坚持以质量和服务开拓市场，赢得信誉。200余项设计作品荣获国家、省、市各级奖励，在全国优秀勘察设计奖、詹天佑土木工程大奖、中国建筑创作奖、空间结构设计奖、建设科学技术进步奖等工程设计领域大奖评选中屡获殊荣。

单位将秉承苛求完美、精益求精的设计宗旨和诚信服务、持续发展的经营理念，充分发挥市场、人才、技术和管理优势，不断提升整体核心竞争力，与社会各界广泛合作，以自身的奋力拼搏和不懈努力领跑行业、奉献精品、服务社会。

办公地址：黑龙江省哈尔滨市南岗区黄河路73号（哈工大二校区）
通信地址：黑龙江省哈尔滨市南岗区海河路202号2545信箱
电　　话：+86-451-86283317
传　　真：+86-451-86283319
邮　　箱：hgdsjy@vip.163.com
网　　址：www.hitadri.cn

Office Address: No. 73, Huanghe Road, Nangang District, Harbin, Heilongjiang (campus II of HIT)
Mail Address: Box 2545, No. 202, Haihe Road, Nangang District, Harbin, Heilongjiang
Tel: +86-451-86283317
Fax: +86-451-86283319
E-mail: hgdsjy@vip.163.com
Http: //www.hitadri.cn

1–2 哈尔滨工业大学主楼

建设地点：黑龙江 哈尔滨
建筑面积：19 928平方米
设计时间：1959年
建成时间：1965年

3–4 抗美援朝纪念馆

建设地点：辽宁 丹东
建筑面积：13 790平方米
设计时间：1984年
建成时间：1993年

5–7 北京亚运会朝阳体育馆

建设地点：北京
建筑面积：8 900平方米
设计时间：1986年
建成时间：1988年

8

10

11

13

9

12

14

15

16

8 沈阳保利心语花园

建设地点：辽宁 沈阳
建筑面积：720 000平方米
设计时间：2007年
建成时间：2008年

9 盘锦市中心医院

建设地点：辽宁 盘锦
建筑面积：180 000平方米
设计时间：2009年
建成时间：在建

10–13 爱建滨江国际社区

建设地点：黑龙江 哈尔滨
建筑面积：2 600 000平方米
设计时间：2003年
建成时间：2008年

14–15 丹东市体育中心

建设地点：辽宁 丹东
建筑面积：80 400平方米
设计时间：2009年
建成时间：在建

16 大庆奥林匹克公园体育馆

建设地点：辽宁 大庆
建筑面积：72 200平方米
设计时间：2009年
建成时间：在建

17–19 哈尔滨国际会展体育中心

建设地点：黑龙江 哈尔滨

建筑面积：370 000平方米
设计时间：2002年
建成时间：2004年

21 黑龙江省图书馆新馆

建设地点：黑龙江 哈尔滨
建筑面积：30 000平方米
设计时间：1999年
建成时间：2003年

21 北京航宇大厦

建设地点：北京
建筑面积：104 458平方米
设计时间：2004年
建成时间：2006年

22

23

24

25

26

27

22–24 东软集团国际软件园

建设地点：辽宁 大连
建筑面积：470 000平方米
设计时间：2006年
建成时间：2008年

25 大连民族学院新校区

建设地点：辽宁 大连
建筑面积：220 000平方米
设计时间：2006年
建成时间：在建

26 北京四季滑雪馆

建设地点：北京
建筑面积：197 000平方米
设计时间：2001年
建成时间：在建

27 黑龙江省博物馆新馆

建设地点：黑龙江 哈尔滨
建筑面积：50 000平方米
设计时间：2008年
建成时间：在建

哈尔滨市建筑设计院
Harbin Architectural Designing Institute

哈尔滨市建筑设计院是具有国家甲级建筑设计和国家首批建筑智能化专项工程资质的设计单位。自1952年创立至今，经过近六十年的奋斗历程，现已发展成为专业设置齐全、设计手段先进、技术实力雄厚、社会信誉卓著的优秀设计单位。全院职工175人，教授级高级技术职称25名，高级技术职称75名，具有国家级执业资格注册人员45名。机构设置为：三个设计分院、建筑规划设计所、工程监理公司、环境艺术设计所、工程技术咨询部、施工图设计审查有限公司、市政分院等。

回顾我院近六十年的发展历程，走过了一条艰苦创业、艰难曲折、再创辉煌的道路。20世纪五六十年代大规模经济建设时期，老一代设计师以艰苦奋斗的创业精神和聪明才智，创作了一批又一批工业与民用建筑。其中，哈尔滨市人民防洪胜利纪念塔、哈尔滨市工人文化宫、哈尔滨友谊宫等建筑，已被我市市民和中外宾客视为哈尔滨市的标志性建筑，并已载入中国建筑史册。进入21世纪，全院职工秉承“科学管理，竭诚服务，精心设计，质量第一”的执业宗旨，加快改革发展的步伐。设计工作在坚持传承哈尔滨市原有历史文脉的同时，不断繁荣建筑创作，强化精品意识，设计理念有了更新的突破。近十几年来先后有80余项工程设计荣获部级、省级、市级等奖励。体现现代城市风貌和时代感的哈市地铁指挥中心，金鼎世纪文化广场、曼哈顿商厦等建筑作品，已成为我市新时期的代表建筑作品及新的城市标志性建筑。哈尔滨人民防洪胜利纪念塔再次荣获中国建筑学会建筑创作大奖，还与圣·索菲亚广场扩建工程同时被广大市民评选为市民心中的“哈尔滨十大名片”。2009年最具影响力的公共建筑观江国际高级商住楼、哈药大厦（哈公馆）荣获全省优秀工程勘察设计三等奖，44层的观江国际商住楼已被誉为东北地区第一超高楼层、最好楼盘、高档住宅社区。2010年我院主持设计的大庆奥林国际公寓A、D区工程在第六届全国优秀建筑结构设计评选中，荣获优秀建筑结构设计二等奖。近六十年来，我院已向社会提供了2万余项设计成果，工程项目不仅遍布哈尔滨、黑龙江，而且拓展到北京、上海、广东、辽宁等省市，以及境外多个国家。

近几年来，我们与境外多家著名科研、设计机构建立了长期合作关系，积累了组织设计大型工程的丰富经验，整体核心竞争能力不断提升。工作中有中国人民财产保险公司工程设计责任保险的诚信保障和ISO9001质量管理体系的严格执行，相信用我们诚实的劳动、聪明的智慧一定会不断创建出更加完美的建筑精品。

With Class-A Architectural Design License, Harbin Architectural Designing Institute is also one of the earliest design institutes which received Architectural Intelligence Special Engineering Qualification. Founded in 1952, after 60 years of development, it has grown into an excellent design institute with complete professional divisions, advanced design tools, technical strength and good social reputation. It has total 175 employees, including 25 Professor Senior Architects, 75 Senior Architects and 45 Registered Architects. The institute is divided into: three branches, architectural planning & design institute, project supervising company, environmental arts design institute, department of engineering technology consulting, construction drawing design & examining Co., Ltd, branch office of Municipal Administration etc.

Looking back at nearly six decades of development, we have gone through a hard, arduous, yet glorious road. In the large-scale economic development period between 1950s and 1960s, the older generation of designers created many industrial and civil buildings by their hard work, entrepreneurial spirit and wisdom. Among which, Harbin Flood Victory Monument, Harbin Workers' Cultural Palace, Harbin Friendship Palace and other buildings have been considered as landmarks by citizens and foreign guests and recorded in China Architectural History. After entering into 21 century, the institute adheres to the practice purpose of "Scientific Management, Heartedly Service, Excellent Design, and Quality Comes First" and accelerates the pace of reform and development. While adhering to the history and culture of Harbin, the design work continuously focuses on architectural creation and strengthens the awareness of competitive products, with updated breakthroughs of design ideas. Over the last decade, there have been more than 80 engineering designs which won the ministerial, provincial and municipal awards. Modern urban style buildings with contemporary features: Harbin Subway Command Center, King Century Culture Square, Manhattan Commercial Building and other architectural works have become the masterpieces of architectural design works of Harbin in the new period and the new city landmarks. Harbin Flood Victory Monument again wins the Architectural Creation Award given by Architectural Society of China, and it is also selected as "Ten Business Cards of Harbin" by citizens together with Expansion Project of St. Sofia Plaza. The most influential public buildings in 2009: Riverview International High-class Commercial and Residential Building, Harbin Pharmaceutical Building (Ha Manor) win the third prize for Excellent Engineering Survey and Design of Heilongjiang. The 44-story Riverview International Commercial and Residential Building is known as the No.1 ultra-tall and best residential community in the Northeast. The Daqing Aolin International Apartment A, D designed by our institute in 2010 wins the second prize for Excellent Architectural Structure Design in the Sixth National Competition of Excellent Architectural Structure Design. In recent six decades, our institute has provided more than 20,000 design works to society, and engineering projects cover not only all over Harbin, Heilongjiang, but also expand to Beijing, Shanghai, Guangdong, Liaoning and other provinces, as well as many foreign countries.

Over recent years, we have established long-term relationships with several well-known foreign science & research and design institutions, and have accumulated rich experience in designing large-scale projects. The overall core competitiveness has been increasing continuously. With integrity of Engineering Design Liability Insurance of PICC and strictly-execution of ISO9001 Quality Management System, we believe that our honest work and wisdom will continuously create more and more perfect architectural buildings.

地址：黑龙江省哈尔滨市道里区友谊路117号
邮编：150010
电话：+86-451-84675544
+86-451-84618448
传真：+86-451-84633760
邮箱：hjzsjy@163.com

Add: No. 117 Youyi Road, Daoli District, Harbin, Heilongjiang
P.C.: 150010
Tel: +86-451-84675544
+86-451-84618448
Fax: +86-451-84633760
E-mail: hjzsjy@163.com

哈尔滨国际汽车城

建设地点：黑龙江 哈尔滨
建筑性质：商业
建筑面积：40 858.58平方米，高度17.40米
功能布局：一、二层为汽车商店，地下室为库房及停车库

新透笼购物广场

建设地点：黑龙江 哈尔滨
建筑性质：商业
建筑面积：总占地14 217.78平方米，总建筑面积108 318.44平方米，
其中地上84 568.44平方米，地下两层23 750平方米
建筑高度及层数：主楼12层，限高为56.6米；裙房7层，限高为36米；地下2层
功能布局：地下二层为人防车库，地下一层至地上7层为商场，
8–12层功能为展厅、办公等附属用房

群力立特商业工程

建设地点：黑龙江 哈尔滨
建筑性质：商业建筑
建筑面积：24 440.79平方米
建筑高度：27米
功能布局：地下一层为停车库及设备用房；地上为商业

哈尔滨市城乡规划展览馆

建设地点：黑龙江 哈尔滨
建筑性质：规划展馆
建筑面积：地上建筑面积20 000平方米
建筑高度及层数：地上1层，地下4层，建筑高度为24.00米
功能布局：地下一、二层均为车库及设备用房
立面为简欧风格，黄色外墙涂料。

地铁控制中心

地铁控制中心位于哈尔滨市南岗区西大直街与和兴路交口处西北角地段清滨公园车站旁。

用地面积为10 561.59平方米，总建筑面积47 855.44平方米；主楼22层，裙楼5层，地下3层，建筑高度99米。结构形式为框架-剪力墙结构。

立面造型以单纯的方格窗为基调，重视对建筑体积感的塑造，打造新颖大方、富有时代特色的标志性建筑。

哈尔滨市跃进小学校教学楼

建设地点：黑龙江 哈尔滨
建筑性质：教育教学
建筑面积：用地面积12 000.00平方米、总建筑面积8 888.87平方米。其中地上8 559.67平方米，局部地下329.20平方米。建筑总高度21.75米
功能布局：一层至五层均为教育教学及相关办公用房
建筑采用框架结构，耐火等级为二级。

眼科医院公寓

建设地点：黑龙江 大庆　　建筑性质：高层公寓
建筑面积：总建筑面积60 123.66平方米
建筑高度及层数：地下2层，地上17层，建筑高度为99.35米
功能布局：地下车库及设备用房，一、二层商业，三层以上跃层公寓

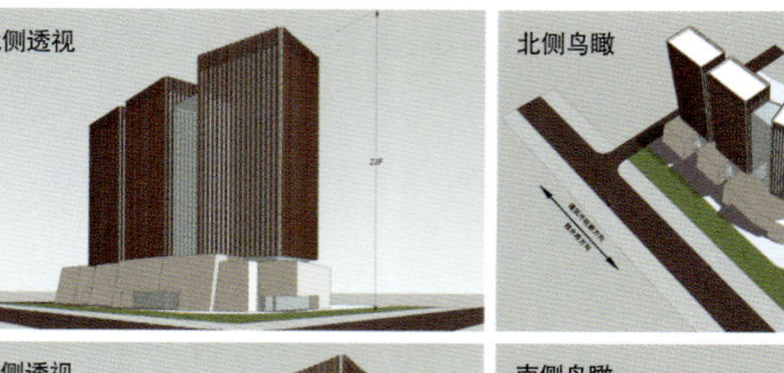

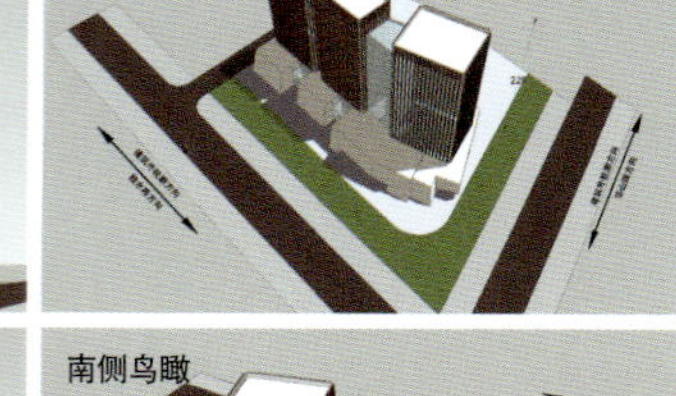

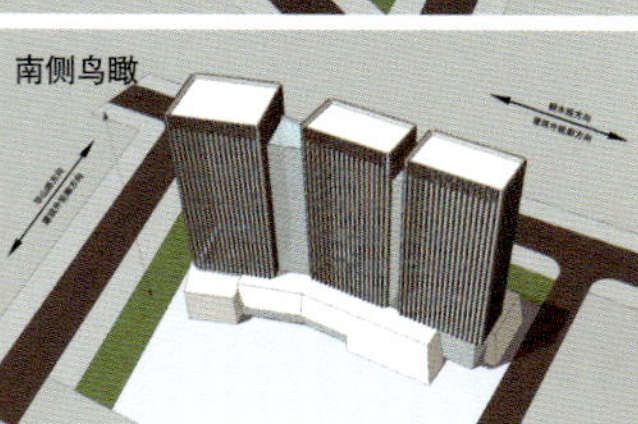

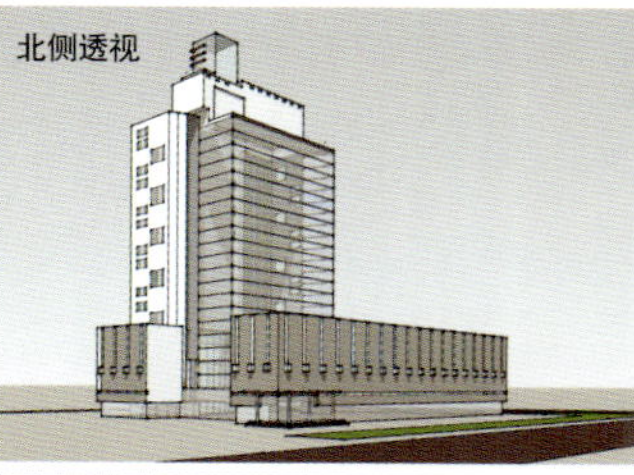

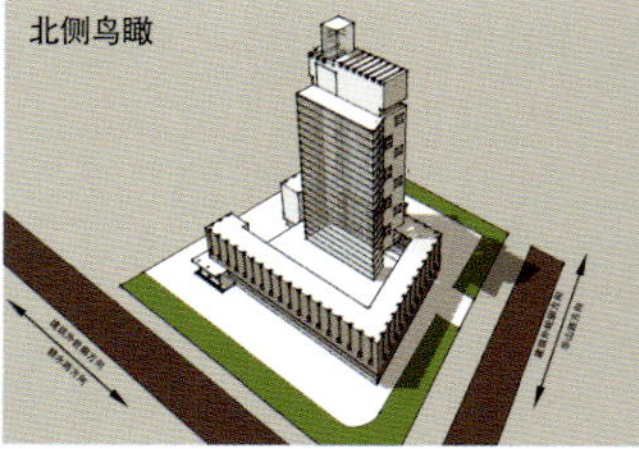

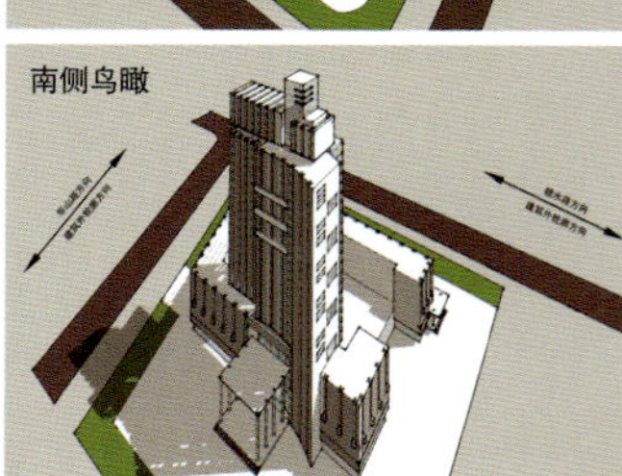

哈尔滨市广播电视局办公楼设计方案

建设地点：黑龙江 哈尔滨
建筑性质：办公用房
建筑面积：地上建筑面积49 887.21平方米
建筑高度及层数：地上22层，建筑高度为92米
功能布局：均为办公用房

防洪纪念塔东侧附属用房工程

建设地点：黑龙江 哈尔滨　　建筑性质：办公用房
建筑面积：地上建筑面积2 740.2平方米
建筑高度及层数：地上3层，地下1层
功能布局：地下一层为车库及设备用房，地上为办公用房

观江国际住宅小区

建设地点：黑龙江 哈尔滨
建筑性质：住宅
建筑面积：总占地2.33公顷，
总建筑面积205 450平方米，
其中公建建筑面积21 500平方米，
住宅建筑面积183 950平方米
建筑高度及层数：主楼42层，限高为150米；
裙房4层，限高为19米；地下2层
功能布局：地下一、二层为人防车库，
地上层为高级住宅

本工程设计总户数895户，居住人口2 685人。

明顺公寓

建设地点：黑龙江 大庆
建筑性质：高层公寓
建筑面积：总建筑面积30 304.8平方米
建筑高度及层数：地下2层，地上30层，
建筑高度为94.35米
功能布局：地下车库及设备用房，
一层接待大堂、小商铺，
二层会馆，三层以上公寓

立面为现代风格，外饰面材质为氟碳漆。

哈尔滨市阿城区海富城居住小区

建设地点：黑龙江 哈尔滨
建筑性质：商业居住混合建筑
建筑面积：781 248.59平方米
建筑高度及层数：地上主要建筑类型有30层、26层、18层、11层、9层、6层；商服建筑类型为3层低层建筑、2层低层建筑、1层裙房；另含地下建筑1层。小区内建筑高度99米。
功能布局：地下一层为车库及设备用房，地上商业建筑临街而建，以靠近上京大道、金涤路为主

立面：以装饰艺术风格为主，屋顶采用混凝土挂瓦，建筑利用苯板造型，进行真石漆喷涂，建筑底部一、二层采用外挂天然大理石的处理方式，重视建筑立面造型变化的丰富性，以塑造极具当代建筑特色的标志性建筑。

哈尔滨地铁一号线电表厂车站（原三角地基坑修复及地铁主变电所、出入口风亭配套工程）

建设地点：黑龙江 哈尔滨
建筑性质：丁类生产厂房
建筑面积：地上建筑面积2 171.58平方米，地下建筑面积9 644.80平方米
建筑高度及层数：地上2层，地下2层，建筑高度为11.70米
功能布局：地下一、二层均为车库及设备用房，地上为变电所

立面为简欧风格，采用黄色外墙涂料。

群力新区中继泵站水处理及监控中心

建设地点：黑龙江 哈尔滨
建筑性质：多层公建和丁类厂房
建筑面积：总建筑面积6 839.14平方米
建筑高度及层数：地上6层，地下1层，建筑高度为22.950米
功能布局：地下1层车库及设备用房，地上6层办公楼，东侧临建

立面为简约欧洲中世纪风格，外饰面材质为砖红色面砖及灰色花岗岩。

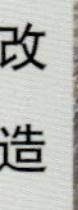

改造前

改造后

圣·索菲亚广场改造工程

建设地点：黑龙江 哈尔滨
主要使用功能：商场
建筑面积：116 954.40平方米
建筑高度及层数：地下1层，地上6层，建筑总高度为23.95米
建筑物结构类型：框架结构

设计说明

本工程为圣·索菲亚广场改造工程，主要内容包括：圣·索菲亚广场周边建筑及广场空间环境的改造；整个街区地下空间的开发；通过对原有地下出入口的整合改善周边交通环境。

本工程新建广场13 000平方米，加上原有广场面积6 600平方米，改造后圣·索菲亚广场总建筑面积19 000平方米。其中，新增绿化面积5 500平方米，加上原有圣·索菲亚广场绿化面积2 000平方米，改造后的圣·索菲亚广场总绿化面积为7 500平方米。本工程中只对华联金太阳做立面改造，为保证广场整体空间一致，在华联和金太阳之间新建一座6层商服楼及三拱门廊，建筑面积4 754平方米，建筑高度23.95米。广场地下为12 200平方米的地下室，其中商场部分6 180平方米，其他地下商业街出口面积1 020平方米，地下车库3 980平方米，停车位99辆。地下商场部分层高为6米，地下车库层高为3.5米。

香格里拉通江亲水广场泵站改造工程

建设地点：黑龙江 哈尔滨
建筑性质：立面改造
建筑面积：用地面积20 000.00平方米，
总建筑面积300平方米
建筑总高度：23.70米
功能布局：在原有泵站的基础上加设挑廊、塔楼、连廊等，用泵站的塔楼控制着广场的空间，同时与公园树丛中的“塔楼”遥相呼应；泵站后侧为“斜廊”，与防洪路及带状公园以调侃式的方式相连，十分有趣；泵站沿友谊路一侧增加较为现代的“空架”，不仅将原泵站的两个独立的房子连接在一起，同时形成第五立面，使得在公路桥一侧的视觉效果焕然一新。

改造前

改造后

医大四院立面改造工程

建设地点：黑龙江 哈尔滨
建筑性质：医疗建筑外墙节能改造
建筑高度及层数：地上10层
立　　面：欧式风格

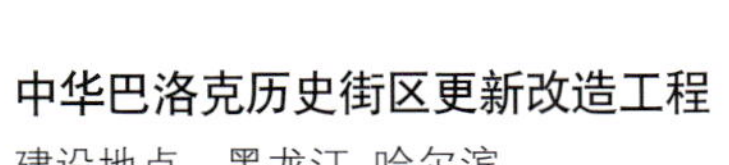

中华巴洛克历史街区更新改造工程

建设地点：黑龙江 哈尔滨
建筑性质：商业建筑
建筑面积：地上建筑面积14.5万平方米，地下建筑面积2.9万平方米
建筑高度及层数：地上2-3层
功能布局：地下为车库及设备用房，地上为商业建筑

立面主要以中华巴洛克历史风貌为主，适当结合现代元素。

哈尔滨天宸建筑设计有限公司

哈尔滨天宸建筑设计有限公司成立于2001年（原名哈尔滨华强建筑设计有限公司），是一家拥有国家甲级建筑设计资质的企业，公司拥有员工40多人，其中国家一级注册建筑师、一级注册结构工程师共8人。

公司自成立以来，一直活跃在东北地区的建筑设计领域，在10年的时间内已完成近百项不同类型的建筑设计任务。公司致力于设计创新，作品囊括了建筑、规划、室内及景观等多种类型。曾多次获得黑龙江地区的设计奖项。公司业务以建筑创新为主，同时提供从前期策划、方案初设到施工图的整体建筑服务。公司以客户需求为服务宗旨，力求创造客户期望的优秀建筑作品。公司同时与本地区政府部门、大型设计院以及相关设计企业都建立友好的合作关系，具有组织与协调大中型建筑项目的工作经验。

公司理念：**创造建筑生命力的设计群体。**

公司目标：公司立志在公共建筑设计领域成为国内领先者，力求形成规范化的管理与经营机制。创建为业主负责、为社会负责、为环境负责的建筑设计企业。

Founded in 2001, Tianchen Architectural Design Co., Ltd. (preceded by Huaqiang Architectural Design Co., Ltd.) is an enterprise with Class-A Architectural Design License. It has over 40 employees, with 8 Class-1 Registered Architects and Class-1 Registered Structural Engineers.

Since foundation, the company has been actively participating in architectural design of the Northeast. In ten years, it has completed nearly a hundred architectural design projects of different types. Company devotes itself to design innovation, with works covering architecture, planning, interior design, landscape and many other fields. It has also won many design awards in Heilongjiang. Emphasizing on architectural innovation, the company also provides integrated architectural services including pre-planning, pre-design and construction drawing. With the purpose of meeting customers' requirements, the company tries to create satisfactory excellent architectural works. Meanwhile, with working experience of organizing and coordinating large-scale and middle-sized architectural projects, the company has also established friendly cooperation relationships with departments of local government, large-scale design institutes and related design enterprises.

Company idea: **Create a design group with architecture spirit.**

Company target: The company determines to be leader in the field of architectural design of China, seeks to establish a standardized management and operation mechanism, and creates an architectural design enterprise that is responsible for owners, society and environment.

地址：黑龙江省哈尔滨市中山路172号常青大厦12层
邮编：150040
电话：+86-451-82695519/712/517
传真：+86-451-82695118
网址：www.had.net.cn
邮箱：2427293@sina.com

Add: Evergreen Building, 12 Floors, No. 172 Zhongshan Road, Harbin, Heilongjiang
P.C.: 150040
Tel: +86-451-82695519/712/517
Fax: +86-451-82695118
Http://www.had.net.cn
E-mail: 2427293@sina.com

大庆公路客运枢纽站

项目地点：黑龙江 大庆
建设单位：大庆市交通局
建筑设计：唐家骏、王葱茏、张 力、及 强
设计时间：2009年
竣工时间：2010年
建筑面积：29 567平方米
摄 影：韦树祥、唐家骏

设计思路

大庆市作为中国最大的石油之城，地处东北高寒地区。由于市区内湖泊湿地纵横，自然景观优美，又被称为"百湖之城"。我们的设计首先力求突出东北冰雪文化特征，同时体现大庆的湿地景观特色。设计运用起伏的曲线形体来演绎东北的雪原地貌，通过建筑细部构件来象征冬季的湿地风光，同时白色表皮材料的使用强化了这一理念。建筑的整体体量简约大方，客运站站房与14层的信息中心形成了良好的体量穿插，做到了形式上的统一。

Daqing Highway Passenger Hub Station

Project Location: Daqing, Heilongjiang
Owner: Dalian Municipal Communications Bureau
Architectural Designers: Tang Jiajun, Wang Conglong, Zhang Li, Ji Qiang
Design Time: 2009
Completion Time: 2010
Floor Area: 29,567 m^2
Photographers: Wei Shuxiang, Tang Jiajun

Design Introduction

Daqing, as the biggest petroleum city of China, is located in the cold area of the Northeast of China. Due to many interlaced lakes and wetlands in the city and beautiful natural landscapes, Daqing is also known as "City of Hundred Lakes". The design emphasizes on the cultural characteristics of ice and snow of the Northeast, while shows the characteristic of wetland landscape of Daqing. Waving lines of the building represent the snowfield landform of the Northeast, while detailed components symbolize the wetland landscape in winter. White surface material emphasizes this idea. The overall appearance of the building is simple and liberal. Hall of passenger station and the information center (14F) intersperse excellently and realize unity of forms.

哈尔滨橄榄山礼拜堂

项目地点：黑龙江 哈尔滨
建设单位：哈尔滨市基督教协会
建筑设计：唐家骏、王葱茏、任 爽
设计时间：2007年
竣工时间：2009年
建筑面积：3 808平方米
摄　　影：唐家骏

设计思路

哈尔滨橄榄山礼拜堂位于哈尔滨市东南部阿城区的山区，距哈尔滨市区约60千米。礼拜堂建筑通过钢结构的运用塑造了新颖的造型和空间。设计突出宗教建筑形体的圣洁感，力争使建筑与自然在室内外有机地结合在一起，同时强调建筑钢结构的美学特征，最终创造出具有时代精神的基督教礼拜堂。

Olives Mountain Church, Harbin

Project Location: Harbin, Heilongjiang
Owner: The Christian Council of Harbin
Architectural Designers: Tang Jiajun, Wang Conglong, Ren Shuang
Design Time: 2007
Completion Time: 2009
Floor Area: 3,808 m^2
Photographers: Tang Jiajun

Design Introduction

Harbin Olives Mountain Church is located in mountainous area of Acheng District, in the southeast of Harbin, which is 60 km away from city center. The use of steel structure of the church creates a new and original appearance and space. The design highlights the sanctification of religious building, and tries to combine the indoor church with outdoor nature organically. Meanwhile, the design emphasizes on the aesthetics characteristics of steel structure, and eventually creates a Christian church with the spirit of time.

镜泊小镇游客咨询集散中心

项目地点：黑龙江 牡丹江
建设单位：镜泊小镇建设指挥部
建筑设计：唐家骏、董广海、及 强、倪小青
设计时间：2011年
预计竣工时间：2012年
建筑面积：11 070平方米

Tourists Consulting and Distribution Center of Jingpo Town

Project Location: Mudanjiang, Heilongjiang
Owner: Construction Headquarter of Jingpo Town
Architectural Designers: Tang Jiajun, Dong Guanghai, Ji Qiang, Ni Xiaoqing
Design Time: 2011
Expected Completion Time: 2012
Floor Area: 11, 070 m^2

HAD

大兴安岭文化体育中心

项目地点：黑龙江 大兴安岭
建设单位：大兴安岭文体局
建筑设计：唐家骏、罗 鹏、及 强、张 力
设计时间：2010年
预计竣工时间：2012年
建筑面积：23 200平方米

Daxinganling Cultural and Sports Center

Project Location: Daxinganling, Heilongjiang
Owner: Daxinganling Sports Bureau
Architectural Designers: Tang Jiajun, Luo Peng, Ji Qiang, Zhang Li
Design Time: 2010
Expected Completion Time: 2012
Floor Area: 23, 200 m^2

Introduction to Harbin Ark Design Co., Ltd.

哈尔滨方舟建筑设计有限公司

哈尔滨方舟建筑设计有限公司坐落于美丽的冰城哈尔滨，是我国建筑设计行业改革大潮中涌现出来的一支新的生力军。多年来，方舟人凭借卓越的团队智慧及拼搏精神始终立于本地区建筑设计行业的不败之地，企业经历了由小到大、由弱到强的深刻变化过程，发展为拥有建筑、规划、景观、结构、给排水、暖通、空调、电气、工程咨询等专业的强大设计企业。目前公司拥有员工138人，其中包括一级注册建筑师、一级注册结构工程师、注册设备工程师与高级工程师48人，公司下设3个综合设计所，1个住宅研究所，1个景观设计所，5个工作室。公司成立十年来，为社会、为城市贡献了许许多多精美的设计作品，得到了社会各界的广泛赞誉，多次被评为省优、部优，并在全国民营设计企业“华彩奖”评选中获奖。今天的方舟凭借着自己的实力与诚信，已经成为黑龙江省内建筑设计行业的顶尖机构，得到了社会各界的充分认可。

我们在《年鉴》中推出的一小部分作品，仅仅反映我们企业的点滴成果，目的在于与全国乃至境外的优秀设计机构创造更多的切磋、交流机会，从而使我们的设计水平有更大的提高。

Harbin Ark Design Co., Ltd. located in the beautiful ice city of Harbin, is a new army emerged in the reform tide of domestic architectural design industry. Over the years, the Ark people has been holding an invincible position in the regional architectural design industry by outstanding virtues of team cooperation and fighting spirit, making the company experience a process of profound change:from small to large and from weak to strong. Now Ark has become a powerful design company with professions including construction, planning, landscape, structure, water supply and drainage, heating, air conditioning and engineering consulting, etc. Currently the company has 128 employees, 48 of whom are Class-A registered architects, Class-A registered structural engineers, registered equipment engineers and senior engineers. Over the past decade since its foundation, Ark has contributed lots of fine designs for the society and city, receiving wide praise from all works of life and winning many prizes in the province, ministry as well as the Nightlife Awards of National Private Design Enterprise. Today's Ark, by its own strength, has become a top institution in the architectural design industry in Heilongjiang receiving full recognition from all social circles.

We will present a small part of our design work in the *Year Book*, which merely exhibit a few achievements of Ark, with an aim to create more opportunities for mutual learning and exchanges with even international excellent design agencies, thus greatly enhancing our design performance.

法定代表人：刘远孝
地址：黑龙江省哈尔滨市道外区红旗大街991号
电话：+86-451-87858515
传真：+86-451-87858504
邮箱：fz2001@vip.163.com
网址：www.fzjz.net

Legal Representative: Liu Yuanxiao
Add: No. 991, Hongqi Street Daowai District, Harbin, Heilongjiang Province
Tel : +86-451-87858515
Fax: +86-451-87858504
E-mail: fz2001@vip.163.com
Http: //www.fzjz.net

大庆奥林匹克体育中心体育场

总建筑面积：49 935.94平方米
层数：4层
设计时间：2010年
竣工时间：建设中
总坐席数：32 031个

Daqing Olympic Sports Centre Stadium

Total Building Area: 49,935.94 m^2
Building Storey: 4
Design Time: 2010
Completion Time: under construction
Seats: 32,031

大庆五星宾馆

建筑面积：233 400平方米
建筑高度：99米
建筑层数：26层
设计时间：2007年
竣工时间：建设中
主要功能：宾馆、商场、
高层公寓、办公

Daqing Five Star Hotel

Building Area: 233,400 m^2
Building Height: 99 m
Building Storey: 26
Design Time: 2007
Completion Time: Under Construction
Mostly Function: Hotel, Commerce,
High-rise Apartment,
Offices

哈尔滨工程大学21B教学实验楼

总用地面积：24 595平方米
总建筑面积：54 767.81平方米
建筑层数：6层
设计时间：2005年
竣工时间：2007年

21B Teaching Laboratory Building of Harbin Engineering University

Total Site Area: 24,595 m^2
Total Building Area: 54,767.81 m^2
Building Storey: 6
Design Time: 2005
Completion Time: 2007

东北农大创业科技园

总用地面积：53 917平方米
总建筑面积：60 176平方米
建筑层数：7层
设计时间：2010年
竣工时间：建设中

Northeast Agricultural University Entrepreneurial Science Park

Total Site Area: 53,917 m^2
Total Building Area: 60,176 m^2
Building Storey: 7
Design Time: 2010
Completion Time: Under Construction

哈尔滨市儿童医院

建筑面积：29 000平方米
建筑高度：45米
建筑层数：10层
设计时间：2005年
竣工时间：2007年

Harbin Children's Hospital

Building Area:29,000 m^2
Building Height: 45 m
Building Storey: 10
Design Time: 2005
Completion Time: 2007

哈尔滨医科大学公共卫生学院教学楼

建筑面积：33 000平方米
建筑高度：30.3米
建筑层数：7层
设计时间：2003年
竣工时间：2005年
主要功能：教学、实验、办公

Teaching Building of Public Health School of Harbin Medical University

Building Area: 33,000 m^2
Building Height: 30.3 m
Building Storey: 7
Design Time: 2003
Completion Time: 2005
Mostly Function: Teaching, Experimental, Offices

哈尔滨医科大学体育馆

建筑面积：20 500平方米
建筑高度：27.6米
设计时间：2004年
竣工时间：2006年

Harbin Medical University Stadium

Building Area: 20,500 m^2
Building Height: 27.6 m
Design Time: 2004
Completion Time: 2006

哈尔滨医科大学图书馆

总用地面积：10 700平方米
建筑面积：26 000平方米
建筑高度：29.8米
建筑层数：5层
设计时间：2005年
竣工时间：2006年

Library of Harbin Medicial University

Total Site Area: 10,700 m^2
Building Area: 26,000 m^2
Building Height: 29.8 m
Building Storey: 5
Design Time: 2005
Completion Time: 2006

黑龙江八一农垦大学图书馆

建筑面积：26 173平方米
建筑高度：30.6米
建筑层数：6层
设计时间：2004年
竣工时间：2005年
藏 书 量：100万册

Library of Heilongjiang Bayi Agricultural University

Building Area:26,173 m^2
Building Height: 30.6 m
Building Storey: 6
Design Time 2004
Completion Time: 2005
Library Measures: 1,000,000

黑龙江八一农垦大学教学主楼

建筑面积：33 900平方米
建筑高度：41.9米
建筑层数：12层
设计时间：2003年
竣工时间：2004年
主要功能：教学、办公

Main Teaching Building of Heilongjiang Bayi Agricultural University

Building Area: 33,900 m^2
Building Height: 41.9 m
Building Storey: 12
Design Time: 2003
Completion Time:2004
Mostly Function: Teaching, Offices

黑龙江中医药大学文体综合楼

建筑面积：80 000平方米
建筑高度：80米
建筑层数：21层
设计时间：2007年
竣工时间：建设中
主要功能：体育训练馆、图书馆、实验

Art & Sports Functioal Building of Heilongjiang University of Chinese Medicine

Building Area: 80,000 m^2
Building Height: 80 m
Building Storey: 21
Design Time: 2007
Completion Time: Under Construction
Mostly Function: Athletics Training Building, Library, Experimental

盟科视界居住社区

总用地面积：78 900平方米
总建筑面积：305 965平方米
建筑层数：32层
设计时间：2007年
竣工时间：2009年

Mengke Sight Community

Total Site Area: 78,900 m^2
Total Building Area: 305,965 m^2
Building Storey: 32
Design Time: 2007
Completion Time: 2009

牡丹江党政中心办公楼

总用地面积：34 194.84平方米
总建筑面积：34 050.31平方米
建筑层数：4层
设计时间：2007年
竣工时间：2009年

Mudanjiang Party-government Center Office Building

Total Site Area: 34,194.84 m^2
Building Area: 34,050.31 m^2
Building Storey: 4
Design Time: 2007
Completion Time: 2009

牡丹江七星公馆

总用地面积：22 506平方米
建筑面积：67 456平方米
建筑高度：18米
建筑层数：6层
设计时间：2009年
竣工时间：2010年

Mudanjiang Seven Star Mansion

Total Site Area: 22,506 m^2
Building Area: 67,456 m^2
Building Height: 18 m
Building Storey: 6
Design Time: 2009
Completion Time: 2010

河南徐辉建筑工程设计事务所是河南省首批通过建设部综合甲级设计资质审批，且是河南省唯一一家以个人名字命名的建筑工程设计事务所，更是中部地区最具综合竞争力的建筑设计机构之一。

徐辉事务所自成立以来，发展稳步，规模不断扩大。其办公设备先进，技术力量雄厚，专业配备齐全，拥有一支国内一流的设计专家队伍和一批具有丰富设计经验的专业人才，员工220多人，其中中级职称以上技术人员80余名，各专业国家注册工程师30余名，各专业专家及技术人员配备齐全，软硬件均达到先进水平，拥有办公场所3 000余平方米，经营效益持续攀升，人均产值年增长迅速，企业品牌、声誉及市场影响力与日俱增。

事务所立足中原，服务全国，经发展现已位居中部地区民营建筑设计企业第一位，尤其在北京、河南、河北、山西、山东、江西、安徽、湖北、海南等地区，凭借事务所雄厚的技术水平、卓越的服务管理理念，赢得了良好的口碑、知名度及美誉度，为中部地区建筑设计行业的发展、创新、突破、崛起发挥了不可替代的作用。

设计范围

☐ 民用建筑设计：教育建筑、办公建筑、科研建筑、文化建筑、商业建筑、服务建筑、体育建筑、医疗建筑、交通建筑、纪念建筑、园林建筑、综合建筑。
☐ 城市规划设计：居住区规划、控制性详细规划。
☐ 景观规划设计：城市景观设计、居住区环境景观设计。
☐ 人防工程设计：公共人防工程、建筑附属人防工程。

"设计天下，建筑人生"是事务所的经营理念，以设计赢得天下，使设计遍及天下，用设计惠及天下；把建筑融入事业人生，用建筑美化大众人生，以建筑构建成功人生。坚持把"为城市负责，为业主负责，为公众负责"的职业精神贯穿始终，把"追求建筑生命真谛，提升人类生活品质"作为企业的使命、最高目标和永远追求，体现出徐辉事务所强烈的社会责任感和历史使命感。

Sunrise Architectural Design Institute is among the first architectural project design firms in Henan licensed by the MOHURD for comprehensive Class-A design, and is the only one named by an individual person, among the most overall competitive architectural design institutes in Central China.

Since its inception, Sunrise grows steadily. With growing scale, advanced office equipment, rich technical resources, and complete specialized fitment, now it has a design team of first-class Chinese experts, and a group of experienced professionals, and more than 220 employees, including over 80 technicians with middle-level titles, and over 30 PRC registered engineers of all specialties. It is equipped with sufficient specialists and technicians, leading software & hardware, office with an area of 3,000 m^2; furthermore, its operating revenue is rising, per capita output value grows fast, corporate brand, goodwill and market influence are increasing daily.

The firm is based on Central China, to serve the whole country. It has developed into the number one private architectural design firm in Central China, especially in Beijing, Henan, Hebei, Shanxi, Shandong, Jiangxi, Anhui, Hubei, Hunan etc. With rich technical resources and excellent service management ideas, it has won goodwill, recognition and reputation, playing an irreplaceable role in the development, innovation, breakthrough and rise of architectural design industry of Central China.

Design Scope

☐ Civil building design: educational building, office building, research building, cultural building, commercial building, service building, sports building, medical building, traffic building, monumental building, landscape building, and complex building
☐ Urban planning design: inhabitancy district planning, regulatory detailed planning
☐ Landscape planning design: urban landscape design, environment and landscape design in residential district
☐ Civil defense design: public civil defense project, architecture attached civil defense project

"Design the world, and build the life" is the management idea of the office. It wants to win the world with its design and makes its design to spread to the whole world and benefits the whole world, integrates the architecture to the career, beautifies people life with architecture, and constructs successful life with architecture. It insists on running through the professionalism of "be responsible for the city, the owners and the public". It takes "pursue the essence of architecture life, and advance human's living quality" as the corporation's mission, the highest object and forever pursuit, which embodies the intense social responsibility and historical mission feeling of Sunrise.

地 址：河南省郑州市农业路22号兴业大厦A座7F-8F
电 话：+86-371-65656666/65336767
传 真：+86-371-65336993
网 址：www.sunrise323.com
邮 箱：sunrise323@126.com

Add: F7-8, Tower A, Xingye Building, 22 Agriculture Road, Zhengzhou, Henan
Tel: +86-371-65656666 / 65336767
Fax: +86-371-65336993
Http: //www.sunrise323.com
E-mail: sunrise323@126.com

公共建筑 Public Buildings

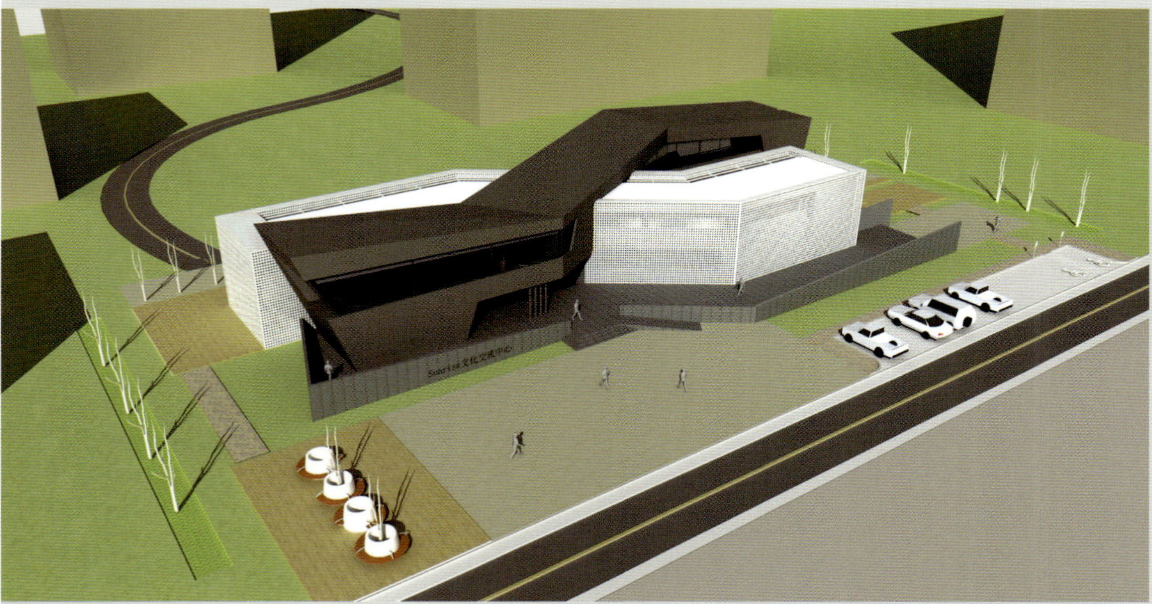

Sunrise文化交流中心

用地规模：2 000平方米
建设规模：1 500平方米
设计时间：2010年
建设地点：河南 郑州
主要设计人员：张朝阳、冯 赋、韩 跃

Sunrise Culture Communication Center

Land Area: 2,000 m^2
Floor Area: 1,500 m^2
Design Time: 2010
Location: Zhengzhou, Henan
Main Designers: Zhang Chaoyang, Feng Fu, Han Yue

方案一

“X房子”方案试图在充满矛盾的设计过程中，通过提炼sunrise企业的文化内涵和文化特点，建立起一种全新的建筑秩序。冷静与热情的哲学关系构成了“X房子”设计的背景与铺垫。

建筑设计从体量入手，深色的扭转体量代表了“热情”，布置公共空间；白色的常规体量象征“冷静”，布置住宿等功能。两者相互穿插交融，通过体量、材质、明暗的对比创造出彼此渗透、相互贯通的空间形态。建筑摆脱单纯的形式标志，既能满足功能上的需要，又能满足文化的展示要求，营造出全新且富有感染力的建筑场所。

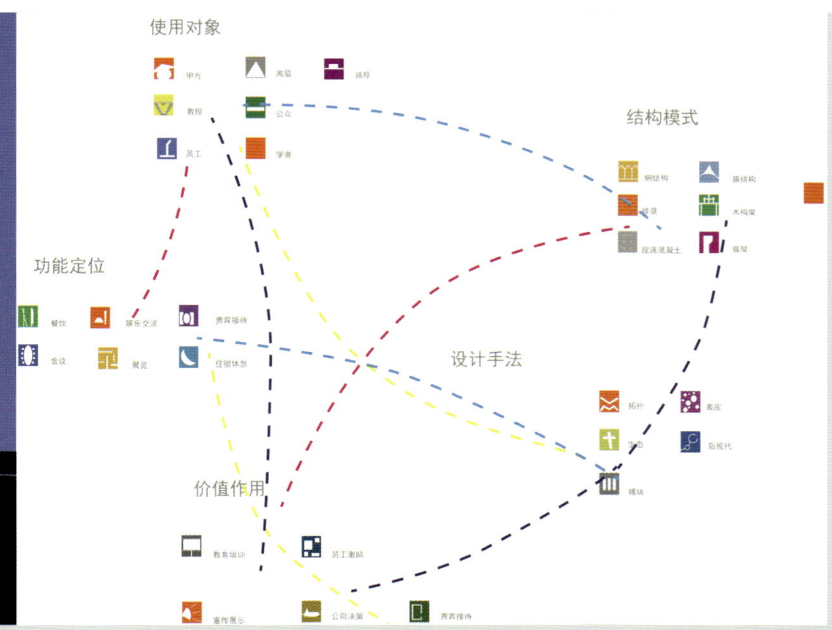

方案二

“自然”设计

建筑是自然的延伸与过渡，从人类最初的穴居树息的时代起，建筑已经不仅仅是一个提供居住等相关活动的场所，建筑本身其实一直在诠释人类的生存方式以及人与环境间的关系。

设计最初的概念出发点就是将原本聚合的、而与自然隔开的空间功能“自然”化，打破原来习以为常的功能聚合体的形式，将各个功能完全剥离，形成一个个小的功能块，然后运用“自然”手法将其随机组合，对比各种形式组合之后优化而成，这样就形成了一个与环境更为一致的新组合形式。在功能使用上分为公共功能和居用功能，公共功能体量在一层随机摆放，然后将居用功能放在其他功能体之上，形成一个既完全开放，又能照顾到个人隐私的功能组合形式，设计手法遵循“自然”，不做人为的限制与要求。

Public Buildings 公共建筑

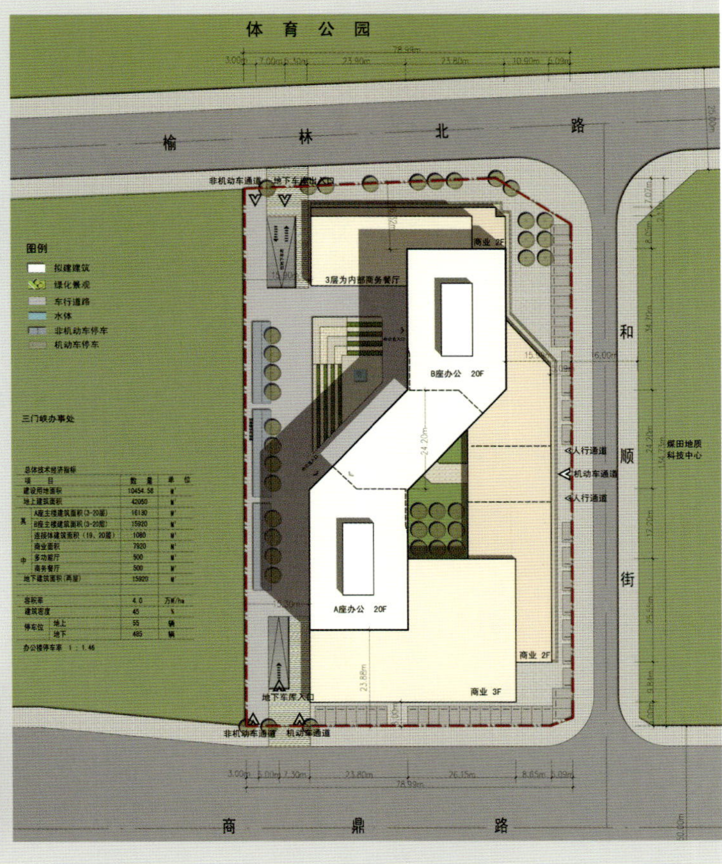

广地商务港

用地规模：10 454.58平方米
建设规模：5.3万平方米
设计时间：2010年
建设地点：河南 郑州
主要设计人员：王 珏、冯 赋

Guangdi Business Harbor

Land Area: 10,454.58 m^2
Floor Area: 53,000 m^2
Design Time: 2010
Location: Zhengzhou, Henan
Main Designers: Wang Jue, Feng Fu

规划设计坚持整体性、文化性、经济性和可操作性的设计原则。规划布局整体划一，结合现代城市文化、办公设计及其细赋的外观，形成强烈并且具有一定品质的外观效果，塑造出具有强烈特色的建筑风格，体现人文与自然、时间与空间的对话。整体布局简洁，在满足规划条件的同时，综合考虑环境效益及社会效益，力争使经济效益最大化。

通过两栋主楼顶部的错位连接，形成完整的建筑体量，一方面保证内部广场有足够舒适度，同时也为城市空间带来了新的活力，通过景观布置使两栋主楼之间有所联系，但又相对独立，这样就通过简单的手法，使得办公的室外空间富有层次和变化，同时也满足承载日常使用和休闲的功能。

绿都德化二期

用地规模：21 825平方米
建设规模：17.5万平方米
设计时间：2009年
建设地点：河南 郑州
主要设计人员：马 磊、王 珏、柴 磊、冯 赋、马海光

Green City Dehua Community Phase-II

Land Area: 21,825 m^2
Floor Area: 175,000 m^2
Design Time: 2009
Location: Zhengzhou, Henan
Main Designers: Ma Lei, Wang Jue, Chai Lei, Feng Fu, Ma Haiguang

绿都德化项目位于郑州市火车站商圈和二七商圈交界处，处于商圈结合部，周边的商品市场经营已形成良好氛围与基础。

设计重点在于解决项目中矛盾性和复杂性的问题：包括城市设计中与周边已建或规划中建筑的群体关系，建筑群体的文化性、艺术性，城市及项目自身的交通，并且打造可持续发展、充满活力的商业空间，以及满足各方面要求的功能规划布局。

建筑群的文化性和艺术性重点表现在群体的主次关系——“和”的概念，与城市环境、商业环境、人文环境之间的有机融合，同时建筑造型中也融入了“史书”的元素，表达其中的文化与和谐。

公共建筑 Public Buildings

许昌候机楼区域规划

用地规模：58万平方米
设计建设规模：155.7万平方米
建设地点：河南 许昌
设计时间：2009年
主要设计人员：徐 辉、马 磊、王 珏、冯 赋、柴 磊、孙 琦

Regional Planning of Airport Terminal, Xuchang

Land Area: 580,000 m^2
Floor Area: 1,557,000 m^2
Location: Xuchang, Henan
Design Time: 2009
Main Designers: Xu Hui, Ma Lei, Wang Jue, Feng Fu, Chai Lei, Sun Qi

本项目是许昌新区的商务商业中心，位于城市“东进”与“北拓”发展方向的交汇面，承接政务新区和产业新城的双重辐射影响。借助高铁站与候机楼便利的交通，影响力将辐射许昌新老城区。形成建筑面积近130万平方米的大型城市综合体。

本方案在进行规划时，便以“城市之光”喻之，其光芒将照耀许昌新老城区，给许昌市带来新的活力与动力。

重要的区域位置，便利的交通，是项目高速健康发展的契机，城市之光的概念同时适用于项目自身。

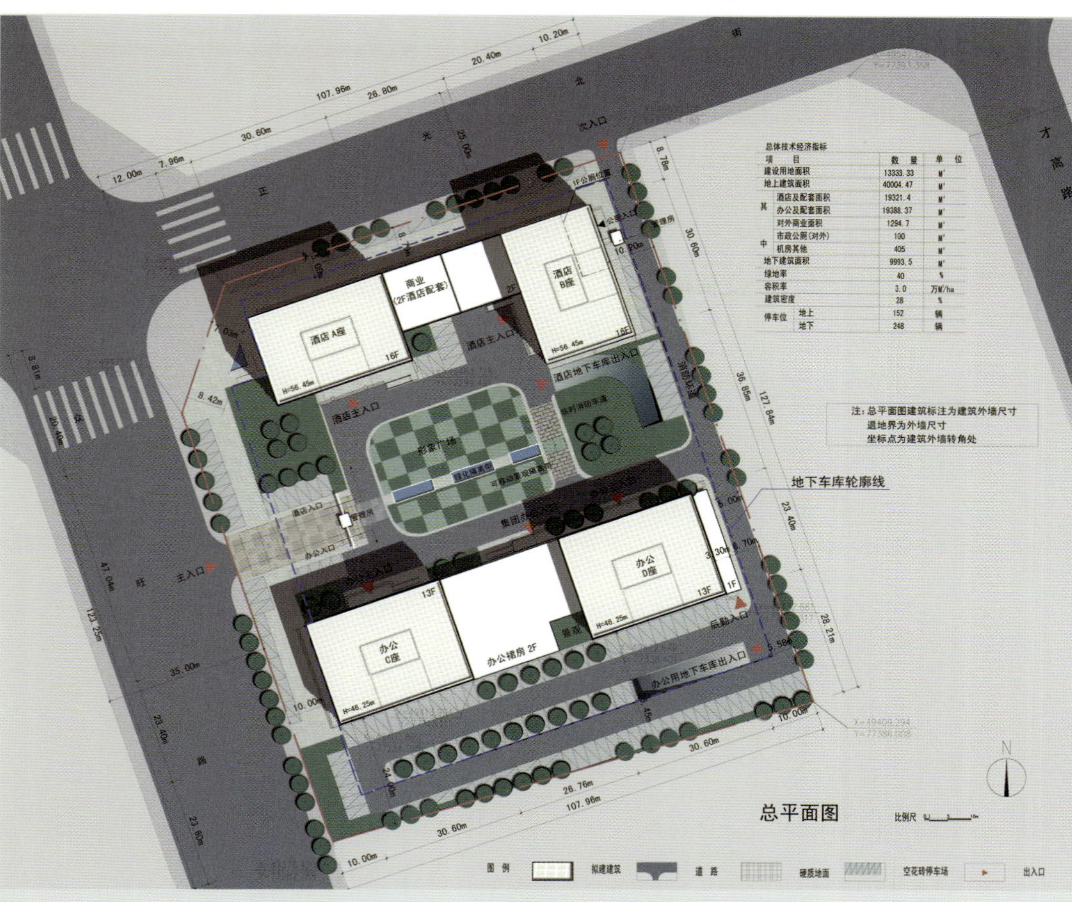

总体技术经济指标

项目		数量	单位
建设用地面积		13333.33	M²
地上建筑面积		40004.47	M²
其中	酒店及配套面积	19321.4	M²
	办公及配套面积	19388.37	M²
	对外商业面积	1294.7	M²
	市政公厕(对外)	100	M²
	机房其他	405	M²
地下建筑面积		9993.5	M²
绿地率		40	%
容积率		3.0	万M²/ha
建筑密度		28	%
停车位	地上	152	辆
	地下	248	辆

天明·锦江商悦酒店

用地规模：13 333平方米
建设规模：40 000平方米
设计时间：2010年
建设地点：河南 郑州
主要设计人员：马 磊、王 珏、柴 磊、王 渊

Tianming-Jinjiang Marvel Hotel

Land Area: 13,333 m^2
Floor Area: 40,000 m^2
Design Time: 2010
Location: Zhengzhou, Henan
Main Designers: Ma Lei, Wang Jue, Chai Lei, Wang Yuan

规划设计尊重城市的发展，满足新区的高品质需求。结合水平线条的立面设计，形成强烈并且具有一定品质的外观效果，融合现代城市文化及酒店经营理念，塑造出全新的建筑风格。通过四栋楼的平行扭转设置，一方面保证内部广场有足够舒适度，同时也给临街商业带来舒适的前广场，使酒店的室外空间富有层次和变化，同时也满足日常使用和休闲的功能。

酒店部分按照“锦江商悦酒店”品牌标准设计，为打造区域高端品质，塑造良好的商务办公环境确定了目标。

住宅建筑 Residential Buildings

安阳元泰清华园

用地规模：48 249平方米
建设规模：11.6万平方米（不含地下）
设计时间：2009年
建设地点：河南 安阳
主要设计人员：王凡召、张喜增、李 彬、王 渊

Yuantai-Tsinghuayuan, Anyang

Land Area: 48,249 m^2
Floor Area: 116,000 m^2 (excl. underground)
Design Time: 2009
Location: Anyang, Henan
Main Designers: Wang Fanzhao, Zhang Xizeng, Li Bin, Wang Yuan

安阳元泰清华园项目位于安阳高新技术开发区，处于安阳工学院南部，与周边建筑形成良好衔接。

设计重点在于提取商都的文化并用现代建筑诠释：寻求城市空间、建筑的有机融合，以文化、空间、环境为主题，打造符合现代人居住的人文社区。

小区一期已基本竣工，在当地具有良好的口碑。

信阳南湾山水

用地规模：31.5万平方米
建设规模：44万平方米
建设地点：河南 信阳
设计时间：2010年
主要设计人员：马 磊、王 珏、齐光辉、应真真、柴 磊、冯 赋、郭红玲

Nanwan Landscape Community, Xinyang

Land Area: 315,000 m^2
Floor Area: 440,000 m^2
Location: Xinyang, Henan
Design Time: 2010
Main Designers: Ma Lei, Wang Jue, Qi Guanghui, Ying Zhenzhen, Chai Lei, Feng Fu, Guo Hongling

信阳金成南湾山水，位于信阳新区南湾湖风景区，距南湾湖仅1.5千米。地块拥有丰富的生态自然山水资源，自然条件得天独厚。项目取材于地中海风情，建筑风格以西班牙建筑形态为主体，结合山地地形、地貌加以创新与变异，形成独具特色的建筑语言。立面、细部尺度、色彩营造均强调"纯美"情趣，以纯粹的建筑气质和丰富的山水资源，将南湾风景区生活带入一种全新的境界。

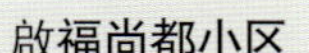

启福尚都小区

用地规模：24.49万平方米
建设规模：100.60万平方米
建设地点：河南 郑州
设计时间：2010年

Qifu Shangdu Community

Land Area: 244,900 m^2
Floor Area: 1,006,000 m^2
Location: Zhengzhou, Henan
Design Time: 2010

主要设计人员：徐 辉、马 磊、孔 波、李健宇、郭红玲、熊兹花、应真真、魏姗姗、刘乃嘉、齐光辉、冯 赋

Main Designers: Xu Hui, Ma Lei, Kong Bo, Li Jianyu, Guo Hongling, Xiong Zihua, Ying Zhenzhen, Wei Shanshan, Liu Naijia, Qi Guanghui, Feng Fu

项目位于郑州西部，东依西流湖公园，南靠郑州一中，与南水北调工程隔中原路相望，南水北调工程规划水面宽100米，两侧各有200米宽的绿化带，生态景观资源优越。项目由四个方正地块组成，基地周边交通网络辐射面广、易达性强，交通非常便利。规划：以景观和谐共生为设计原则，以"空间·景观·环境"为主题，遵循"花园里面种房子"的理念，让景观不再是建筑的配角，充满风景、人情，让家在绿色中生长，让居者体验"天人合一"的景观感受。

机械工业第六设计研究院
No. 6 Institute of Project Planning & Research of Machinery Industry

机械工业第六设计研究院（简称中机六院）创建于1951年，是国家大型综合设计研究院，全国勘察设计行业综合实力百强单位，隶属中央大型企业集团——中国机械工业集团有限公司。

现有8个工程所、3个分院、3个子公司、2500多名员工，其中中国工程院院士1人、中国工程设计大师2人、英国皇家特许建筑设备注册工程师协会荣誉资深会员1人、享受政府特殊津贴专家29人、研究员级高级工程师110人、高级工程师386人、各类国家注册工程师500多人。

拥有工程咨询设计甲级资质8个，乙级资质7个和压力容器、压力管道设计专业资质共17个，工程监理综合资质，工程总承包甲级资质，建筑企业施工资质二级，具有国家商务部颁发的对外工程承包经营资格证书。

业务涵盖工业、民用、市政和环境工程领域的咨询、设计、监理、管理、代建和总承包。建院以来，完成工程项目10 000余项；主编、参编国家和行业标准、规范21项；荣获国家科技发明二等奖1项，中国土木工程创新最高奖詹天佑奖1项，鲁班奖6项，国家科技进步及优秀工程设计金、银、铜奖25项，省部级奖300余项。

工业工程涵盖机床工具、无机非金属材料、重矿机械、轻工烟草、机车车辆、石化机械、风电机械、工程机械、轻纺机械、通用机械、农用机械、电工电器、仪器仪表、标准件、汽车及汽车零部件、军工等16大类机械行业。

民用工程涵盖大型公建、会展、文化、体育、交通、办公、商业、金融、医疗、教育、宾馆、酒店、住宅以及城镇、景观规划等。

市政与环境工程涵盖城市集中供热、城市供水、污水、工业废水治理、生活垃圾处理及危险固体废物处理等。

中机六院是国内机床工具和无机非金属材料两个行业唯一的专业设计院，是烟草、铸造、重矿、工程机械、民用建筑五大行业设计强院，在大型工厂和园区规划、企业生产流程再造、高难度结构、暖通空调、工业除尘、信息智能化、绿色工业建筑、市政和环境工程等许多方面具有国内一流的工程技术。

我院秉承"服务是立院之本、创新是兴院之道、人才是强院之基"的理念，竭诚以一流的技术、一流的队伍、一流的管理为国内外客户提供工程建设领域的全过程、全方位服务！

No.6 Institute of Project Planning & Research of Machinery Industry (Abbreviated as SIPPR) is a large-sized national comprehensive design institute and one of the top 100 design institutes in China survey & design industry subordinated to central enterprise group --- China National Machinery Industry Corporation established in 1951.

SIPPR has 8 engineering institutes, 3 branches, 3 subsidiary companies and over 2500 staff, of which there are 1 academician of Chinese Engineering Academy, 2 Chinese engineering design masters, 28 experts enjoying special subsidy from State Department, 110 senior engineers at researcher level, 386 senior engineers and more than 500 all kinds of national certified engineers.

Besides, SIPPR has 8 Grade A qualifications, 7 Grade B qualifications, 17 pressure vessel & other qualifications, comprehensive supervision qualification, Grade A general contract qualification, Grade B architecture construction qualification and direct foreign business right.

Business scope of SIPPR includes consultation, design, supervision, management, construction agent & general contract of industrial, civil, municipal & environmental protection projects. It has finished more than 10,000 projects since its establishment. What's more, SIPPR attended the compilation of 21 national, local & industry standards and won 1 2nd Prize of National Scien-tech Invention, 1 Zhan Tianyou Prize---the highest prize of China civil engineering projects, 6 Lu Ban Prizes, 25 National Gold, Silver & Copper Prizes & more than 300 Ministerial & Provincial Prizes.

Industry project includes such 16 machinery industries as machine tool, inorganic nonmetal material, light industry, tobacco, rolling stock, petroleum machinery, wind power machinery, engineering machinery, textile machinery, general machinery, agricultural machinery, electrical engineering and electric apparatus, instruments and meters, standard parts, automobile and auto parts and military, etc.

Civil projects includes large-sized public buildings, exhibition center, cultural buildings, sports buildings, transportation buildings, office buildings, commercial buildings, financial buildings, hospital buildings, educational buildings, hotels, restaurants, residential buildings, urban & landscape planning, etc.

Municipal projects include urban central heat supply, municipal water supply & drainage, industrial sewage treatment, domestic refuse treatment and hazardous solid waste treatment, etc.

SIPPR is the only professional machine tool & inorganic nonmetal material design institute in China and strong design institute of tobacco, casting, heavy mining machinery, engineering machinery & civil projects with advanced technology in large-sized plant & industrial park planning, enterprise flow reconstruction, high-difficult structure, industrial dust remove, information intelligence, green industrial buildings, municipal and environmental protection projects.

SIPPR sticks to the concept of "Service is the base of existence, innovation is the way to prosper and talents is the foundation to make SIPPR stronger " and sincerely looks forward to cooperating with friends both home and abroad by our powerful comprehensive strength, rich design experience, advanced design concept and best service.

地址：河南省郑州市中原中路191号
邮编：450007
电话：+86-371-67606088 / 6005 / 6087
371-67606016
传真：+86-371-67639571
371-67628091
网址：www.sippr.com.cn

Add: No.191, Zhongyuan Road, Zhengzhou, Henan Province, P. R. C
P.C.: 450007
Tel：+86-371-67606088 / 6005 / 6087
371-6760 6016
Fax: +86-371-6763 9571
371-6762 8091
Http://www.sippr.com.cn

沈阳机床（集团）有限责任公司发展数控机床及成套装备整体改造项目

该项目是中机六院承担“设计、监理、项目管理”业务的典型代表，总投资22亿，占地76万平方米，建筑面积58万平方米，被誉为“东北振兴的样板和典范”、“中国最大的数控机床装备制造基地”。

Numerical Controlled Machine Tool and Outfit Development Transformation Project of Shenyang Machine Tool (Group) Co., Ltd.

SIPPR undertakes design, supervision & project management of this project. Its total investment is 2,200 million, floor space 760,000 m^2 and building area 580,000 m^2. It's regarded as "the biggest intelligence and net factory" and "model and sample factory in machine tool industry".

大连重工·起重集团重组搬迁改造项目

该项目包括泉水、中革、旅顺、双D港四个生产基地，占地面积105万平方米，建筑面积45万平方米，建设投资37亿元。大重、大起强强联合的战略性重组的成功实施，对全国重机行业大型骨干企业重组以及东北老工业基地改造具有重要的示范意义。该项目荣获机械工业科技进步三等奖、机械工业优秀工程咨询成果二等奖。

Dalian Heavy-duty Industry & Crane Group Removing & Transforming Project

This project includes such four manufacturing bases as Quanshui, Zhongge, Lushun & Double-D Port with the floor space 1,050,000 m^2, building area 40,000 m^2 and total investment 3.7 billion Yuan. It plays important role in the regroup & transformation of Northeast China industrial base and won 3rd Prize of Scien-tech Improvement of Machinery Industry and 2nd Prize of Excellent Consultation Achievement of Machinery Industry.

中国北车集团大连机车车辆有限公司旅顺基地建设项目

该项目占地188.6万平方米，建筑面积73.5万平方米，建设投资50亿元，年生产各类机车1 000台，城市轨道车辆1 000辆，中、高速柴油机1 000台，建成世界级、国内最大、技术最先进的轨道交通装备和通用动力机械研发和制造基地。

CNR Group Dalian Rolling Stock Co., Ltd. Lushun Base Construction Project

Floor space of this project is 1,886,000 m^2, building area 735,000 m^2, total investment more than 5 billion Yuan and annual output 1,000 all kinds of rolling stocks, 1,000 urban rail cars and 1,000 middle & high speed diesel engines. When it's finished, it will be the world level and the largest & most advanced railway transportation equipments & general motive power machine research & manufacturing base in China.

郑州煤矿机械集团有限责任公司高端液压支架生产基地建设项目

本项目占地32万平方米，建筑面积24万平方米，建设投资11.4亿元，年产各类高端液压支架15 000架。该项目的实施进一步巩固郑煤机集团在国内煤机行业的领先地位，形成亚洲最大、世界一流的液压支架研发、生产基地。

High-end Hydraulic Support Manufacturing Base of Zhengzhou Coal Mine Machinery Group Co., Ltd

Floor space of this project is 320,000 m^2, building area 240,000 m^2, total investment 1.14 billion Yuan and annual output 15,000 all kinds of hydraulic supports. This project makes Zhengzhou Coal Mine Machinery Group the largest & most advanced hydraulic supports research, development & production base in Asia & even in the world.

中煤张家口煤矿机械有限责任公司煤矿装备产业园

该项目占地171万平方米，一期建筑面积36万平方米，建设投资21亿元。企业通过搬迁改造，实现企业流程再造和产品升级，建成国内最大的煤机装备产业园。

China Coal Group Zhengzhou Coal Mine Machinery Co., Ltd Coal Mine Machinery Equipment Industrial Park

Floor space of this project is 1,710,000 m^2, building area 240,000 m^2 and total investment 2.1 billion Yuan. The company will be the largest coal mine machinery industrial park by the removing, transformation, production process reconstruction & product up-gradation.

宝鸡石油机械有限责任公司搬迁改造建设项目

该项目占地94.8万平方米，建筑面积38.5万平方米，建设投资30亿元，建设内容包括渭滨新厂区建设及原有北厂区改造两部分，建成我国最大的石油钻采装备制造基地和石油钻采装备研发制造中心，全球最大的陆地钻机和泥浆泵研发制造基地。

Removing & Transforming Project of Baoji Petroleum Machinery Co., Ltd.

Floor space of this project is 948,000 m^2, building area 385,000 m^2 and total investment 3 billion Yuan. The construction mainly includes Weibin New Plant & transformation of original north plant. When it's finished, it will be the largest petroleum drilling equipment manufacturing base & research & manufacturing center in China and largest land rig & slurry pumps researcher, developer & manufacturer in the world.

中国第一重型机械集团公司铸钢车间工程总承包

该项目生产纲领为年产铸钢件80 000吨，最大铸钢件重达500吨，是国内铸钢单件最大的生产企业；厂房最大吊车起重量为500吨，是亚洲最大的吊车吨位铸钢车间。

General Contract of China No.1 Heavy-duty Machinery Group Metallurgy Branch Cast Steel Workshop

Output of this project is 80,000 t cast steel products per year and heaviest cast steel product is 500 t which is the largest single cast steel product in China. Maximum crane hoisting capacity is 500 t which is the largest in Asia.

沈阳铸锻工业园

该项目占地面积70万平方米，建筑面积31万平方米，年产铸锻件30万吨，由沈阳重型机械集团公司铸锻公司、沈阳鼓风机集团铸造公司、沈阳矿山集团铸锻公司等近十家企业组成，是中国已建成的最大的铸锻工业园，已成为国内第一个集中建设铸锻企业的典范，对国内城市中铸锻件生产企业的搬迁、改造、重组起到示范作用。

Shenyang Casting & Forging Industrial Garden

Floor space of this project is 700,000 m^2, building area of 310,000 m^2 and annual output 300,000 t casting & forging products. It's consisted of about 10 companies, such as Shenyang Heavy-duty Machinery Group Corporation Casting & Forging Company, Shenyang Fan Group Casting Company, and Shenyang Mine Group Casting & Forging Company, etc. It's the largest casting & forging industrial park finished in China and first successful model of centralized-built casting & forging company which plays an important role in removing, transformation & recombination of casting & forging products manufacturers.

国电联合动力（保定）技术有限公司项目

该项目由国电集团公司投资建设，位于保定市国家高新技术产业园区，占地20.6万平方米，建筑面积7.1万平方米，由总装车间、叶片生产车间、打磨车间、电控车间、原料仓库、综合楼、公用站房等组成，年产800台（套）风力发电整机装备和800台（套）电控及风机叶片，是中国第一、亚洲最大的风电设备生产基地。

China Power Synergy Drive (Baoding) Technology Co., Ltd.

This project locates at Baoding National High-tech Industrial Park and is invested by China Power Group Company with the floor space 206,000 m^2, building area 71,000 m^2 and annual output 800 wind power generation whole-set equipments and 800 electric control & fan blades. It's consisted of assembly workshop, blade production workshop, polishing workshop, electric control workshop, raw material warehouse, comprehensive building & public station and is the largest wind power equipment manufacturing base in China & Asia.

湖北中烟工业园区总体规划及武汉卷烟厂易地技改项目

该项目建设用地面积115.5万平方米，一期工程总建筑面积38.8万平方米，建设投资50亿元，年产卷烟150万箱，是迄今为止我国烟草行业最大建设规模的易地技术改造项目。

General Planning of Hubei China Tobacco Industrial Garden & Technical Innovation Project of Wuhan Tobacco Company

Floor space of this project is 1,150,000 m^2, building area of the Phase I 388,000 m^2, total investment about 5,000 million Yuan and annual output 1.5 million boxes of tobacco. It's the largest removing & technical innovation project in domestic tobacco industry.

杭州卷烟厂“十一五”易地技术改造项目

该项目位于杭州市西湖区科技经济工业园，占地40.5万平方米，年产卷烟100万箱以上，总投资20.6亿元。其建筑面积近19万平方米的联合工房是国内烟草行业第一家按照最高标准“三星级”设计的“绿色工房”，对我国烟草行业建设“环境友好型、资源节约型”企业具有重要的示范作用。

“11th Five-Year-Plan” Technical Innovation Project of Hangzhou Tobacco Company

Floor space of this project is about 405,000 m^2, total investment RMB 2,060 million Yuan, total building area 620,000 m^2 and annual production capacity is more than 1 million box of cigarette. United workshop of this project with the building area 190,000 m^2 is a three-star green workshop which adopts such over ten technologies as roof greening, waste regeneration, after heat recycling & utilization and rain water utilization, etc.

将军集团济南卷烟厂易地技术改造项目

该项目占地约7公顷，建筑面积47.4万平方米，建设投资20亿元，是国家烟草专卖局批准立项的我国烟草行业第一家占地47.5万平方米、年产100万箱企业的易地技术改造项目，项目荣获机械工业科技进步二等奖。

Removing & Technical Innovation Project of Jinan Tobacco Company

This is the 1st technical innovation project of a tobacco company with the floor space about 70,000 m^2, building area 474,000 m^2, total investment about 2,000 million Yuan and the annual output of 1 million boxes approved by State Tobacco Monopoly Bureau. It won 2nd Prize of Scien-tech Improvement of Machinery Industry.

大观·国际住宅小区

时　　间：2009年
地　　点：河南 郑州
建筑面积：330 000平方米

Daguan International Residential District

Time: 2009
Place: Zhengzhou, Henan
Building Area: 330,000 m^2

郑州白庄城中村改造——甲天下西湖新城

时　　间：2006年
地　　点：河南 郑州
建筑面积：613 000平方米

Zhengzhou Baizhuang Urban Village Reconstruction – Jiatianxia Xihuxincheng Residential District

Time: 2006
Place: Zhengzhou, Henan
Building Area: 613,000 m^2

北京原乡美利坚别墅

时　　间：2006年
地　　点：北京
占地面积：66 700平方米

Beijing Yuanxiang USA Villa

Time: 2006
Place: Beijing
Floor Space: 66,700 m^2

郑州国际会展中心

Zhengzhou International Conference & Exhibition Center

郑州国际会展中心由中机六院和日本黑川纪章建筑都市事务所联合设计，是一个集展览、会议、交流、信息发布、观光等于一体的多功能、现代化、智能化的特大型公共建筑。总投资23亿元，建筑面积22.6万平方米，展览中心展厅跨度102米，3 560个国际标准展位；会议中心由5 000人多功能厅、1 200人国际会议厅、两个400人会议厅和十几个中小型会议室等组成；无线局域网覆盖会展室内外所有空间，30余个智能化子系统高度集成。

中机六院在设计中大胆突破、勇于创新，创下了国内会展中心八项设计之最：7万平方米的钛锌金属屋面，6万平方米原浆混凝土外墙饰面，360° 旋转电子大屏幕，大面积室外透水混凝土，3.4万平方米无柱展览大厅，十余米高的超重型活动隔断，单件重34吨的铸钢件节点，30米大跨度预应力梁板结构等。

该项目工程设计先进、科技含量高，荣获中国建筑工程鲁班奖和第七届中国土木工程詹天佑奖。

Zhengzhou International Conference & Exhibition Center is an oversize multi-functional, modern and intelligence public building combined with the functions of exhibition, conference, communication, information and visiting. Its total investment is 2,300 million Yuan, building area 226,000 m^2, span of the exhibition hall is 102 m and international standard exhibition booths number is 3,560. The Conference Center is consisted of a 5,000-people multi-functional hall, 1,200-people international conference hall, two 400-people meeting halls and more than 10 middle or small-sized meeting rooms; Besides, wireless LAN integrated by more than 30 subsystems covers all the space in and out of exhibition hall.

SIPPR creates eight "the first" in the design of Zhengzhou International Exhibition & Conference Center, that is: 70,000 m^2 titanium zincum metal roof, 60,000 m^2 virgin pulp concrete exterior wall finish, 360° rotary electric screen, large area exterior pervious concrete, 34,000 m^2 exhibition hall without column, over 10 m overweight movable partition, cast steel node with the single weight of 34 t and 30 m long-span pre-stressing beam & slab structure. All these mentioned above reflect the comprehensive design ability of SIPPR in oversize public building design.

This project won Lu Ban Prize & 7th China Civil Engineering Zhan Tianyou Prize.

金科国际大厦

时　　间：2009年
地　　点：河南 郑州
建筑面积：7.5万平方米

Jinke International Mansion

Time: 2009
Place: Zhengzhou, Henan
Building Area: 75,000 m^2

国家棉花及纺织服装产品质检中心综合楼

时　　间：2007年
地　　点：河南 郑州
建筑面积：4.2万平方米

Comprehensive Building of National Cotton, Textile & Clothes Product Quality Control Center

Time: 2007
Place: Zhengzhou, Henan
Building Area: 42,000 m^2

青岛颐和·国际

时　　间：2004年
地　　点：山东 青岛
建筑面积：12万平方米

Qingdao Yihe International Mansion

Time: 2004
Place: Qingdao, Shangdong
Building Area: 120,000 m^2

重庆龙头寺片区保利花园

时　　间：2008年
地　　点：重庆
建筑面积：1.93万平方米

Chongqing Longtousi Area Baoli Garden

Time: 2008
Place: Chongqing
Building Area: 19,300 m^2

吉林土木风设计集团 JILIN TUMUFENG DESIGN GROUP

土木风设计是由吉林土木风建筑工程设计有限公司、吉林土木风环境艺术设计有限公司、吉林土木风旅游规划设计有限公司、吉林土木风苗木培育有限公司、上海土木风设计有限公司组成的集规划设计、建筑设计、景观设计、室内设计为一体的集团化设计公司。持有建设部认证的建筑工程设计、装饰工程设计甲级资质，规划设计、风景园林设计乙级资质。

公司现有各级专业技术人员138人，其中国家一级注册建筑师8人，一级注册结构工程师10人，国家注册电气工程师、国家注册公用设备工程师8人，国家注册规划师7人，省优秀青年建筑师1人。公司通过设计实践不断吸收、学习国内外优秀设计公司的成功经验和先进的经营模式，努力将公司建设成为省内一流，国内有影响的集规划、建筑、景观、装饰四位一体的设计公司。

土木风的长期发展目标：成为具有高度社会责任感和市场化的，以创造具有持久价值的建筑产品为根本目的，具有强大的研究、咨询、设计和管理能力，服务于城市建设的综合技术服务公司。公司通过项目负责制，使公司服务于社会和业主的管理制度得到有效执行。本着诚信、守法、全过程、全面服务的原则，全心全意为客户服务的经营理念，严格执行国家制定的有关建筑法规，通过有价值的创造性的工作，努力为业主和社会创造优质的设计产品。

Tumufeng Design, is a group corporation integrating planning and design, architectural design, landscape design, interior design, consisting of Jilin Tumufeng Architectural Engineering Design Co.,Ltd., Jilin Tumufeng Environment Art Design Co.,Ltd., Jilin Tumufeng Tourism Planning and Design Co.,Ltd., Jilin Tumufeng Seedling Cultivation Co.,Ltd., Shanghai Tumufeng Design Co.,Ltd., which has Grade A qualification certificate of architectural engineering design and decoration design, Grade B qualification certificate of planning and landscape design, issued by the Ministry of Housing and Urban-Rural Development.

At present, there are 138 professional and technical personnel at all levels, among them, eight PRC Class-A registered architects, ten PRC Class-A registered structure engineers, eight PRC registered electric and public equipment engineers, seven registered planners, one provincial outstanding young architects. Through the design practice, the company absorbs and learns the successful experience and advanced business model from the excellent companies at home and abroad constantly, trying its best to be a top company in Jilin province, and an influential domestic design company with planning, architecture, landscape, decoration all in one.

The long-term goal of Tumufeng is to be a comprehensive technical service company with highly social responsibility and market-oriented, creating the enduring value for building products as fundamental purpose, with strong research, consulting, design and management to service the urban construction. Project responsibility system of the company makes sure the management system for serving socials and clients are implemented effectively. In the principle of good faith, law-abiding, the whole process, comprehensive service, and dedication to customer service philosophy, strict implementation of the relevant laws for buildings, through valuable creative work, it will do the best to create high-quality design products for the owners and the community.

法定代表人：石铁军
地址：吉林省长春市东民主大街519号
电话：+86-431-88591360
传真：+86-431-88591360
邮箱：tumufeng@126.com
网址：www.tumufeng.com

Legal Representative: Shi Tiejun
Add: No.519 East Minzhu Street, Changchun, Jilin Province
Tel: +86-431-88591360
Fax: +86-431-88591360
E-mail: tumufeng@126.com
Http://www.tumufeng.com

土木風設計
TUMUFENG DESIGN

1–2 长春老年大学

建设地点：吉林 长春
项目类别：综合教学楼
建筑规模：45 000平方米
设计时间：2011年

1-2 Changchun University of Senior Citizen

Construction Location: Changchun, Jilin
Project Category: Comprehensive Teaching Building
Building Scale: 45,000 m^2
Design Time: 2006

3–5 长春青年城商业广场

建设地点：吉林 长春
项目类别：商业综合体
建筑规模：160 900平方米
设计时间：2010年

3-5 Changchun Youth City Plaza

Construction Location: Changchun, Jilin
Project Category: Commercial Complex
Building Scale: 160,900 m^2
Design Time: 2010

1 融大・天玺
建设地点：吉林 长春
项目类别：综合办公楼
建筑规模：49 000平方米
设计时间：2010年

1 Rongda • Tianxi

Construction Location: Changchun, Jilin
Project Category: Comprehensive Office Building
Building Scale: 49,000 m^2
Design Time: 2010

2 长春市担保大厦
建设地点：吉林 长春
项目类别：写字楼、公寓
建筑规模：120 000平方米
设计时间：2011年

2 Changchun Guaranty Tower

Construction Location: Changchun, Jilin
Project Category: Office Building, Apartment
Building Scale: 120,000 m^2
Design Time: 2011

1

2

3

3 华镡国际
建设地点：吉林 长春
项目类别：高级公寓
建筑规模：23 000平方米
设计时间：2011年

3 Huaxin International

Construction Location: Changchun, Jilin
Project Category: Senior Apartments
Building Scale: 23,000 m^2
Design Time: 2011

土木風設計
TUMUFENG DESIGN

1 天伦·中央区
建设地点：吉林 长春
项目类别：商业综合体
建筑规模：171 000平方米
设计时间：2010年

1 Tianlun • Central District
Construction Location: Changchun, Jilin
Project Category: Commercial Complex
Building Scale: 171,000 m^2
Design Time: 2010

2 成·悦天下
建设地点：吉林 长春
项目类别：住宅小区
建筑规模：500 000平方米
设计时间：2011年

2 Success • Happy World
Construction Location: Changchun, Jilin
Project Category: Residence Community
Building Scale: 500,000 m^2
Design Time: 2011

3–4 新星宇·和邑
建设地点：吉林 长春
项目类别：住宅小区
建筑规模：40 000平方米
设计时间：2010年

3-4 New Star • Heyi
Construction Location: Changchun, Jilin
Project Category: Residence Community
Building Scale: 40,000 m^2
Design Time: 2010

土木風設計
TUMUFENG DESIGN

1 万晟・金地公馆B区

建设地点：吉林 长春
项目类别：商业建筑
建筑规模：17 600平方米
设计时间：2011年

1 Wansheng • Jindi Residence Area B

Construction Location: Changchun, Jilin
Project Category: Business Building
Building Scale: 17,600 m^2
Design Time: 2011

2–3 博润・一品江城商业步行街

建设地点：吉林 松原
项目类别：商业步行街
建筑规模：50 000平方米
设计时间：2011年

2-3 Borun• Yipinjiangcheng Commercial Pedestrian Street

Construction Location: Songyuan, Jilin
Project Category: Commercial Pedestrian Street
Building Scale: 50,000 m^2
Design Time: 2011

2

3

東南大學建築設計研究院

东南大学建筑设计研究院始建于1965年，是国家批准的具有独立法人资格（1993工商注册）的建筑行业建筑工程设计甲级、公路行业（公路）甲级、市政公用行业（道路、桥梁）甲级、风景园林专项工程设计甲级、文物保护规划与设计甲级、电力行业（火力发电）乙级、市政行业（环境卫生、热力）乙级等工程设计资质。

该院现拥有国家一级注册建筑师46人，一级注册结构工程师40人，注册城市规划师4人，其他注册工程师等81人，是一支综合技术实力雄厚的设计单位。对承接的每一项工程都本着"用户第一，质量第一"的原则精心设计，取得了很好的信誉和社会效益。该院在历届国家、部委和省优秀设计评选中，均取得了优异的成绩，先后荣获各级优秀设计奖400多项，其中国家级银质奖6项、铜质奖3项，属国内一流的著名设计院。

The Architectural Design & Research Institute of Southeast University was established in 1965, as an independent corporate entity (incorporated in 1993) approved by the government. It is qualified for Building Construction Design Class-A, Highway Industry (Highway) Class-A, Municipal Utilities (Road, Bridge) Class-A, Landscaping Special Project Design Class-A, Antiquity Protection Planning & Design Class-A, Power Generation (Thermal Power Generation) Class-B, Municipal Industry (Environmental Hygiene, Heating Power) Class-B and other design licenses.

This Institute is a comprehensive design team with rich technical resources, including 46 Class-1 Registered Architects, 40 Class-1 Structural Engineers, 4 Registered Municipal Planners, and other 81 registered engineers. It elaborates every project in the principle of "User First, Quality First", winning good faith and social benefits. It has good results from national, ministerial and provincial design appraisals, winning more than 400 excellent design prizes at all levels, including 6 silver prizes and 3 bronze prizes at national level. It is among the first-class design institutes in China.

地址: 江苏省南京市玄武区四牌楼2号
电话: +86-25-83793178
传真: +86-25-83793176
邮箱: ad@adriseu.com
网址: http://adri.seu.edu.cn/

Add: 2 Fourth Archway, Xuanwu District, Nanjing City, Jiangsu
Tel :+86-25-83793178
Fax:+86-25-83793176
E-mail: ad@adriseu.com
Http://adri.seu.edu.cn/

九江市文化艺术中心

建筑地点：江西 九江
建筑面积：27 600平方米
设计时间：2010–2011年
竣工时间：建设中
主要用途：表演、会议、电影、培训

项目始于九江这座有着典型山水城市特色的八里湖新区内。设计充分考虑地段周边的八里湖、七里湖、胜利岛等自然地貌特征，以一个水平向舒展而连绵起伏的连续屋顶轮廓线表达着对场地地景的呼应，成为基地周边优美的自然生态环境得以延伸的一部分，将大剧场、多功能剧场以及培训办公区三大功能主体统一在其下，各自相对独立，又以打开的半室外灰空间和大平台——"城市客厅"相联系，形成面对胜利岛、胜利碑，以及老城区方向的两个视线通廊，这个公共区域就像一个全天候对市民开放的"城市客厅"，使该建筑成为真正意义的公共建筑。那婉转起伏的屋顶既像音乐流动的跌宕起伏，又恰似庐山连绵的山峰和九江地域"襟江带湖"的水体地貌，故而谓之"山水一脉"。

设计充分考虑建造的技术与经济可实施性，满足建筑内部功能、空间、结构与外部造型的统一。由于建筑外墙与屋面形体为不规则的三维曲面形式，故设计采用了专业三维软件RHINO（犀牛）、MAYA（玛雅）进行施工图设计。屋顶采用小直径无缝钢管的双层钢结构网壳结构与集构造、排水、保温、隔热为一体的直立锁边铝镁锰板屋面系统，外帷幕墙分为透明与不透明两种系统，其中透明部分采用钢结构索绗架与肋玻点支式玻璃幕墙，不透明部分采用玻璃纤维增强水泥板（GRC）系统，可进行各种曲度的三维面浇注，便于工厂预制、现场安装。

目前本项目施工图设计已结束，正处于施工过程中。

中国电子人才基地

建筑地点：北京
建筑面积：301 954平方米
设计时间：2011年
竣工时间：2012年
主要用途：办公、研发

一谷：规划利用基地东西向较长的特点引入一条长近500米的绿色之谷，整个园区内绿意盎然，生机勃勃，两期用地通过绿谷组合成为一个完整的有机生命体。

两城：规划将建筑尽量沿四周道路的退让线设置，形成外紧内松的空间布局。同时将"L"形建筑分别设于基地四个角部，建筑对内空间围合，对外形象完整。

形体：外围建筑为六层，内部采用两层，自外向内退层的手法消解了建筑轮廓线的生硬感。设计从自然界谷道外形中获得灵感，将绿谷两侧的裙房处理成自然流动的线性形态。

地下空间：将共用功能如餐饮、会所、研发配套设于中心绿谷地下，并通过设置下沉庭院和采光井来改善地下空间的使用舒适度。

江苏运河文化城“国际会议中心”

建筑地点：江苏 宿迁
建筑面积：74 237平方米
设计时间：2011年
主要用途：住宿、餐饮、会议、娱乐

江苏运河文化城“国际会议中心”基地选址于骆马湖南岸、通湖大道与女贞路交会口南侧，规划建设用地规模约29公顷，工程总建筑面积为91 707平方米，包括配置340间客房的临湖酒店主楼一座、可举办1 500人会议的宴会餐饮中心一座、娱乐健身会所一座、15栋大小不一的国宾酒店、一座室内网球馆及设备用房。

基地北侧越过通湖大道即为骆马湖，湖体宽阔，风景秀丽，位置重要。因此，北侧为基地最为优质的景观方向，景观条件优越。

基地东侧的女贞路将作为未来宿迁运河文化城的滨湖景观大道，具有较强的礼仪性和公共性。基地西侧及南侧的规划路等级相对较低，公共性较弱，车流量较小。场地东高西低，落差约4米，基地西侧标高与通湖大道25米左右的标高落差也约4米。

拟建大楼为江苏运河文化城“国际会议中心”——主楼，总建筑面积为74 237平方米。另有贵宾楼A（总统楼）7 500平方米，贵宾楼B（部长楼）3 376平方米，贵宾楼C 3 828平方米，贵宾楼D 966平方米，室内网球馆含独立设备用房1800平方米。

主楼为地上六层，建筑高度23.90米（从室外地坪到主楼屋面），多功能宴会区及会议娱乐部分两层，底层架空（局部半地下为车库）。

建筑的主要功能是提供住宿、会议、餐饮、娱乐及其完善配套功能。

东南大学苏州研究院

建筑地点：江苏 苏州
建筑面积：41 880平方米
设计时间：2007年
竣工时间：2008年
主要用途：教学、科研、实验、研发

东南大学苏州研究院的设计亮点在以一组现代主义风格来表达苏州特色地域主义内涵的教育建筑群，从而形成一个功能实用、造价适度、建造合理、空间与形式美观而不虚夸的“新苏州·新东大”的空间场所。

规划总体上采用“造城”理念——“造一座紧凑、节地、联系紧密而容纳教学、科研与办公不同功能区域的教育之城”。整体布局沿南北方向拉开成三条建筑，北侧建筑与南侧二期建筑平行于道路和用地边界形成围合的街墙；西侧庭院向西朝体育公园打开；东侧面临林泉街形成主入口广场和西庭院；而中部信息中心的底层无柱架空空间，则将东入口广场与西后庭院以及西侧的城市道路整体串联在一起，形成校园的东西主轴线。

设计采用白色为主，搭配适度灰色的建筑外墙，使人联想到粉墙黛瓦的古城苏州特色，主要使用在东西长向的使用空间领域中，而南北长向的联系空间则以U形玻璃、玻璃幕墙、聚碳酸酯板等透明的材料形成较空透的界面，实现了从不透明到透明的连续转换。

南京市妇女儿童活动中心

建筑地点：江苏 南京
建筑面积：14 326平方米
设计时间：2007—2008年
竣工时间：2009年
主要用途：活动、培训、办公

妇女儿童活动中心的设计亮点在于城市公共空间的整体营造和建筑性格的恰当表达。

南京市妇女儿童活动中心选址在南京河西新城文化中心区内，与先期建成的金陵图书馆相毗邻，并与同期规划设计的基督教圣恩堂共占用同一地块。活动中心的基本形体来自其西侧教堂基座的延伸，以此将相邻建筑强化一体，与图书馆建筑呼应。

一道东西向的"峡谷"穿破基本形体，将活动中心分成南北两部分，使城市人流可以从四周不同的方向汇集到这个文化区域，同时形成活动中心的半公共室外活动场所。连接南楼与北楼的空中过道在不同的方位和高程上穿插，在"峡谷"中创造出随机的动感与活力。这个"峡谷"在视觉上与其西侧的基督教钟塔构成一种自然的联系与张力。

活动中心的外墙以两种不同灰度的千思板(trespa)为主体饰面材料，由主形体凹入的平台空间饰以不同色彩的穿孔铝板。"峡谷"两侧为透明的玻璃明框幕墙，使其内外之间形成连续的空间体验。幕墙的双曲面构形及其竖向明框上点缀的彩色铝构件的随机分布方式通过参数化设计的程序得到有效的控制。流动的幕墙界面、悦动的色彩恰如其分地再现了妇女、儿童柔美活泼的特点。

人民日报社报刊综合业务楼

建筑地点：北京
建筑面积：137 883平方米
设计时间：2011年
竣工时间：建设中
主要用途：办公、会议、图书馆

人民日报社报刊综合业务楼是人民日报社建社以来最重大的建设项目之一，由国际会议中心、图书馆和办公主楼组成，总建筑面积137 883平方米，主楼32层，建筑高度180米。本设计方案通过象征的手法与建筑形态上的动感，体现着建筑的标志性和现代感，是时代特色同传统文化内涵的融合。

以人为本的设计思想

以人字形的平面表达人民日报社的主题，创造尺度宜人的空间关系；自然柔和的曲线造型，形成良好的视觉效果。

与自然环境和谐统一

对自然环境进行最小的干预，保留原有基地良好的植被条件；对道路的退让处理，使高层建筑形体在视觉上弱化，减小对周边建筑日照条件的影响。

对时代精神的积极回应

以独特的建筑造型理性的表达结构受力的逻辑，在稳定的结构之中寻求视觉上的动态感，利用奔腾的动势，反映出发展与进步的观念。

苏州工业园区设计研究院有限责任公司

Suzhou Industrial Park Design & Research Institute Co.,Ltd.

苏州工业园区设计研究院成立于1995年，是依照国外设计事务所运作模式创立的新型甲级设计研究院，服务于园区开发建设。园区设计院已于2002年9月由国有公司成功转制为由员工持股的有限责任公司。

苏州工业园区设计院采用国际通行的项目经理制，提供高效的、全方位的设计和服务。

苏州工业园区设计院已通过了ISO9001：2000质量体系认证，取得了中国质量体系认证（CCQS）证书和英国尤卡斯（UKAS）质量体系认证证书。

苏州工业园区设计院拥有一支由500余人组成具有年富力强、学有所长、精通业务，有实际经验，并通晓国际惯例的规划师、建筑师、景观设计师、结构工程师、电气工程师、暖通空调工程师、给排水工程师和概预算工程师的优秀团队。

苏州工业园区设计院成立16年来，已完成各类工程设计项目数百项，其中包括文教卫生类建筑、居住类建筑、办公类建筑、商业类建筑、城市规划类、景观类、工业建筑类等。同时，园区设计院还为众多外资项目提供项目管理或设计施工总承包服务。

苏州工业园区设计院重视学习国外的先进管理思想、设计技术与运作模式，已与多家国外公司建立了良好的合作关系，熟知国际最新设计理念，并积累了丰富的设计经验。

苏州工业园区设计院的设计作品获得了社会广泛好评，并获得了多项国家级、部级和省级、市级优秀工程设计奖。

苏州工业园区设计院目前有合资企业和子公司：

苏州工业园区设计研究院工程公司、苏州工业园区建设监理有限责任公司、苏州工业园区设计研究院装饰工程有限公司、苏州工业园区方元置业有限公司、境群国际规划设计顾问（苏州）有限公司、苏州工业园区赛普装饰工程有限公司、苏州工业园区设计研究院上海分院。

Suzhou Industrial Park Design and Research Institute, was established in 1995, a new Mode and Class A Research Institute in accordance with the mode of operation of foreign design firms, serving on park development and construction. Park Institute in September 2002 was successfully transformed from state-owned to the employee-owned limited liability company.

Suzhou Industrial Park Design Institute uses the international accepted system for project managers to provide efficient, comprehensive design and service.

Suzhou Industrial Park Design Institute has passed the ISO9001:2000 quality system certification, received the Quality System Certification of China (CCQS) and the UKAS Quality System Certification.

Suzhou Industrial Park Design Institute has an excellent team of more than 500 people, with prime of life, deep learning, proficiency, practical experience, knowledge of international customs. They are planners, architects, landscape architects, structure engineers, electric engineers, HVAC engineers, drainage engineers and budget estimate engineers.

Suzhou Industrial Park Design Institute has completed hundreds of types of engineering projects since its foundation for 16 years, including cultural and educational construction, housing construction, office buildings, business architecture, urban planning, landscape, industrial buildings etc. At the same time, the Park Institute also provides project management services or general contracting for a number of foreign projects.

Suzhou Industrial Park Design Institute focuses on learning foreign advanced management ideas, design and mode of operation, has a good working relationship with a number of foreign companies, knowledge of the latest international design concept, and has accumulated rich experience in the design.

The design works of Suzhou Industrial Park Design Institute with widespread praise, has won several national, ministerial and provincial, municipal engineering design excellence awards.

Suzhou Industrial Park Design Institute currently has joint ventures and subsidiaries:

Suzhou Industrial Park Design Institute Engineering Company, Suzhou Industrial Park Construction Supervision Co., Ltd., Suzhou Industrial Park Design Institute Decorative Engineering Co., Ltd., Suzhou Industrial Park, Fang Yuan Properties Ltd., Environmental Group International Planning and Design Consultants (Suzhou) Co., Ltd., Suzhou Industrial Park Sapp Decoration Engineering Co., Ltd., Suzhou Industrial Park Design Institute Shanghai Branch.

地址：江苏省苏州工业园区苏虹中路393号
邮编：215021
传真：+86-512-62586259
电话：+86-512-62586258
网址：WWW.SIPDRI.COM

Add: 393 Suhong Middle Road,
Suzhou Industrial Park, Jiangsu
P.C.: 215021
Tel: +86-512-62586258
Fax: +86-512-62586259
Http://WWW.SIPDRI.COM

苏州工业园区设计研究院新办公楼

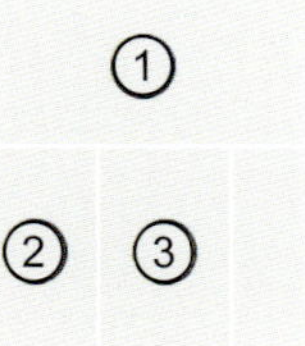

1 中国桑蚕丝绸博物馆
Chinese Silkworm Silk Museum

2–3 苏州广播电视总台现代传媒广场
合作设计：日建设计
Suzhou Radio and Television Station Modern Media Plaza
Co-design: Nikken Sekkei

1 四川绵竹体育中心
Sichuan Mianzhu Sports Center

2–3 四川绵竹历史博物馆
Sichuan Mianzhu History Museum

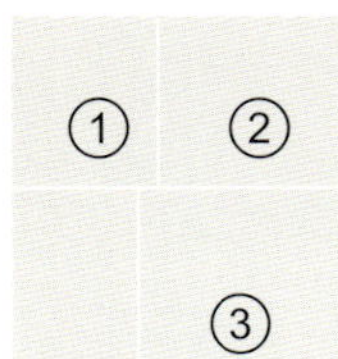

1 吴江应急指挥中心大厦方案
Wujiang Emergency Command Center Building Program

2 苏州工业园区科技园七期纳米技术孵化基地方案投标
Phase Seven Nanotechnology Incubator Bid Program
Technology Park, Suzhou Industrial Park

3 苏州工业园区档案管理中心
合作设计：JPW
Suzhou Industrial Park File Management Center
Co-design: JPW

1 博世中国总部大楼
Bosch China Headquarters Building

2 博世汽车柴油系统股份有限公司
Bosch Automotive Diesel Systems Co., Ltd.

3 布林顿斯地毯制造（苏州）有限公司
Brintons Adams Carpet Manufacturing (Suzhou) Co., Ltd.

4 卡特彼勒（苏州）有限公司
Caterpillar (Suzhou) Co., Ltd.

5 艾利（苏州）有限公司
Avery (Suzhou) Co., Ltd.

①	② ③
④	⑤

1 中国矿业大学国家大学科技园总部基地
National University Science Park Headquarters Base, China University of Mining and Technology

2 南京莱斯大型电子系统工程科研生产基地
Nanjing Royce Large Electronic Systems Engineering Research and Production Base

3 苏州工业园区金姬顿邻里中心
Jin Jidun Neighborhood Centers, Suzhou Industrial Park

4 阳澄湖澜庭度假酒店
Yangcheng Lake Lanting Resort Hotel

5 吴江东太湖温泉度假酒店
Wujiang East Taihu Lake Spa & Resort Hotel

江苏华海建筑设计有限公司
Jiangsu Huahai Architectural Design Co., Ltd.

江苏华海建筑设计有限公司前身为徐州市第三建筑设计研究院，成立于1984年，为国家甲级建筑工程设计单位，同时具有甲级工程总承包及甲级工程监理资质，并于2004年9月在徐州市事业单位中率先改企转制成功。现有职工78人，其中，中高级技术职称42人、国家一级注册建筑师6人、一级注册结构工程师6人、其他专业各级注册工程师12人。历经20多年的艰苦努力，发展成为专业设置配套齐全、技术力量雄厚、管理方法科学的建筑设计公司。公司下设建筑设计所、结构设计所、设备设计所。

多年来，公司的工程设计项目遍及苏、鲁、豫、皖、冀、闽、浙、鄂、京、沪、内蒙古等十几个省市自治区。40多项工程获省（部）、市优秀设计奖。近年来，每年完成数百项工程项目。完成了御景湾、山水湾、清水湾、碧水湾、湖光山色、兴隆大厦、新亚大厦、石家庄雅清小区、淮北青岛花园、万宁华府、润敏香槟城、鼓楼生态园、山水康桥、贾汪传世经典、新沂城市花园、邳州汇龙国际、富贵第一城、南京水月秦淮、连云港爱丁堡公寓、宿州明丽大厦、宿州检察院、久隆凤凰城等一批高层建筑或大型住宅小区；完成了安居工程、风华园、徐州科技城、淮海食品城、新城区惠民花园等一批省（部）、市重点工程。其中，凤凰山康居小区被列为国家重大科技产业工程项目，并荣获"国家小康住宅示范小区"称号；久隆凤凰城、天润花园、徐州人家、华美·沁园、公园100、嘉慧园等小区被《徐州日报》评为2004年读者最喜爱的十大精品楼盘，占入选楼盘的50%以上；山水华美、苏商·御景湾分别荣获2006年度、2007年度江苏省优秀住宅奖。

公司设计的云龙湖水上世界、徐州国际会展中心、徐州公交大厦、汇源置地广场等，建成后已成为徐州市具有一定代表意义的标志性建筑，取得了良好的经济效益和社会效益，为该市城乡建设作出了显著的贡献。公司承接的上海世博会城市人馆与中国国家馆的内馆设计，于2009年7月顺利通过了世博局专家组的评审。目前，公司正在设计徐州鼓楼广场和菏泽豪庭绿洲两个大型中央商务区及高层住宅综合体。

公司的钢结构工程设计所成立于1995年，是徐州市最早成立的钢结构工程设计研究机构，该所设计的轻钢网架工程遍及全国各地，自成立以来完成了包括中央电视台1号演播厅在内的上千个网架工程，为徐州市获得"网架之乡"的美誉提供了有力的技术支持。

多年来，公司坚持设计与研究相结合，走技术发展和人才发展的道路。公司与中国矿业大学建立产学研联办关系，每年派多人参加市、省及全国性的学术研讨与交流，每年有十几篇学术论文在省级、国家级乃至全国核心期刊上发表；还参编《粉煤灰小型空心砌块填充墙》、《太阳能热水系统与建筑一体化设计标准图集》等多套徐州市标准图集。

全体职工积极倡导"团结向上，爱岗敬业，精心设计，求实创新"的企业精神，坚持改革，致力发展，连续多年荣获省、市"两个文明建设先进单位"、"先进集体"等光荣称号；被徐州市工商行政管理局评为"AA级重合同守信用单位"；并顺利通过ISO9001：2000质量认证。

自2005年起，公司在永安财产保险公司投保了810万元的"建设工程设计责任保险"，成为徐州市及淮海经济区首家投保的设计公司，该保险不仅增强了公司的抗风险能力，也为建设项目提供了设计质量的保证。

Jiangsu Huahai Architectural Design Co., Ltd., formerly known as Xuzhou the Third Institute of Architectural Design, founded in 1984, is a design unit with the National Class A qualification, and has Class A General Contracting and Engineering supervision qualification. In September 2004, it first completed the successful reformation in Xuzhou City public institutions into enterprises. At present, there are 78 employees, of which 42 senior and intermediate titles, six are the First Class registered architects of China, six are the First Class Registered Structural Engineers, and 12 other registered professional engineers at all levels. After 20 years of hard work, it has developed into an architectural design company with professional setting, strong technical force and scientific management methods. Company consists of architectural design studio, structural design studio, equipment design studio.

Over the years, our company's engineering design projects are located throughout Jiangsu, Shandong, Henan, Anhui, Hebei, Fujian, Zhejiang, Hubei, Beijing, Shanghai, Inner Mongolia and other provinces. More than 40 projects won the excellent design awards of provincial (ministry). In recent years, hundreds of projects are completed each year. Including Royal Lagoon, Mountain and Water City, Clear Water Bay, Green Water Bay, Lakes and Mountains, Xinglong Building, New Asia Building, Shijiazhuang Yaqing Community, Huaibei Qingdao Garden, Wanning Hua Fu, Runmin Champagne City, Drum Tower Ecological Park, Mountain and Water Cambridge, Jia Wang Chuanshi Classic, Xinyi City Garden, Pizhou Huilong International, Rich the First City, Nanjing Shuiyue Qinhuai, Lianyungang Edinburgh Apartments, Suzhou Mingli Building, Suzhou Procuratorate, Jiulong Phoenix City and a number of high-rise buildings or large-scale residential communities; completed housing projects, Feng Hua Park, Xuzhou Science and Technology City, Huaihai Food Town, New City Huimin Garden and a number of provinces (Ministry) and municipal key projects. Among them, the Phoenix Mountain Healthy Community is classified as national key industrial projects, and won the "Well-off Residential National Demonstration District" title; Jiulong Phoenix City, Tianrun Garden, Xuzhou People, Beauty Qinyuan, Park 100, Jiahui Garden and other communities are entitled by Xuzhou Daily as 2004's Top Ten Readers Favorite Boutique Real Estate, accounted for more than 50% real estate selected; Mountain and Water Beauty, Union were Sushang • Royal Lagoon, was respectively won Jiangsu Province Outstanding Residential Award in 2006 and 2007.

Our company designed the World on the Yunlong Lake, Xuzhou International Convention and Exhibition Center, Xuzhou Bus Building, Huiyuan Landmark Plaza, etc., after the completion they have become a significant landmark of certain representatives in the city, achieved good economic and social benefits, and made a significant contribution for the Urban and Rural Construction of the city. The company took the design of Shanghai World Expo Urban Museum and China National Museum, in July 2009 has passed the expert assessment of the Expo Bureau. At present, our company is designing Xuzhou Drum Tower Square and Heze Royal Oasis, two large CBD and high-rise residential complex.

Steel structure engineering design studio established in 1995, is the oldest engineering research institute of steel construction in Xuzhou city, the studio designed the light steel grid engineering throughout the country, since its foundation, has completed thousands of grid engineering including CCTV No.1 studio Hall, it provided strong technical support for the reputation of Grid Xuzhou City.

Over the years, our company insists on the combination of design and research, technological development and takes the path of human development. Our company established the relationship of production, learning and research with China University of Mining, sending people attend the annual municipal, provincial and national academic discussions and exchanges, each year more than a dozen papers published in provincial, national and even the national core journals; also supervised, *Small Hollow Block Wall Filled with Fly Ash, Solar Water Heating Systems and Building-integrated Design Standards Atlas and Sets of the Standard Atlas* of Xuzhou City.

All workers actively promote the enterprise spirit, "unity, love and dedication, well-designed, realistic and innovative" and insist on reformation, efforts to develop, for many years won the provincial and municipal "Advanced Unit in the two Civilizations", "Advanced Group" and other titles of honor, rated "AA-level contract and keeping promises units" by Xuzhou City Administration for Industry and Commerce; and passed ISO9001: 2000 quality certification.

Since 2005, our company has insured 810 million Yuan in Yong An Property Insurance Company for "Construction and Engineering Design Liability Insurance", the first insurance design company of the Huaihai Economic Zone and Xuzhou City. The insurance not only enhances the anti-risk capacity of our company but also affords the assurance of design quality for building projects.

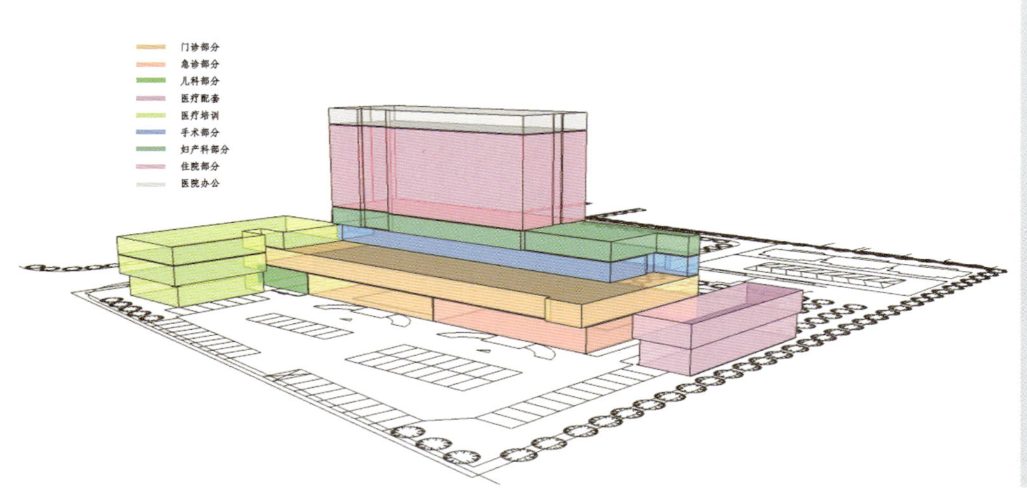

砀山县红十字医院

项目地点：安徽 砀山
总占地面积：17 944平方米
总建筑面积：31 635.4平方米
容 积 率：1.6
绿 化 率：36%

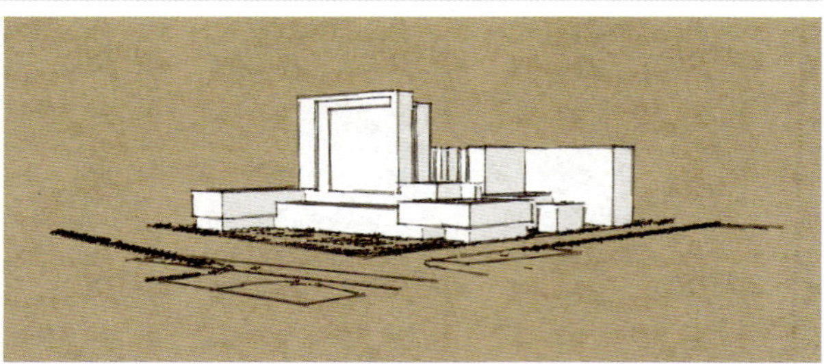

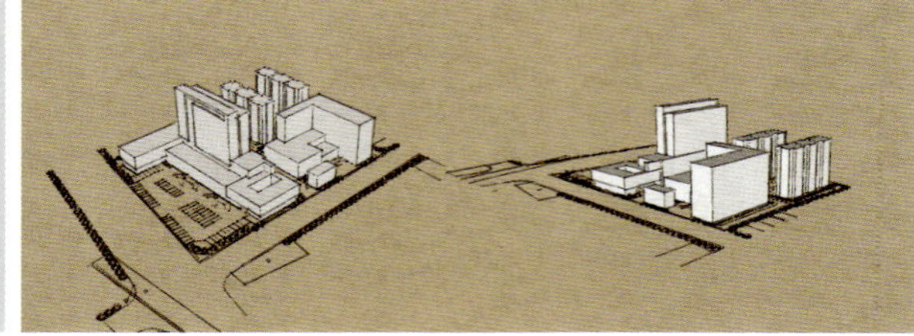

新沂阳光水岸绿洲规划建筑设计方案

项目地点：江苏 新沂
规划总用地面积：294 668.1平方米
规划总建筑面积：1 040 526.0平方米
绿 地 率：20.1%
容 积 率：3.0

其中：
A地块
规划总用地面积：119 049.7平方米
规划总建筑面积：305 581.3平方米
B地块
规划总用地面积：80 530.0平方米
规划总建筑面积：360 320.5平方米
C地块
规划总用地面积：99 748.0平方米
规划总建筑面积：374 624.2平方米

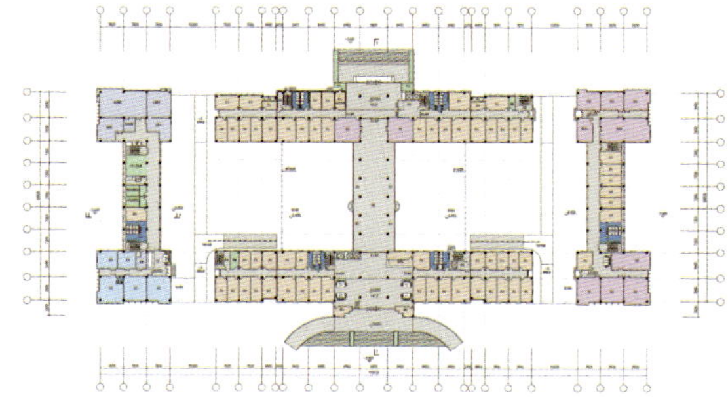

一层平面图

徐州市贾汪新城商务中心

项目地点：江苏 徐州
建筑面积：40 000平方米
建筑高度：38.10米
建筑层数：地上9层，地下1层

本工程建成后作为县级人民政府的施政中心，庄严、壮观，体现出“开放”、“公正”、“民主”等特征。

徐州月星世纪城商业中心住宅部分规划建筑方案

该项目位于江苏省徐州市高铁商务区，建设用地面积82 758平方米，本项目设计总建筑面积约为40万平方米（含地下室）。商业面积为30万平方米，住宅为10万平方米，建筑方案为地下层、地上为五层商业、酒店、办公及公寓等。该方案则为该工程的住宅部分设计。

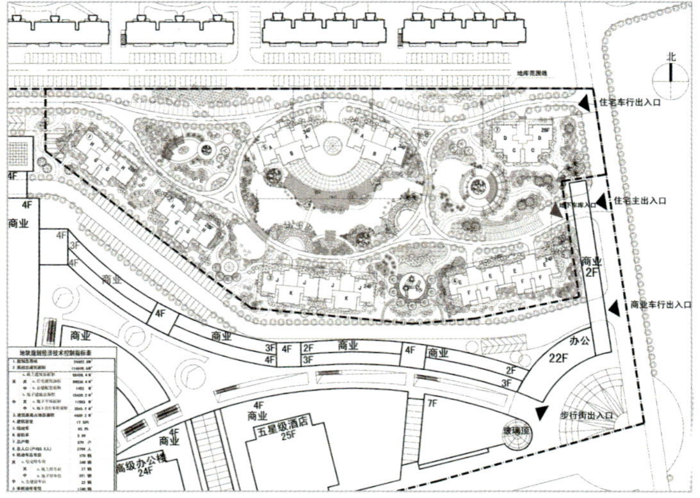

淮北市烈山区人民法院审判综合楼

项目地点：安徽 淮北
总用地面积：15 022.83平方米
总建筑面积：11 682.2平方米
建筑密度：18.2%
容 积 率：0.78
绿 地 率：41.3%
建筑高度：43.20米
层　　数：11层

本法院共设计刑事庭1个，中法庭1个，小法庭7个。

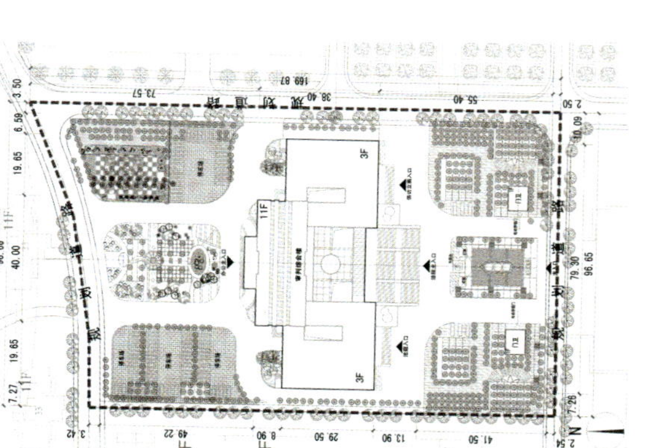

总平面图

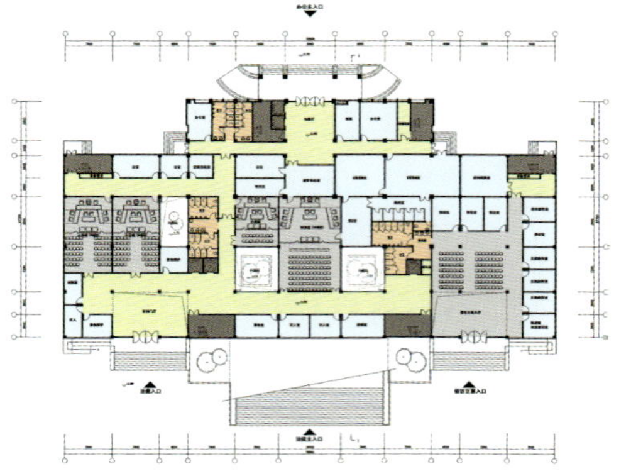

一层平面图

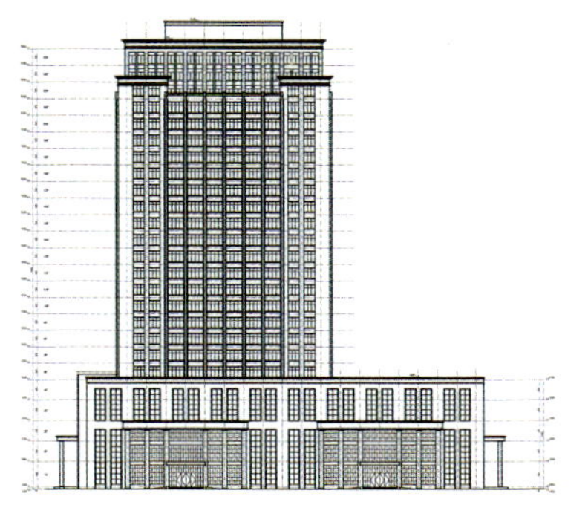
西立面图

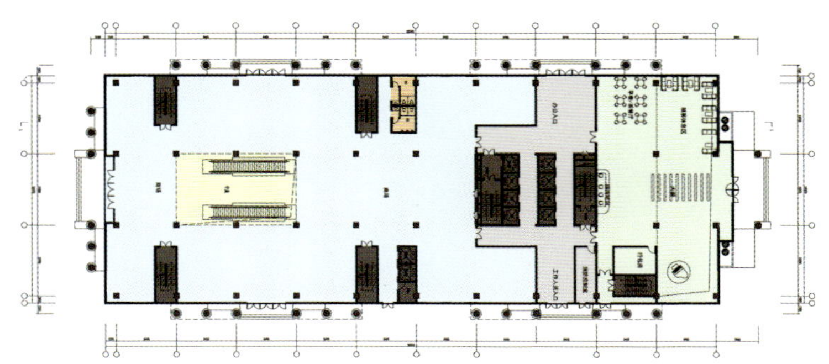
裙房一层平面图

徐州市国际东方商务中心

项目地点：江苏 徐州
基地面积：5 142.2平方米
总建筑面积：46 600平方米

容 积 率：7.51
建筑密度：53.8%
绿 化 率：15.2%

苏商御景湾规划建筑设计

项目地点：江苏 徐州
总用地面积：31 667平方米
总建筑面积：491 828平方米

容 积 率：3.0
绿 地 率：30.2%
建筑密度：43%

2010—2011
主要建筑设计作品
Main Architectural Design Works
Annual Review of Chinese Architectural Design Works

ATELIER ZHOULING
南京大学建筑与城市规划学院周凌工作室

地址: 江苏省南京市汉口路22号，南京大学费彝民楼1508
Add: Room 1508, Fei Yimin Building, Nanjing University, No. 22 Hankou Road, Nanjing, Jiangsu
联系人: 周 凌
Contact: Zhou Ling
邮箱 / E-mail: zl7070@163.com, archidot@163.com
电话 / Tel: +86–25–83621621, +86–25–83621622, 13915971665
传真 / Fax: +86–25–83595673
邮编 / P.C.: 210096

周凌建筑工作室成立于2005年，完成政府、学校、办公、展览及各类文化等建筑项目，多次获得省部级建筑设计奖项。主创建筑师周凌从东南大学建筑学院博士毕业，现为南京大学建筑与城市规划学院副教授、硕士生导师，建筑系副系主任。主要代表作品有宁通高速公路收费站房、兴化档案馆、南京大学百年纪念堂、南京大学教学综合楼、南京理工大学学生生活区、上海世博公园建筑、武进西太湖中心广场展览建筑、绿地21世纪企业公园、绿地21世纪影城等。

工作室作品2001年获得“江苏省中青年建筑师建筑创作奖”二等奖；2003年获“中国青年建筑师奖”设计竞赛奖；2004年获“第五届中国青年建筑师奖”；2005年入选欧洲首本介绍中国当代建筑师的书籍《中国制造——新中国建筑》（慕尼黑DVA出版社）；2005年，作品应邀参加上海国际建筑设计规划展；2007年参加上海“40x40华人青年建筑师展”；2008年参加“深圳艺术三年展”。2004—2008年，周凌工作室作品被中国各大建筑杂志，如《建筑学报》、《世界建筑》、《时代建筑》刊载报道；文章多次发表于《建筑师》、《城市建筑》、《香港建筑业导报》、《建筑文化》等刊物；作品被《建筑实录》、《中国中青年建筑师画像》等书收录。2005年以来，工作室作品被国内外各网站杂志报道，如德国著名建筑网站“世界建筑”网站http://www.world–architects.com/、意大利建筑网站http://www.europaconcorsi.com等。

Founded in 2005, Atelier Zhou Ling has completed government, school, office, exhibition and all kinds of cultural projects. It has won several provincial awards for architectural design. Chief Creative Architect – Zhou Ling, who graduated from Architecture Department of Southeast University with a Doctor's Degree, is now the Associate Professor, Master Instructor and Vice-director of Architecture Department of School of Architecture and Urban Planning of Nanjing University. Main representative works are: Toll House of Nanjng Expressway; Xinghua Archives; Centennial Hall of Nanjing University; Comprehensive Teaching Building of Nanjing University; Students Living Area of Nanjing University of Science and Technology; Shanghai World Expo Park; Central Square Exhibition Hall on the West of Taihu Lake, Wujin; Green Land 21st Century Enterprise Park; Green Land 21st Century Movie Theater etc.

Works of Atelier Zhouling won the Second Prize of "Architectural Creation Award for Young Architects of Jiangsu" (2001); "Chinese Young Architects Award" of Design Competition (2003); the "Fifth China Young Architects Award" (2004). It was collected in *Made in China – New Chinese Architecture* which is the first European book to introduce contemporary architects of China (Munich DVA Press) (2005). In 2005, its works were invited to participate in Shanghai International Exhibition of Architectural Design and Planning; in 2007, Atelier Zhouling attended the "Exhibition of 40x40 Chinese Young Architects"; in 2008, it participated in the "Shenzhen Free Art Base Triennial". During 2004 to 2008, works of Atelier Zhouling were reported by famous architectural magazines of China, such as *Architectural Journal*, *World Architecture* and *Time Architecture*, published in *Architects*, *Urban Architecture*, *Hong Kong Building Review*, *Architectural Culture* and other publications, collected by *Architectural Record*, *Portrait of Chinese Young Architects* and other books. Since 2005, Atelier Zhoulingn has been reported by all magazines and websites at home and abroad, such as the famous architectural website of Germany – "*World Architecture*" (http://www.world-architects.com/), as well as famous architectural website of Italy (http://www.europaconcorsi.com) etc.

宁德图书馆
NINGDE LIBRARY

设计单位/Design Company：南京大学建筑与城市规划学院周凌工作室/ATELIER ZHOU LING
设计团队/Design Team：周 凌、应 超/Zhou Ling, Ying Chao
建筑面积/Floor Area：12 000 m^2
设计时间/Design Time：2011年
项目地点/Location：福建 宁德
项目状态/Project Status：建设中/Under construction

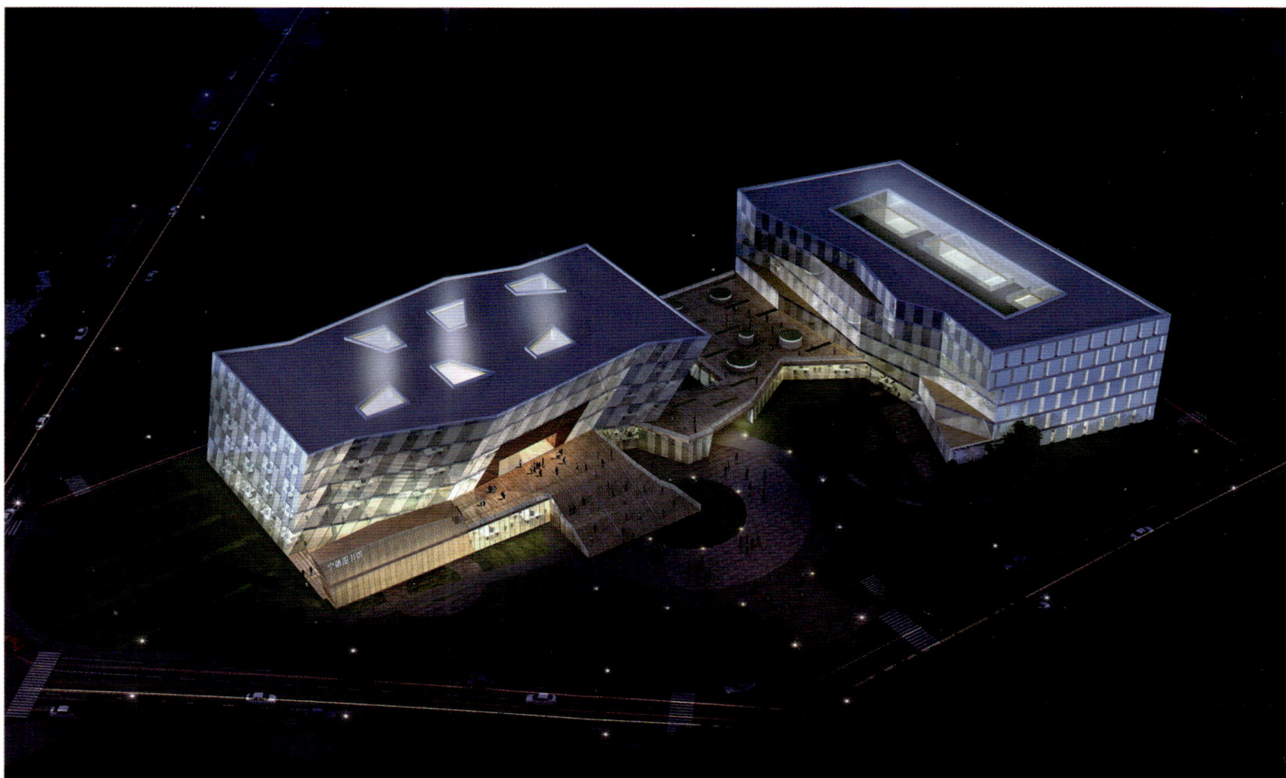

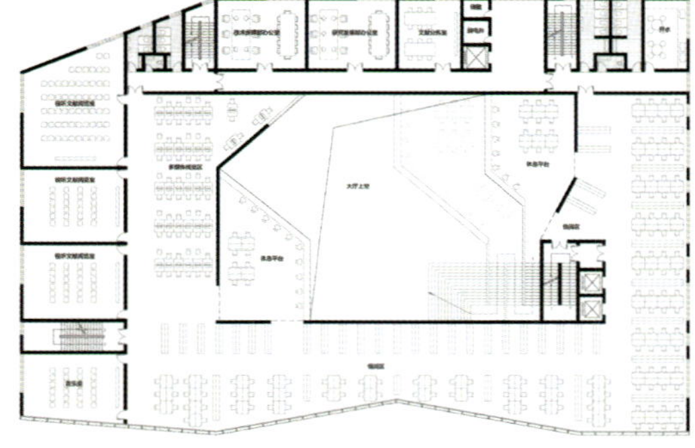

四层平面图

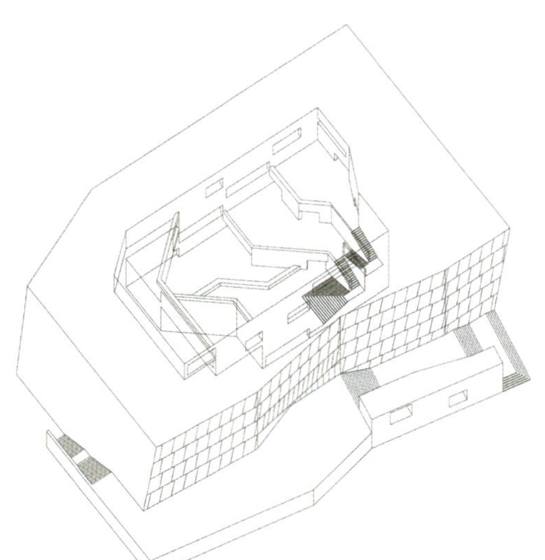

吴江新城中心广场
THE SQUARE OF WUJIANG CITY CENTER

设计单位/Design Company：南京大学建筑与城市规划学院周凌工作室/ATELIER ZHOU LING
设计团队/Design Team：周 凌、张 雷、胡 昕、夏 芸/Zhou Ling, Zhang Lei, Hu Xin, Xia Yun
建筑面积/Floor Area：88 000 m^2
设计时间/Design Time：2011年
项目地点/Location：江苏 苏州
项目状态/Project Status：建设中/Under construction

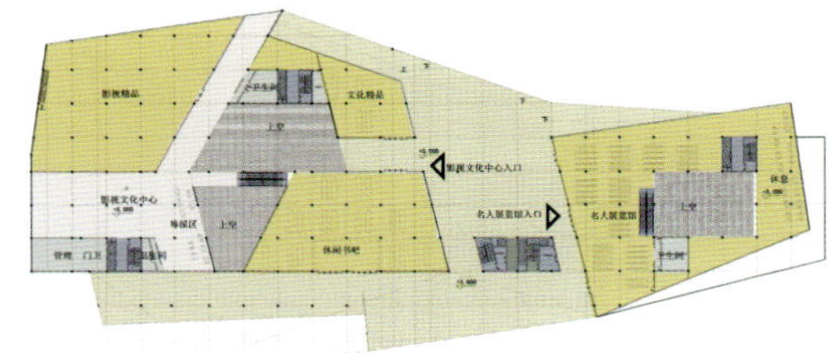

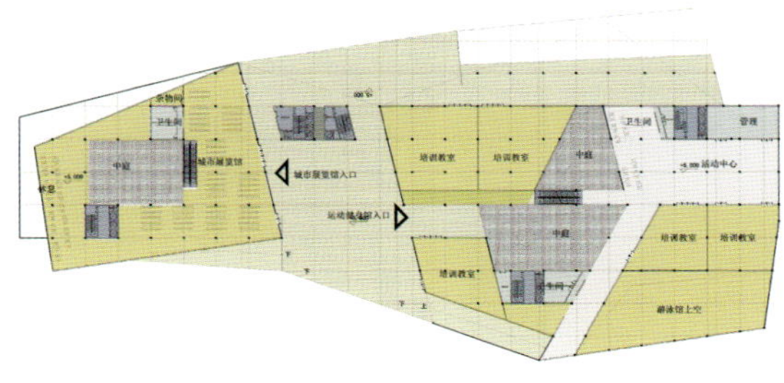

赭石门会所
OCHER DOOR CLUB

设计单位/Design Company：南京大学建筑与城市规划学院周凌工作室/ATELIER ZHOU LING
设计团队/Design Team：周 凌、刘琨鹏/Zhou Ling, Liu Kunpeng
建筑面积/Floor Area：2 430 m^2
设计时间/Design Time：2011年
项目地点/Location：四川 成都
项目状态/Project Status：建设中/Under construction

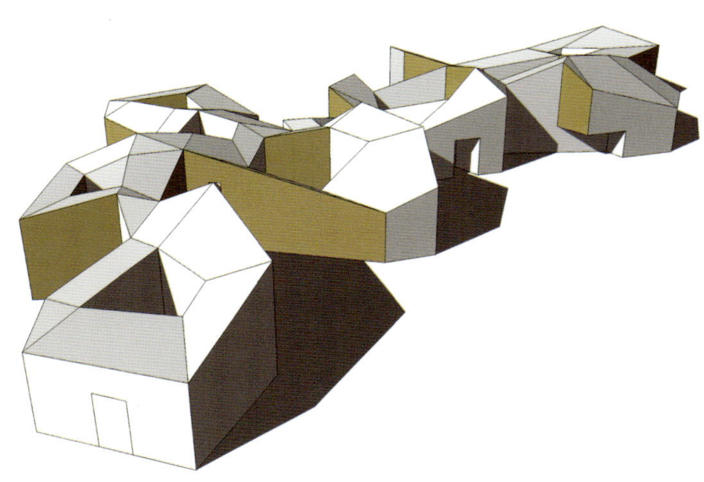

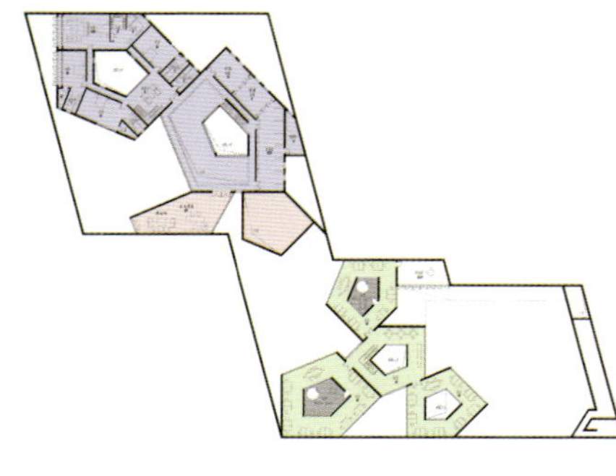

岱山十五班幼儿园
DAISHAN 15 CLASS NURSERY

设计单位/Design Company：南京大学建筑与城市规划学院周凌工作室/ATELIER ZHOU LING
设计团队/Design Team：周 凌、张 茹/Zhou Ling, Zhang Ru
建筑面积/Floor Area：4 279.6 m^2
设计时间/Design Time：2011年
项目地点/Location：江苏 南京
项目状态/Project Status：建设中/Under construction

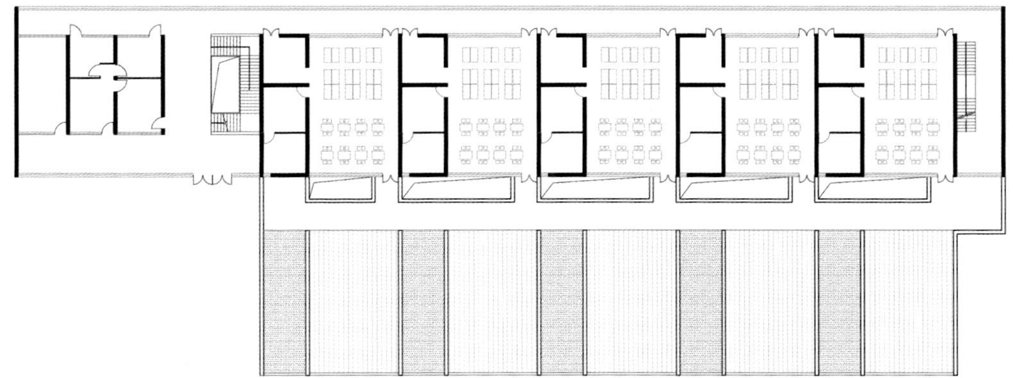

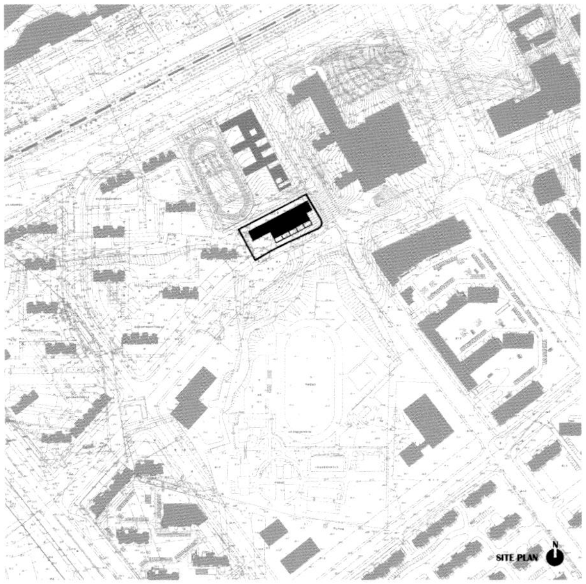

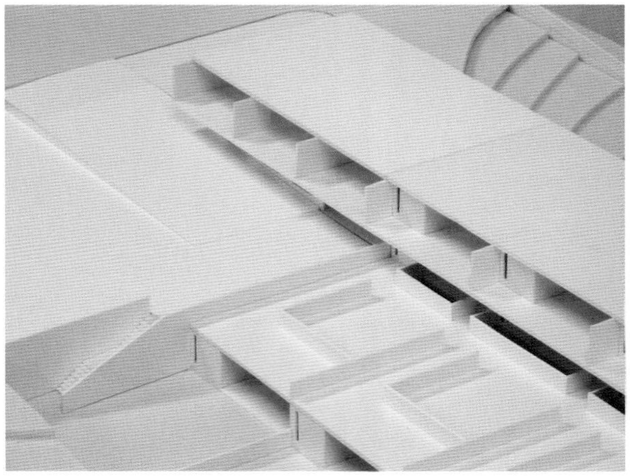

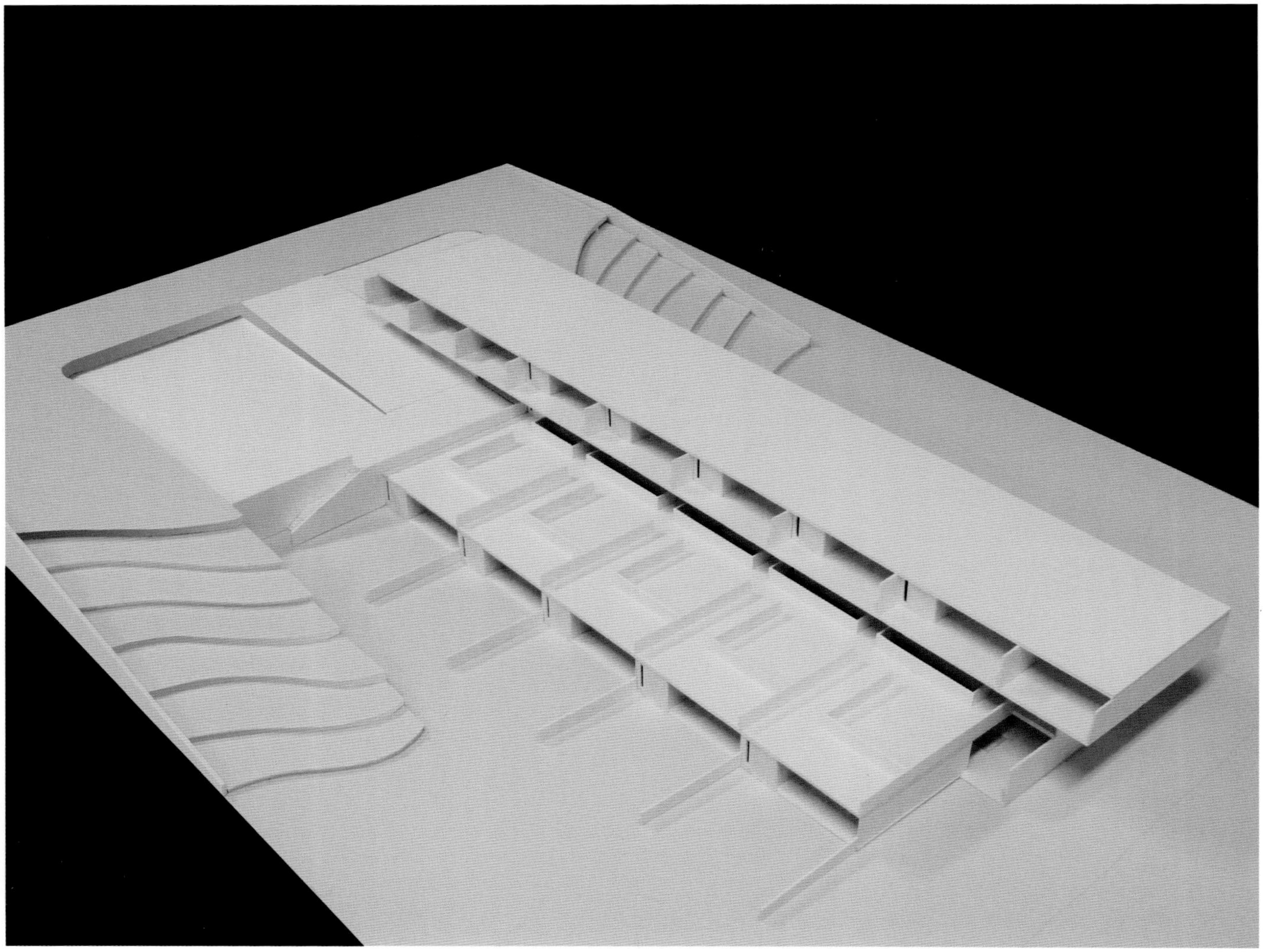

辽宁省建筑设计研究院
LIAONING PROVINCIAL BUILDING DESIGN & RESEARCH INSTITUTE

辽宁省建筑设计研究院（LDI）是国家甲级建筑设计和甲级工程勘察单位，也是国家首批具有建筑智能化专项工程设计资质的单位，沈阳市高新技术企业。本院创建于1956年1月，由沈阳、大连、锦州三市的设计人才汇集而成，至今已发展成为能够为工程建设提供全方位综合服务的专业设置齐全、设计手段先进、技术力量雄厚、社会信誉卓著的优秀勘察、设计和咨询单位。全院职工305人，其中教授级高级建筑师和教授级高级工程师25人，高级建筑师和高级工程师67人，建筑师和工程师103人；具有国家级执业资格注册人员82人，享受国务院特殊津贴专家6人。本院下设4个综合性建筑设计研究所、建筑方案研究所、规划设计研究所、市政设计研究所，以及林立岩大师工作室、环境艺术研究所、建筑经济所、地源热泵应用技术研究所等部门，院属公司有岩土工程公司、项目管理咨询公司、施工图审查咨询有限公司等实体。2006年成立了大连分院。

建院50多年来，已经向社会提供了1.3万余项设计成果，工程项目不仅遍布辽宁，而且拓展到北京、上海、广州、深圳、杭州、宁波、无锡、西宁、威海、大庆、吉林、梅河口、通辽、赤峰等城市，以及也门、布基纳法索、俄罗斯、喀麦隆、塞舌尔、巴基斯坦等国家。近年来，我院与德国、澳大利亚、美国、英国、日本、韩国等国家，以及我国香港、台湾的建筑师事务所有着广泛的合作，共同投标、合作设计了国内的多个项目。

20世纪80年代以来，共获国家、部、省、市级优秀工程设计（勘察）奖179项和科技进步奖72项，并荣获辽宁建设系统质量管理奖、重合同守信用单位、辽宁省工程勘察设计行业CAD应用先进单位、档案管理国家一级单位、辽宁省“八五”期间建设科技工作先进单位、全国建筑节能先进单位、沈阳市节能设计先进集体、全国建设技术创新工作先进单位等奖励和称号。

2000年12月，我院总工程师林立岩同志被建设部授予“中国工程设计大师”称号；2008年院长杨晔、副院长兼总工师李庆钢被授予首批“辽宁省工程设计大师”称号。院长杨晔同志1990年获联合国教科文组织奖（UNESCO PRIZE）；1998年获中国建筑学会青年建筑师奖；2004年当选“沈阳市十大中青年建筑师”、“沈阳市十大杰出青年”；2006年被评为“辽宁省劳动模范”、“辽宁省优秀共产党员”，被中国勘察设计协会授予“优秀企业家”称号；2008年被授予“辽宁省十大杰出青年”称号，并当选2008奥运火炬手；2009年荣获“全国建筑设计行业国庆60周年管理创新奖”。副院长兼总工程师李庆钢2005年荣获“第五届辽宁青年科技奖十大英才”称号。2007年总建筑师曹辉获“辽宁五一劳动奖章”，2009年获“辽宁省直属机关十大杰出青年”称号。我院及所属岩土工程公司、项目管理咨询公司于2000年12月通过ISO9000（1994版）质量体系认证；2003年通过GB/T 19001－2000 idt ISO 9001:2000标准质量体系认证。

The institute (LDI) is a national Class-A building design and project exploration firm. As one of the intelligent building special item project designing and one of the first groups of institutes in the country;it has qualification of intelligent building special item project design. Founded in January, 1956, the institute has collected design talents from Shenyang, Dalian and Jinzhou. Now it has become an excellent building design and exploration firm with complete specialties, advanced design method, great technological power and excellent social reputation. The institute has 305 staff members, including 25 professorial architects and engineers, 67 advanced architects and engineers, 103 architects and engineers, 82 national class registered architects and engineers and 6 experts enjoying special state allowance. The institute comprises four complex Architecture Design and Research Section, Master Lin Liyan Studio, Environmental Art Institute and Building Economy Section as well as IAT on GSHP. Affiliated companies include geotechnical engineering company, project managing consultation company, construction drawing review consultation Co., Ltd. and so on.

More than 50 years since the foundation of the institute, it has provided more than 13,000 achievements to the society. These engineering projects can be found not only in Liaoning but also in many other cities such as Beijing, Shanghai, Guangzhou, Shenzhen, Hangzhou, Ningbo, Wuxi, Xining, Weihai, Daqing,Jilin, Meihekou, Tongliao, Chifeng as well as some foreign countries like Yemen, Burkina Faso, Russia, Cameroon, Seychelles. Recent years, the institute has broad cooperation with some foreign architects firms including Germany, Australia, America, Britain, Japan and Korea. We work together to win the bid and design a number of domestic projects.

Since 1980s, the institute has won 179 excellent project design (exploration) prizes and 72 technological progress prizes with national, ministerial, provincial and municipal level. Moreover, it has been awarded series of honors and titles such as Liaoning Provincial Construction System Quality Management Prize, Unit Focusing on Contract and Credit, Liaoning Provincial Exploration and Design Industry CAD Applying Advanced Unit, National 1st Class File Management Unit, Advanced Unit for Construction Technology Work in Liaoning Provincial "Eighth Five-Year Plan" Period, National Building Energy-Saving Advanced Unit, Shenyang Energy-Saving Design Advanced Group, and National Construction Technology Creation Work Advanced Unit.

In December 2000, Mr. Lin Liyan, Chief Engineer of the institute, was awarded the honorary title of "Master of China engineering design" by the Ministry of Construction of People's Republic of China. In 2008, President Yang Ye and Vice President Li Qinggang were awarded the first batch of title for Master of Liaoning Engineering Design. Mr. Yang Ye, the president, was awarded the UNESO PRIZE in 1990, the Youth Architect Prize of CAA in 1998, and he was also selected into "Top 10 Youth Architects of Shenyang" and "Top 10 Excellent Youths of Shenyang" in 2004. In 2006 he was selected as Liaoning Province Working Model, Liaoning Excellent Communist, and was awarded the title of Excellent Entrepreneur by China Exploration Design Association. In 2008 he was awarded the title of Top Ten Youths of Liaoning Province, and was selected as the 2008 Olympic torch bearer. In 2009 Yang Ye was awarded "Franchise Innovation Award of the 60th National Day" by CEDA. Vice President Li Qinggang were awarded "Liaoning Youth Science and Technology Award for Top Ten of excellence" in 2005. Cao Hui, as the chief architect, was awarded May 1st Labor Medal of Liaoning in 2007, and also awarded the title of Top Ten Excellent Youths of directly subordinate organization under Liaoning Province in 2009. Geological Engineering Co., Ltd. and Project Management Consultant Co., Ltd, subordinated by institute and the institute obtained the certificates of ISO9000 Quality System approval (1994 version) in December 2000, and GB/T 19001-2000 idt ISO 9001:2000 Standard Quality System approval in 2003.

联系方式 / Contact Us

网址 / Http://www.ldi.com.cn 邮箱 / E-mail: ldi@ldi.com.cn

沈阳
地址 / Add: 辽宁省沈阳市和平区和平南大街84号
No.84, Heping Nan Street, Heping District, Shenyang, Liaoning
电话 / Tel: +86-24-23389443 传真 / Fax: +86-24-23389612 邮编 / P.C.: 110005

大连
地址 / Add: 辽宁省大连市高新园区黄浦路624号
No.624, Huangpu Street, Hi-tech Zone, Dalian, Liaoning
电话 / Tel: +86-411-39750590 传真 / Fax: +86-411-39750590 邮编 / P.C.: 116000

辽宁省建筑设计研究院
LIAONING PROVINCIAL BUILDING DESIGN & RESEARCH INSTITUTE

规划设计研究所
PLANNING AND DESIGN INSTITUTE

随着城市化进程的加快，城市规划在日趋规范的房地产市场中备受瞩目。为了更好地顺应市场需求，也使得院设计部门更加完善，该院于2010年成立规划设计研究所。规划设计研究所现有设计人员5人，其中一级注册规划师1人，具有高级以上技术职称2人，中级技术职称3人。团队年轻而充满活力，严谨又不失创新，设计思路新颖独特。在探索中不断前行，在磨炼中自我完善。

规划设计研究所主要参与设计了沈抚新城总部基地及五号大街西侧用地修建性详细规划、沈北新区新城子镇规划、沈北新区清水台镇规划、仙人岛能源产业区总规调整、黄河北大街城市设计、桓仁满族自治县西江居住区规划、新宾满族自治县新区城市设计、新宾满族自治县南杂木镇城市设计、抚顺汽车城、沈阳亿达房地产城市设计、沈阳国际汽车城概念规划、天湖大桥南4S店规划、抚顺市经济开发区初级中学规划设计、抚顺经济开发区李石站南安置区规划设计。其中，沈北新区财落中心镇控制性详细规划、修建性详细规划荣获辽宁省优秀工程勘察设计城市规划设计项目一等奖，沈阳新世界花园居住区规划荣获辽宁省优秀工程勘察设计城市规划工程类三等奖，沈阳靠山屯新农村建设方案设计荣获新农村建设方案设计竞赛三等奖。

在工作和实践中，大家逐渐认识到城市规划是对城市未来建设蓝图的描绘，是对城市经济快速稳定发展的战略指导，更是实现人与自然和谐共生的科学手段。大家时刻牢记作为一名规划师的职责，要坚持一流的标准，瞄准国际前沿，努力使规划设计经得起实践和历史的检验。

孙建凯　规划设计研究所副所长
Sun Jiankai Deputy Director

1992年毕业于沈阳建筑工程学院建工系。现任规划设计研究所副所长，是注册城市规划师、国家二级注册建筑师、高级规划师。

主要代表作品：沈阳新世界花园居住区规划（荣获省厅三等奖）、沈抚新城总部基地规划、沈北新区财落镇控制性及修建性详细规划（荣获省厅一等奖）、沈阳国际汽车城概念规划、新宾满族自治县南杂木镇城市设计、雅温得（喀麦隆）居住区修建性详细规划、本溪桓仁西江居住区规划、沈阳市黄河北大街地块城市设计、辽宁省十二运前期选址及筹备规划、沈阳辽中靠山屯新农村规划设计（荣获省厅三等奖）等。

感言：规划设计作品只有平民化、通俗化、人性化才有亲和力，才能更加便于公众的积极参与和使用，所以设计师要用自己的心灵去真切地体验生活、去感受万物。

孙洋　规划设计研究所副所长
Sun Yang Deputy Director

2001年毕业于沈阳建筑大学城市规划专业，2004年获沈阳建筑大学城市规划专业硕士学位，研究方向是城市更新与城市设计。

主要代表作品：西藏那曲地区城市总体规划、朝阳市建平县城市总体规划、四川省绵阳市安县晓坝镇总体规划、瓦房店市驼山滨海旅游度假区总体规划、凌源市南河新区控制性详细规划及城市设计、阜新玉龙新城概念规划及起步区城市设计、沈阳道义碧桂园修建性详细规划设计、锦州开发区顺发阳光海岸修建性详细规划设计、朝阳城市商业网点布局规划、大连市长兴岛户外广告总体规划等。

获奖情况：获省优秀设计奖9项、院优秀设计4项。项目内容涉及城市总体规划、控制性详细规划、修建性详细规划、城市设计、城市更新及改造、城市商业网点规划、旅游度假区总体规划、新农村建设规划等规划设计的各个领域。

辽宁省残疾人中等职业技术学校
Liaoning Handicapped Secondary Vocational & Technical School

设计单位：辽宁省建筑设计研究院
建设地点：辽宁 沈阳
建筑面积：98 700平方米

通过多方案的对比分析，探求最适宜、最合理的解决方案。最终形成了对于新校区规划建筑的指导原则，重点落实在如何针对特殊的环境肌理、特殊的场地地形地势、特殊的使用人群、特殊的发展需求，做出全面、合理和最适宜的回应。

1. 延续周边肌理，建筑主要以东西方向水平展开。结合场地地势的竖向变化，将水平建筑架空于两侧较高地势之上。

2. 最大限度地保持基地原始风貌，建筑仿佛从地面生长而出，与周边环境浑然一体。

3. 结合场地中间地势低且平坦的现状，创造方便、安全、宜人的步行和残疾人通行空间。其他场地依坡就势，既有情趣又有经济性。

4. 在用地中部和西侧预留兼顾近远期发展的空间，既能满足当前实训教学之用，又能形成优美环境，并为未来发展留有充分余地。

俯视校园的建筑布局和沿主要城市道路的建筑轮廓，仿佛一艘巨轮扬帆远航，也象征着省残疾人中等职业技术学校和这里的莘莘学子的顽强精神和美好明天。

沈抚新城总部基地
Shenfu New City CBD

设计单位：辽宁省建筑设计研究院
建设地点：辽宁沈阳市与抚顺市交界
规划建设用地：75.15公顷

东北电网电力调度交易中心大楼
GRID Power Dispatching Trading Center

设计单位：辽宁省建筑设计研究院
建设地点：辽宁 沈阳
规划建设用地：85 713平方米

建筑的总体布局首先来自基地周边环境，基地周边有大尺度的高速路与河流，周边空地面积较大，只有圆形平面的新华社和方形平面的汽车展厅，因此采用圆形平面与未来的不确定的周边建筑和高速路、河流形成友好对话，与此同时在形式上与东北电网有限公司的标志和中国传统的天圆地方的朴素哲学相吻合。

主楼、辅楼和副楼分别占据基地的三个角落，东北角设计入口广场作为与周边环境之间的缓冲空间，成为城市的客厅，通过弧形边界水面设计在总平面上形成完整圆形，并与传统风水理论产生关系，功能单元之间通过中庭和架空连廊相接，形成建筑内部的对话。

架设通廊与浮桥直通浑河河面，搭建人与自然之间的桥梁。

抚顺主题假日酒店
Shenfu Resort & Spa Hotel

设计单位：辽宁省建筑设计研究院
合作设计：德国奥尔韦伯建筑师事务所
建设地点：辽宁 抚顺
建筑面积：49 000平方米

辽宁省国际接待中心
Liaoning International Reception Center

设计单位：辽宁省建筑设计研究院
建设地点：辽宁 沈阳
设计时间：2011年
建成时间：2012年

辽宁省图书馆、档案馆
Liaoning Libraries & Archives

辽宁省图书馆

设计单位：辽宁省建筑设计研究院
合作设计：法国VP建筑事务所
建设地点：辽宁 沈阳

辽宁省档案馆

设计单位：辽宁省建筑设计研究院
合作设计：法国VP建筑事务所
建设地点：辽宁 沈阳

大连天工建筑设计有限公司
Dalian Tech-go Architectural Design Co., Ltd.

大连天工建筑设计有限公司具备国家建筑设计甲级资质，拥有一流的办公场所和先进的技术装备，在采用国家领先的工程设计软件的同时，已实现网络化设计与管理，办公面积3 000平方米。现有设计人员80多人，由多名建筑师、规划师、结构工程师、设备工程师及景观设计师共同组成的拥有丰富的大型项目运作经验、众多的成功合作客户及专业合作公司的综合性设计咨询业团队，在城市综合开发项目、公共建筑项目、大型工业项目、居住社区项目、商业及休闲项目、大型景观环境和室内环境项目设计中所展现的创造性构想能力、领先的设计理念和坚持不懈的探索精神，得到了社会公众的广泛认可。

本企业的优势来源于独特的多专业、多文化的设计师团队，在这里云集了一批不同年龄、背景经历各异的设计师，带来了创新设计概念、先进的设计管理经验，凝聚成了崭新的特色企业文化——技术、效率、合作、愉悦、感恩、创新。

本企业作为一支国际化、品牌化的实战团队，以最优异的策略解决项目中的审美问题、技术问题、经济成本与回报方案，关注从产品定位、工程设计到项目竣工过程中的每一个细节，并坚持每个作品都能体现设计价值的终极标准，提升客户的品牌价值。

With Grade A License of Architectural Design, Dalian Tech-go Architectural Design Co., Ltd. has first-class office building and advanced technology and equipment, and has realized network design and management while adopting advanced engineering software in China. The office area is 3,000 m^2. Now it has 80 designers. It is an integrated team of design consulting composed of a number of architects, planners, structural engineers, equipment engineers and landscape designers, which has rich experience in operating large projects and has many successfully-cooperated clients and professional cooperation companies. In the designs of urban comprehensive development projects, public construction projects, large-scale industrial projects, residential community projects, commercial and leisure projects, large-scale landscape projects and interior environment projects, creative vision capability, advanced design idea and persistent spirit of exploration of the team have been widely recognized by the public.

The company's advantage comes from its unique multi-disciplinary, multi-cultural team of designers. A team gathered by a number of designers with different ages, backgrounds and experiences brings innovative design concepts, advanced design management experience, which are cohered into a brand-new characteristic of enterprise culture -- technology, efficiency, cooperation, joy, gratitude and innovation.

As an international and branded design team, it uses the best strategies to solve aesthetic issues and technical issues of projects, as well as issues of economic costs and returns; it pays attention to product positioning, engineering design and every detail of the process before project completion; it insists on the ultimate design standard, showing the design value of each design work, and enhancing customers' brand value.

地址：辽宁省大连市沙河口区星海中龙园3号
邮编：110021
电话：+86-411-84395555
传真：+86-411-84395888
邮箱：Tech-go@vip.sina.com

Add: No.3 Xinghai Zhonglongyuan, Shahekou District, Dalian, Liaoning
P.C.: 110021
Tel: +86-411-84395555
Fax: +86-411-84395888
E-mail: Tech-go@vip.sina.com

1

2

3

4

1. 通化亿城 · 伯瑞斯花园
2. 大商沈阳新玛特
3. 大连中华名城
4. 大连市中心医院
5. 乳山市长城文化苑
6. 大连市第四十四中学
7. 赤峰国际会展中心
8. 大连友谊商城
9. 大连长兴岛临港工业园区规划展示馆
10. 鹤岗市大润发
11. 旅顺世达购物广场
12. 阜新公安局

Shenyang Wanchen Architecture & Plan Design Co., Ltd.

沈阳万宸建筑规划设计有限公司

沈阳万宸建筑规划设计有限公司为股份制有限责任企业，成立于2005年9月，具有国家住房和城乡建设部颁发的建筑工程甲级设计资质、规划乙级设计资质。

公司现有员工90人，工程技术人员84人，其中高级工程师11人，中级技术职称人员17人，国家一级注册建筑师4人，国家一级注册结构工程师3人，国家注册设备工程师4人，是一支既有理论知识，又有实践经验的高素质、高水准的设计团队。公司一直本着以人为本的管理理念，坚信员工的智慧和激情永远是公司发展的源动力，员工的团队精神永远是公司发展的最有力保障。

公司主要承揽建筑工程设计业务，自成立以来，坚持“以设计质量求生存，以诚信服务求信誉，以技术创新求振兴，以科学管理求发展”的宗旨，已先后完成了数百项、近600万平方米的工程设计项目，作品遍及辽宁省各主要城市以及吉林、黑龙江两省。万宸的每一项作品都凝聚着设计人员的聪明才智和创作激情以及对品牌追求的执著，正是员工的不懈努力，使业主和社会认识了万宸、认可了万宸。

努力争创辽沈地区第一建筑设计品牌一直是万宸不懈的追求，为社会和业主回报优秀的作品和优质的服务是万宸一贯的追求。目前，全公司员工正以饱满的工作热情、精湛的业务水平期待着与您的合作。

Shenyang Wanchen Architecture & Plan Design Co., Ltd. was established in September 2005, as a joint-stock company with limited liability licensed for class-A construction engineering and class-B planning design by the Ministry of Housing and Urban-Rural Development of the PRC. Among the ninety employees in the company, eighty four of which are technicians, composed of eleven senior engineers, seventeen engineers with the middle-level technical title, four first-class registered architects, three first-class registered structure engineers and four registered utility engineers, and has formed a high-standard, good-quality design team with both theoretical knowledge and practical experience. With the management idea of “based on people-oriented”, the company firmly believes that the wisdom and passion of employees is the motivation for development, and teamwork will always be the strongest security in the future.

The company mainly engages in construction design. Since its establishment, under the spirit of “survive in the competition of design quality, win reputation by integrity service, improve by technology innovation and develop by scientific management”, the company has completed hundreds of projects covering an area of almost six million square meters in main cities of Liaoning province, Jilin province and Heilongjiang province. Each project is embodied in designers' ingenuity and creative passion. With our dedication to the pursuit of brand and unremitting efforts, owners and the society start to know and accept Wanchen.

Striving to become the No.1 architectural design brand in Liaoning and Shenyang area has been the relentless pursuit of Wanchen, as well as providing outstanding works and good-quality service to the society and owners. At present, with excellent skills, all employees are full of enthusiasm and looking forward for your cooperation.

地址：辽宁省沈阳市和平区南京南街80号
1-26-1、1-26-2、1-26-3
邮编： 110005
电话： +86–24–23524311
传真： +86–24–23508881
邮箱： wanchensj@sina.com

Add: 1-26-1, 1-26-2, 1-26-3, 80 Nanjing South Street, Heping District, Shenyang City, Liaoning Province
Tel : +86–24–23524311
Fax: +86–24–23508881
E-mail: wanchensj@sina.com

沈阳长途客运中心

建设地点：辽宁 沈阳
建筑面积：306 600平方米
设计时间：2009年

Shenyang Long-distance Passenger Transportation Center
Location: Shenyang, Liaoning
Floor Area: 306,600 m²
Design Time: 2009

沈阳黄金交易中心

建设地点：辽宁 沈阳
建筑面积：302 000平方米
设计时间：2010年

Shenyang Gold Trading Center

Location: Shenyang, Liaoning
Floor Area: 302,000 m²
Design Time: 2010

本项目以高耸的金色双塔的形式将黄金交易及办公等功能配套垂直结合在两个对称的建筑体中。裙房部分为金色通透的黄金交易大厅，简洁大气，极具标志性，主体双塔采用菱形作为立面符号，大气的线条，细腻的窗墙渐变，形成富有冲击力的视觉震撼，形成具有象征意义的城市地标。

In the shape of a high-rise golden double-tower, the building functions as both gold trading and office, equally designed into two symmetrical parts. The skirt building is golden, transparent, simple and iconic, which is used as the gold trading hall. The front of main double-tower part is in the shape of rhombus, with simply-cut lines and delicate gradual changes of the window walls. With a strong visual shock of its appearance, the building has become a symbolic city landmark.

玉龙新区核心区规划设计

建设地点：辽宁 阜新
建筑面积：265 000平方米
设计时间：2010年

Core Area Planning and Design of Yulong New District

Location: Fuxin, Liaoning
Floor area: 265,000 m^2
Design Time: 2010

阜新市玉龙新区规划设计理念体现由水源起始生命的规划理念，沿中心水域缓缓生长展开，布置自然、生动，与周围景观水体相辅相成，并形成规划展示馆、会展中心和会议中心三馆环抱博物馆的形态，象征着"玉龙抱珠"的美好寓意。

行政区由外向内、由高到低错落布置，对外展开，象征玉龙新区向城市敞开的新区形象，展现玉龙新区行政区的稳重、大气、现代的建筑形象。

The planning and design idea of Fuxin Yulong New District embodies the concept that life originates from water. The construction starts slowly around the central water area. The layout is natural and vivid, complementing with the surrounding water landscape. Standing in the center, the museum is surrounded by Planning Exhibition Hall, Exhibition Center and Convention Center, which represents the blessing meaning that "Dragon embraces the bead".
From the outside to the inside, the administration area is well-arranged from the tall buildings to the low ones. Opening to the outside symbolizes the image of Yulong New District -- opening to the city. Representing the image of Yulong New District, the architecture is steady, generous and modern.

昌图博物馆

建设地点：辽宁 昌图
建筑面积：9 700平方米
设计时间：2010年

Changtu Museum

Location: Changtu, Liaoning
Floor Area: 9,700 m^2
Design Time: 2010

辽河左岸小区规划设计

Planning and Design of Liao River Left Bank Community

建设地点：辽宁 盘锦
建筑面积：1 896 000平方米
设计时间：2011年

Location: Panjin, Liaoning
Floor Area: 1,896,000 m^2
Design Time: 2011

盘锦辽河左岸文化城，是一个综合性地产文化城项目，在开发理念、功能价值、生活体现、系统工程、开发模式、资源整合六方面均进行高标准、高视野、高品位规划，力求在盘锦打造一个生态之城、文化之城、魅力之城。

公建部分建筑立面为现代风格，结合中式传统文化图案，体现了大气且不失文化底蕴的建筑形象。住宅部分为新古典风格，对称的立面格局、深色的墙面，使建筑显得更加稳重。立面通过细腻的拼砖、精致的线脚、丰富的墙面凹凸再现了高档住宅的公馆风范。

The Liao River Left Bank Culture City of Panjin city is a comprehensive real estate culture city project. Based on high-standard, far-sighted and high-quality planning of development concept, functional value, living environment, system engineering, development pattern and integration of resources, we strive to build an eco-city, culture-city and charming city.

Combined with Chinese traditional culture patterns, the façade of public building is of modern style, showing a generous image with rich cultural features. The residential part is of neo-classical style with a symmetrical facade pattern. With dark walls, the building looks more stable. Through delicate tile-matching, exquisite moldings, rich concave-convex of the wall, the façade reappears the mansion style of high-end residence.

世界温泉部落——加拿大郡

World Spring Tribe – the Canada Town

建设地点：辽宁 沈阳
建筑面积：468 000平方米
设计时间：2011年

Location: Shenyang, Liaoning
Floor Area: 468,000 m^2
Design Time: 2011

项目定位为加拿大风情小镇，具有欧洲小镇别具风情的宜居生活小区。北侧商业地块设有浓郁法式风情的高档温泉度假商业中心。

项目的景观体系犹如一片枫叶，由中心向四周延伸，形成景观轴、文化轴、中心湖面、塔楼，最终形成特色完整的景观系统。景观系统与温泉酒店、会所共同为业主提供高品质生活空间。建筑典雅稳重、层次丰富，充分展现出该项目整体的尊贵品质。

The project is aimed at building a Canadian-style town, with characteristics of European towns; a comfortable living area with exotic features. At the commercial area on the north, there is a luxury French-style spa resort business center.

The landscape system of this project is like a maple leaf, extending from the center to the edges, and with landscape axis, culture axis, central lake and the tower, it has formed a complete and special landscape system. Together with spa club, the landscape system provides owners a high-quality living space. With rich levels, the whole architecture is steady and elegant, and has fully demonstrated the dignity and quality of the overall project.

京邦铭城

建设地点：辽宁 盘锦
建筑面积：3 702 000平方米
设计时间：2011年

Jing Bang Ming Cheng

Location: Panjin, Liaoning
Floor Area: 3,702,000 m^2
Design Time: 2011

金域华城规划设计

建设地点：辽宁 葫芦岛
建筑面积：310 000平方米
设计时间：2010年

Jinyuhuacheng Planning and Design

Location: Huludao, Liaoning
Floor Area: 310,000 m^2
Design Time: 2010

本项目定位为东南亚风情。建筑通过现代简约的建筑风格、建筑色彩和材质变化的结合，营造出极具东南亚风情的园区特征。利用现代的廊架、景墙、叠水等景观处理手法和科学环保的绿植方式，把居住区内部相对封闭的生活环境空间和居住区外部开放城市空间改造成别具特色的宜居城市空间。

The project is to build a community with Southeast Asia style, and create a highly distinctive feature of Southeast Asia through modern minimalist architectural style, combinations of building colors and material changes. Through landscape approaches such as modern gallery frame, landscape wall and water landscape etc. and scientific, environmental-friendly planting methods, we will make the relatively closed living space inside the community and the open space outside the community into a comfortable urban space with distinctive features.

1 中园书苑

建设地点：辽宁 朝阳
建筑面积：13 000平方米
设计时间：2010年

1 Zhong Yuan Shu Yuan

Location: Chaoyang, Liaoning
Floor Area: 13,000 m^2
Design Time: 2010

2 东风湖宾馆

建设地点：辽宁 本溪
建筑面积：15 000平方米
设计时间：2009年

2 Dongfeng Lake Hotel

Location: Benxi, Liaoning
Floor Area: 15,000 m^2
Design Time: 2009

3 沈阳鑫丰西海岸

建设地点：辽宁 沈阳
建筑面积：76 500平方米
设计时间：2010年

3 Shenyang Xinfeng West-coast

Location: Shenyang, Liaoning
Floor Area: 76,500 m^2
Design Time: 2010

4 沈阳兴隆女人街

建设地点：辽宁 沈阳
建筑面积：322 000平方米
设计时间：2010年

4 Shenyang Xinglong Ladies' Street

Location: Shenyang, Liaoning
Floor Area: 322,000 m^2
Design Time: 2010

5 新民凯帝宾馆

建设地点：辽宁 沈阳
建筑面积：37 000平方米
设计时间：2009年

5 Xinmin Kaidi Hotel

Location: Shenyang, Liaoning
Floor Area: 37,000 m^2
Design Time: 2009

6 本溪市中医院综合门诊楼

建设地点：辽宁 本溪
建筑面积：12 600平方米
设计时间：2009年

6 Out-patient Building of Benxi Hospital of TCM

Location: Benxi, Liaoning
Floor Area: 12,600 m^2
Design Time: 2009

OTHERS STUDIO

别人建筑工作室

地址：辽宁省沈阳市沈河区五爱街2-3号地王国际花园白金公寓C座706
邮编：110015
电话：+86-24-31306708
传真：+86-24-31306708
网址：www.others.cn
邮箱：others815808@sina.com

Add: 706, Royal Plaza Platinum Tower C, 2-3 Wuai Street, Shenhe District, Shenyang, Liaoning
P.C.: 110015
Tel: +86-24-31306708
Fax: +86-24-31306708
Http://www.others.cn
E-mail: others815808@sina.com

我们总是在找不到自我的时候，想到别人，在别人的节日里寻找快乐。“在别人的思想中思想，在别人视线里生存”。直到我们找到了自己的节日。

那正是工作室成立之初的一种状态，所以将工作室命名为“别人”，以纪念当时的心态。“别人”不是理论或口号，它是繁杂地交织在一起的建筑认识论和始终影响着创作心态的一个关键词，它时而是“地狱”，时而又成了“天堂”，它的被解释的多样性正好说明了生活与建筑本身的多样性。

We always think about others and look for the happiness in others' festivals when we can't find ourself. "Think in others' thought, and live in others' sight" until we find our own festivals.

That was our state exactly when we just built our studio. So we named it "Others" to commemorate this mentality. "Others" is not a theory or a slogan. It is a complex architecture epistemology and a keyword to affect our creation. Sometimes it is "hell", while sometimes it is "heaven". It can be explained in a lot of ways, which just illustrates the life's and the architecture's diversity.

圣泰能源总部基地

项目地点：内蒙古 通辽
设计时间：2010年
建筑面积：20 030平方米
设 计 师：杨 胤、高德战、倪伟宁

圣泰能源总部基地主要由总部会所及办公两部分组成。在最初的设计任务中也是要求分别建会所和办公两栋楼。结合项目的特点及地段的特征，将两个部分通过室内花园连接起来，从而形成一个完整的“大建筑”。

会所、办公、花园这三个部分并不是像切片一样平行独立，而是通过体积的咬合、空间的渗透成为一个浑然的整体。室内花园向两侧扩散，在会所及办公两部分内部形成若干个缝隙，这些缝隙分割出餐饮、会议、客房、办公、娱乐等多个功能体。人在缝隙及花园内的行走、空气的流动、树影的摇曳，将各个分离的体积结合为有机的生命体。

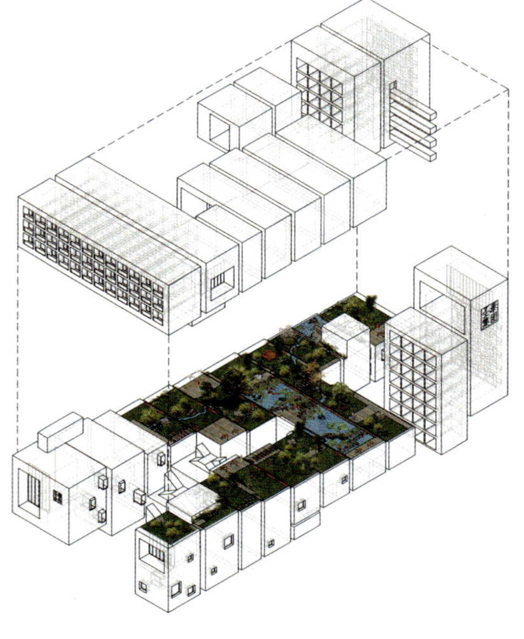

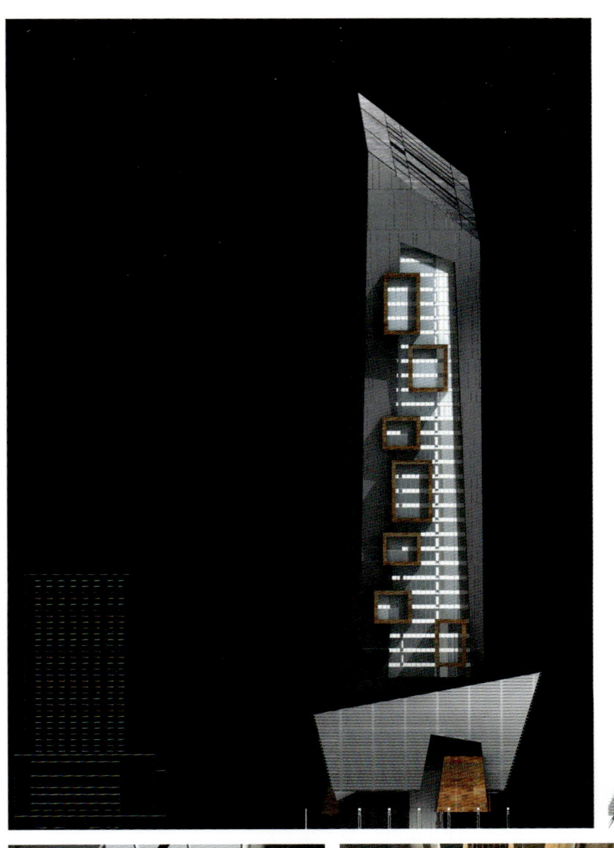

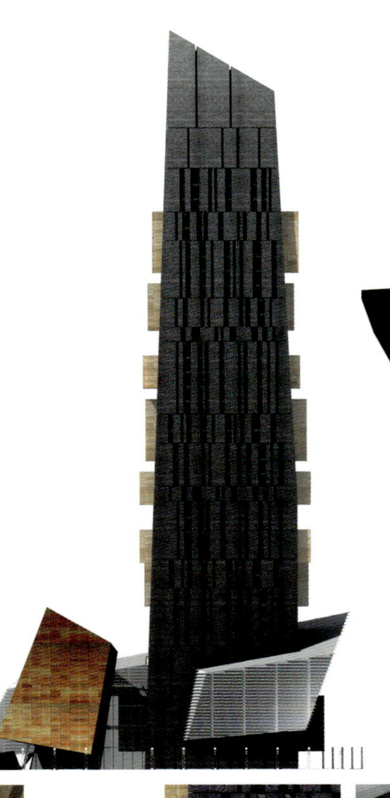

YG艺术中心
YG Art Center

用地面积：0.47公顷
地上建筑面积：50 233平方米
地下建筑面积：14 910平方米
总建筑面积：64 423平方米
容积率：10.7
其中，艺术中心15 880平方米，写字楼17 123平方米，商住17 233平方米
设计时间：2009年
设计师：杨 胤、张亦宁、倪伟宁

艺术展示类建筑可以有以下两种方式：

1.建筑以背景方式呈现，用以容纳、烘托展品。

2.建筑与展品共同表演，建筑融入展品中，即具有强烈的个性、标志性、戏剧性，呈现出一种形体与空间的变奏，建筑是容器的同时也是展品。

此方案选择了后面的一种方式，它产生于这个喧闹的年代、躁动的都市，它不选择隐身，而是积极地参与，它要激活的不仅是一个区域、一个城市，更应该是一种创造性的心态和真实的人性的反思。

墨，在砚中游动，每次都沿着不同的轨迹，不经意的，不可预料的。

人，在生活中，每个人都有不同的生活轨迹，所谓记录，其实都是片段的。

这里的建筑，整体中涵盖了些许片段与不经意，这也映射了我们对这种世界的理解。建筑以开敞的空间形态与原生态景观呼应，在原有植被全部保留的基础上，在树木间巧妙、谨慎地设置铺装，水景、草地灯、座椅、雕塑等，使其具有足够的可进入、可停留性。

这难得的宁静，不应该因为建筑的融入而被打破。艺术中心入口空间采用雕塑手法，彰显空间张力的同时，又有效地把基地西侧的原生态景观收纳其中，这也是建筑在设计之初的主要概念之一。

静逸的内部展示空间，建筑的外部形态与内部的空间感受完全契合，磨砂玻璃外墙均匀稳定了随时变动的日光，金属百叶进一步调整了照度，随墙而上的楼梯既提供了垂直动线，又活跃了空间本身。

非匀质空间——建筑力求摆脱整齐划一的标准层模式，想法来自于人的生活本身就不应该是标准化的，在主体中，嵌入了尺寸、位置都不同的盒子空间，用这种方式打造不可预知的尺度变化，形成个性化的场所空间。

“建构”不仅显示结果，它更是一个过程、一个动态。建筑突出体现了一个过程，强化了一种体块间的搭接、穿插的关系，还原建筑的本意。我们认为，过程永远重于结果。

关于景观：除了建筑给城市带来全方位的景观价值外，其内部空间同样运用景观设计手法，达成内与外、内与内的全景交融，建筑不仅是功能与雕塑形体调和的结果，同时也是各种具有空间张力的场景以空间序列的方式合理组织的结果。

关于流线：三种主要功能、五种流线相对独立，互不干扰，不仅是使用的需要，更是管理的需要。

艺术中心的展示空间分为上下两部分，顶部三层为永久展区，底部六层为临时展区，可根据面积、性质的不同需求灵活布展。

另外，在主体的1至2层设置了常年对外营业的画廊，在保障日常人气的同时，又可以分担一部分艺术中心的运营成本。

工大设计
GRANDDESIGN

内蒙古工大建筑设计有限责任公司
INNER MONGOLIA GRAND ARCHITECTURE DESIGN CO.,LTD.

法定代表人、董事长：张鹏举
地址：内蒙古呼和浩特市爱民路49号
邮编：010051
电话：+86–471–6576005 / 6575347 / 6577170
传真：+86–471–6576322
邮箱：Gdsijy@yahoo.com.cn
网址：www.Nmggdsj.com.cn

Legal Representative, the Chairman: Zhang Pengju
Add: No. 49 Aimin Road, Hohhot City, Inner Mongolia
P.C.: 010051
Tel: +86–471–6576005 / 6575347 / 6577170
Fax: +86–471–6576322
E-mail: Gdsijy@yahoo.com.cn
Http://www.Nmggdsj.com.cn

内蒙古工大建筑设计有限责任公司前身为成立于1979年的内蒙古工业大学建筑设计研究院，2002年2月改制为有限责任公司，2004年通过ISO 9001：2000质量体系认证，并逐步建立起完善的现代企业制度，经过20余年的发展，公司取得的成就令人瞩目，现已持有国家授予的建筑工程甲级设计证书。

公司现有职工64人，其中教授4人，副教授8人，高级工程师12人，工程师10人，一级注册建筑师5人，一级注册结构工程师5人，是一支具有高素质的精干的设计队伍，同时公司以高校学术力量为依托，充分发挥广大专业教师的设计潜能，在内蒙古自治区树立了自身独特的行业地位。公司业务范围涵盖建筑设计、规划设计、室内设计、景观设计、岩土工程地质勘探、工程技术咨询、预应力结构设计及施工等内容。

Inner Mongolia Grand Architecture Design Co.,Ltd. was first established as Inner Mongolia University of Technology Architecture Design Institute in the year of 1979, and has been awarded the National Class One Design Qualification Certificate for Architecture Design Engineering. The company has finished incorporation in February 2002, and passed the authentication of ISO9001:2000 quality system in December 2004.

The company now has 64 people,including 4 professors, 8 associate professors, 12 senior engineers, 10 engineers, 5 Class One Registered Architect, 5 Class One Registered Structure Engineers, and is a high-efficient design team. By making use of the academic strength of the university, the company has established solid position in the industry in Inner Mongolia. In recent years, the company has worked in Inner Mongolia and other provinces in China, as well as in the Republic of Mongolia, and has won many awards for excellent design works granted by the Ministry of Construction of China and Inner Mongolia, being widely-praised in society.

恩格贝沙漠科学馆

项目地点：内蒙古 鄂尔多斯
项目功能：展览、学术研究
建筑面积：5 771平方米
设计时间：2010年
竣工时间：2011年
项目建筑师：张鹏举、郭 彦、张 恒

Engebei Science Museum

Venus: Ordos, Inner Mongolia
Function: Exhibition, Research
Scale: 5,771 m^2
Design Time: 2010
Completion Time: 2011
Architects: Zhang Pengju, Guo Yan, Zhang Heng

作品简介

项目位于库不齐沙漠的边缘。方案汲取当地民居的建造智慧和形式特征，将建筑平伏在大地上，并嵌入在背景大青山轮廓线与自然坡地线之间，最大限度地保留了基地的地域特征。同时，建筑外墙材料采用土黄色传统的水刷石做法，进一步与基地环境相融合，力求表现建筑的独特特征。

Introduction

The project located at the edge of Hobq Desert. The design learned from the local residential construction feature and form characteristic, made the building flat on the ground, integrated into the natural background of mountain and natural sloping fields, got much more retention for the base of the geographical features. Meanwhile, the exterior walls use wheat granitic plaster to combine with the base environment displaying the character of the building further.

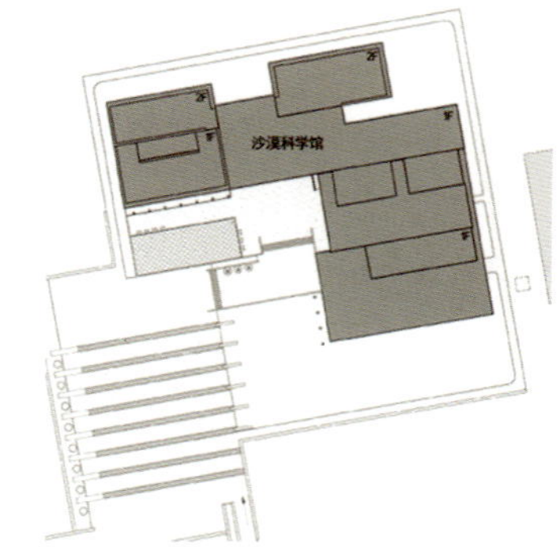

乌兰察布博物馆、图书馆

项目地点：内蒙古 乌兰察布
项目功能：展览、学术研究、图书
建筑规模：5 920平方米
设计时间：2008年
竣工时间：2009年
项目建筑师：张鹏举

Ulanqab Museum, Library

Venus: Ulanchap, Inner Mongolia
Function: Exhibition, Research, Library
Scale: 5,920 m^2
Design Time: 2008
Completion Time: 2009
Architect: Zhang Pengju

作品简介

项目由市博物馆、市图书馆以及市文化局办公楼三部分组成，建筑采用化整为零、平直而方、七方联体、底层架空等策略将建筑置于方台之上，增强了识别性，加强了整体感，以较少的土方解决场地低的现状，内部空间获得了自然采光及通风的可能性，具备了日后灵活发展的潜力。同时，立面肌理的提取以及源于地域文化元素色彩的选择增加了建筑的地域认同感。

Introduction

The project consists of three buildings— museum, library and culture bureau of the city. The buildings are on the square construction, and not integrate but straight and square, seven parties conjoined, the bottom overhead space, all the forms enhanced to identify, strengthened the integrity, and solved the low status of earth with less earthwork. The Inner space also has the possibility of the natural lighting and ventilation. At the same time, facade texture and colors learned from the regional culture elements, increasing the building's geographic identity.

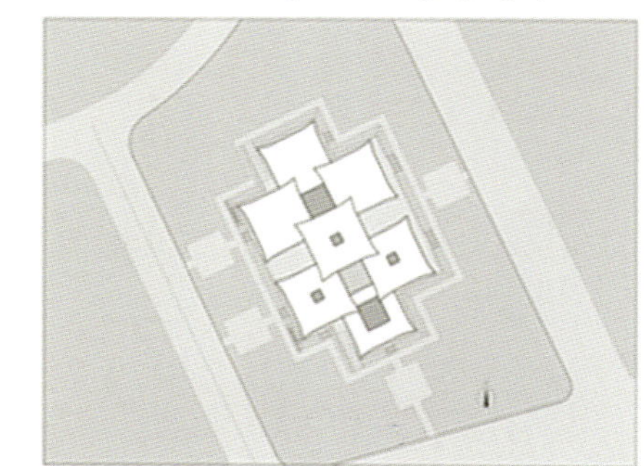

Inner Mongolia Youbang Architecture Design Institute

内蒙古友邦建筑设计事务所有限公司

地址：内蒙古包头市昆都仑区钢铁大街54号民主大楼409室
Add: Room 409, Democracy Building, No.54 Gangtie street, Kundulun district, Baotou, Inner Mongolia
法定代表人：马玉斌 / Legal Representative: Ma Yubin
电话/Tel: +86–472–5130567 +86–472–5130564
E-mail: STYG123456@126.com MYB_BT2001@163.COM
Http://www.btjzsj.com

内蒙古友邦建筑设计事务所有限公司（甲级）成立于2001年，原名为包头新成建筑设计事务所，可承揽建设工程业务以及项目管理和小区规划、工业与民用建筑设计及相关技术咨询与管理服务。

事务所拥有40余名专业设计人员，其中国家一级注册建筑师5人，中高级职称的专业设计人员20余人。先后完成了各类设计百余项，如呼和浩特明泽家苑小区、包头鼎太风华小区、乌海居家苑小区、包头金荣装饰建材城、鄂尔多斯市康巴什财税大楼、伊金霍洛旗医院、包头市教育局办公楼、包头市滨河大厦、包头市胜源滨河新城、呼和浩特满族风情城、鄂尔多斯达拉特旗胜达广场、通辽蒙泰大厦等。

我们倡导“与时俱进、共创辉煌”的企业精神，坚守“不断创新、追求完美”的设计理念，以健全的质量保证体系维护企业信誉，不断打造出一流的设计精品。

Founded in 2001, Inner Mongolia Youbang Architecture Design Institute was originally called Baotou Xincheng Architecture Design Institute, undertaking industrial and civil architectural design as well as the corresponding technical consultation.

There are more than 40 professional designers in our company, including 5 national First Class registered architects and over 20 junior and senior professional designers. For these years we have completed more than fifty different types of design projects, such as the Hohhot Mingze Jiayuan, Baotou Dingtai Fenhua, Wuhai Jujiayuan, Jinrong Jiancaicheng in Baotou, Ordos Kangbashi Finance and Revenue Building, Yijinhuoluo Qi Hospital, Baotou Education Bureau Office Building, and so on.

Advocating to the enterprise spirit "keeping up with the times, creating the brilliant prospect", and being in pursuit of "innovation and perfection", we maintain good credibility by improving the quality assurance system, creating the first class exquisite designs.

1

2

3

1–2 鄂尔多斯市达拉特旗胜达广场 总建筑面积：22万平方米
Shengda Plaza, Talat Banner, Ordos Total Building Area: 220,000 m²

3 包头市滨河大厦 总建筑面积：2.9万平方米
Riverside Building, Baotou Total Building Area: 29,000 m²

4–7 包头市昆区哈业脑包中学 总建筑面积：1.5万平方米
Hayenaobao High School, Kun District, Baotou Total Building Area: 15,000 m²

8 包头市建筑职工教育培训中心二期 总建筑面积：4万平方米
Building Workers Education & Training Center Phase II, Baotou Total Building Area: 40,000 m²

4

5

6

7

8

9

10

9 包头市云龙骨科医院住院楼 总建筑面积：2.6万平方米
Hospitalization Building, Yunlong Orthopaedics Hospital, Baotou Total Building Area: 26,000 m^2

10 包头市贵花苑 总建筑面积：5万平方米
Guihua Garden, Baotou City Total Building Area: 50,000 m^2

11–12 呼和浩特满族风情城 总建筑面积：45万平方米
Manchu Exotic Town, Hohhot Total Building Area: 450,000 m^2

13 通辽市蒙泰大厦 总建筑面积：5.3万平方米
Mengtai Building, Tongliao Total Building Area: 53,000 m^2

14 鄂尔多斯市达拉特旗精品移民小区 总建筑面积：14万平方米
Elite Migrants' Quarter, Dalat Banner, Ordos Total Building Area: 140,000 m^2

15 包头市胜源滨河新城 总建筑面积：52万平方米
Shengyuan Riverside New Town, Baotou Total Building Area: 520,000 m^2

16 包头市东河区现代化高级中学 总建筑面积：6.9万平方米
Donghe District Modern Senior High School, Baotou Total Building Area: 69,000 m^2

13

11

12

14

15

16

Inner Mongolia Zhihui Architectural Design and Consultant Co., Ltd.

内蒙古智汇工程设计咨询有限责任公司

内蒙古智汇工程设计咨询有限责任公司始建于2000年，经过十多年的发展，公司综合实力不断增强。搭建了坚实的企业发展平台，取得了良好的社会信誉，汇集了一批资深专家和业内精英，其中包括高级职称17人、国家一级注册建筑师4人、注册城市规划师2人、国家一级注册结构师3人、国家各类注册人员18人，为客户提供规划编制、建筑设计、市政设计、环境设计和设计咨询等全方位的设计服务。拥有建筑工程甲级设计资质和城市规划编制乙级设计资质。

公司始终秉承："集众人之智、汇行业精英、创时代精品"的企业核心理念，遵循质量第一、服务至上的原则，强调质量是设计工作的灵魂，服务是设计工作的保证。十多年来，公司完成了一系列包括大中型公共建筑、居住小区及城市规划等具有社会知名度和影响力的作品。多项工程获国家级和省级奖励。智汇人将始终坚持自己的核心理念，继续为广大客户提供优质服务。让我们携起手来，共创美好的城市环境与符合时代特征的建筑作品。

Inner Mongolia Zhihui Architectural Design and Consultant Co., Ltd. was established in 2000. After more than ten years of development, The company's comprehensive strength unceasingly strengthens, Build a solid enterprise development platform, Achieved good social reputation. And thus collects a group of senior experts and industry elites.Including 17 senior engineers, 4 Class One Registered Architect, 2 Registered urban planners, 3 Class One Registered Structure Engineers, and 18 registered engineers in other relevant areas. Include many types of construction design engineers, such as urban planning, architectural design, municipal design, environment design and design consulting,etc. It has Grade-A qualification of architecture, Grade-B qualification of urban planning.

The company always adheres to the crowd: "gather all wisdom, collect industry elite, genera the high-quality goods "the enterprise core idea, Follow the quality first, service is supreme principle, emphasize the quality is the soul of the design work, service is the guarantee of the design work. For over ten years, the company has completed a series ofsocial fame and influence works, including large and medium-sized public buildings, residential area and urban planning and design projects, of which a number of engineering were awarded as national and provincial awards. Zhihui people will always adhere to the own core idea, continue to provide quality services to our customers. Let us join hands to create beautiful city environment and conform to the time features of building works.

1

2

地址：内蒙古呼和浩特市大学东街巨海商厦九层
法定代表人：董剑钧
电话(传真)：+86-471-5291147
邮编：100020
邮箱：zhihuigongsi@vip.sina.com
网址：www.zhedc.com

Address: Floor 9, Juhai Mall,university East Street, Hohhot, Inner Mongolia
Legal Representative: Dong Jianjun
P.C.:100020
Tel (Fax): +86-471-5291147
E-mail: zhihuigongsi@vip.sina.com
Http://www.zhedc.com

3

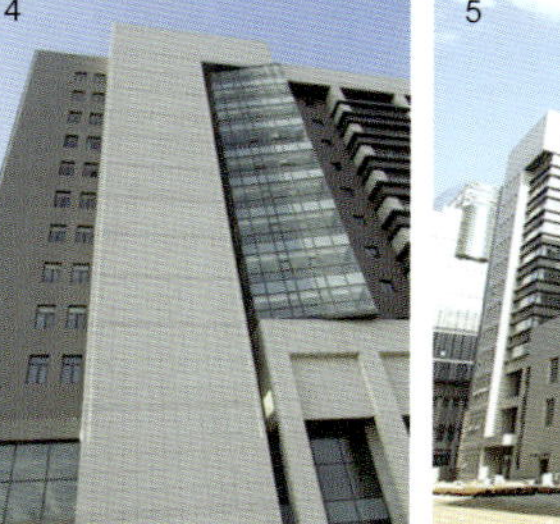
4

5

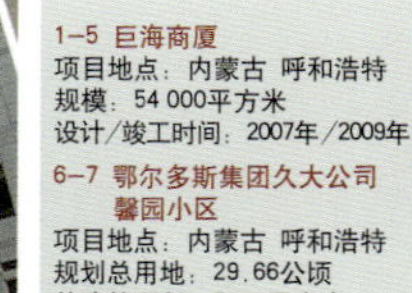

1–5 巨海商厦
项目地点：内蒙古 呼和浩特
规模：54 000平方米
设计/竣工时间：2007年/2009年

6–7 鄂尔多斯集团久大公司馨园小区
项目地点：内蒙古 呼和浩特
规划总用地：29.66公顷
总建筑面积：52万平方米
设计时间：2009年

8 内蒙古师范大学鸿德学院图书馆
项目地点：内蒙古 呼和浩特
规模：23 000平方米/100万册
设计/竣工时间：2008年/2009年

9–10 内蒙古师范大学鸿德学院教学综合楼
项目地点：内蒙古 呼和浩特
规　　模：9 600平方米
设计/竣工时间：2006年/2007年
获奖情况：
自治区优秀设计二等奖

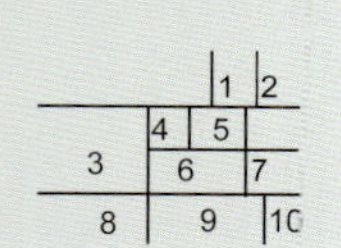

6

7

8

9

10

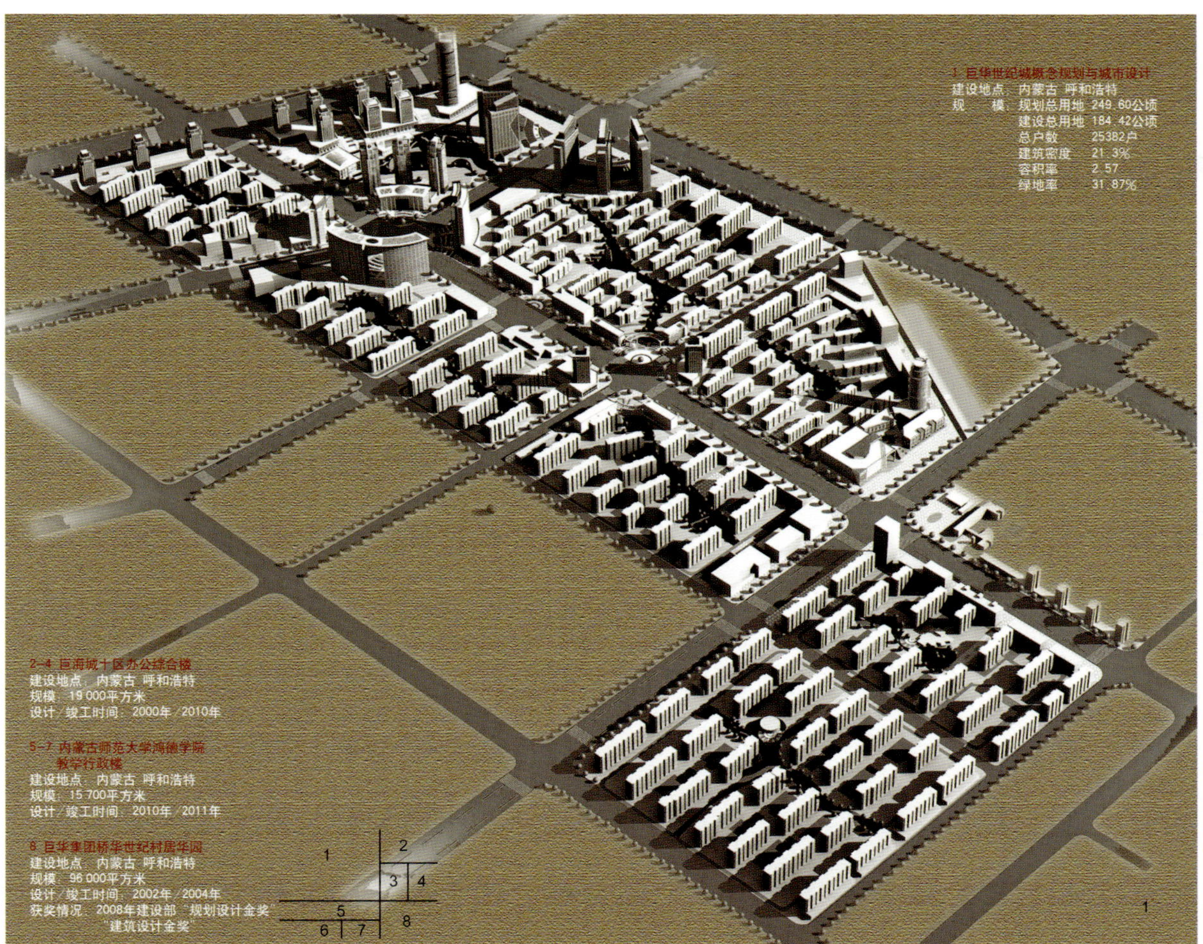

1 巨华世纪城概念规划与城市设计
建设地点：内蒙古 呼和浩特
规　　模：规划总用地 249.60公顷
建设总用地 184.42公顷
总户数 25382户
建筑密度 21.3%
容积率 2.57
绿地率 31.87%

2-4 巨海城十区办公综合楼
建设地点：内蒙古 呼和浩特
规模：19 000平方米
设计/竣工时间：2000年/2010年

5-7 内蒙古师范大学鸿德学院教学行政楼
建设地点：内蒙古 呼和浩特
规模：15 700平方米
设计/竣工时间：2010年/2011年

8 巨华集团桥华世纪村居华园
建设地点：内蒙古 呼和浩特
规模：96 000平方米
设计/竣工时间：2002年/2004年
获奖情况：2008年建设部"规划设计金奖"
"建筑设计金奖"

Inner Mongolia Zhuyou Architectural Design & Consulting Co., Ltd.

内蒙古筑友建筑设计咨询有限责任公司

内蒙古筑友建筑设计咨询有限责任公司，前身为内蒙古新千年建筑设计有限责任公司。2009年6月注册资金增加至300万元人民币，重组更名成立了内蒙古筑友建筑设计咨询有限责任公司。通过建设部考核和审批，于2010年2月获建筑工程甲级设计资质，证书编号A115002169-6/5。各专业设计骨干增加至88人，其中高级专业技术职称22人，中级专业技术职称15人，初级专业技术职称51人。

公司自成立以来，先后承接了数百项建设项目，从方案设计、初步设计、施工图设计到后期设计服务，涵盖了办公、商业、酒店、学校、住宅、公寓、别墅等建设项目。现已建成并投入使用的办公楼有内蒙古党委联合接访中心、内蒙古建设银行档案库、内蒙古军区、呼和浩特市赛罕区人民法院、阿巴嘎旗国家税务局、镶黄旗审计局、临河区农村信用社、集宁区农村信用社、桃花信用社、巴彦淖尔市邮政局、准格尔建设银行、通辽市建设银行等；商业设计项目有富邦商务中心、白音华工人文体中心、白音华医院、金利达酒店、东方御景酒店、呼市巴彦淖尔南路小学、集宁八中、集宁五中等；住宅设计项目有昕嘉园、风华园、现代华城、阳光名座、一鑫康乐苑、鄂尔多斯市凤凰新城和宏源一品、准格尔南苑小区B区、乌海阳光景园、内蒙古高院天河公寓和海东公寓等。

“设计创造价值”的核心理念始终贯穿在我们的作品中，因此这些设计项目为客户提供了经济、实用、安全、可靠的设计产品。“质量为本，技术为先，求持续发展；进取为能，创意为源，树知名品牌”是我们一贯坚持的质量方针。今后，筑友将始终坚持自己的设计理念，以激情和梦想打造一流设计品牌而再接再厉！

Inner Mongolia Zhuyou Architectural Design & Consulting Co., Ltd. is preceded by Inner Mongolia New Millennium Architectural Design Co., Ltd. It is reorganized with present name in June 2009, with registered capital increased to RMB3,000,000. By examination and approval of the Ministry of Construction, in February 2010, it was qualified for Construction Project Class-A Design, with Qualification Certificate No. A115002169-6/5. The core team increases to 88 designers, of whom, 22 designers hold senior technical titles, 15 designers hold medium technical titles, and the rest 51 designers hold preliminary technical titles.

Since it inception, this company has undertook hundreds of construction projects, from conceptual design, preliminary design, construction drawing design and later-phase design services, covering office, commercial, hotel, school, residential, apartment, villa and other building projects. Now it has built up and put into operation such office buildings as Inner Mongolia CCP Committee Joint Complaint Reception Center, Inner Mongolia CCB Archives, Inner Mongolia Military Command, Saihan District People's Court (Hohhot), Abag Banner State Administration for Industry & Commerce, Bordered Yellow Banner Auditing Bureau, Linhe District Agricultural Credit Cooperative, Jining District Agricultural Credit Cooperative, Taohua Credit Cooperative, Bayannaoer Post Office, Junggar CCB, Tongliao CCB; and its commercial projects include Fubang Business Center, Baiyinhua Worker's Cultural & Sport Center, Baiyinhua Hospital, Jinlida Hotel, Oriental Royal View Hotel, Bayannaoer South Road Elementary School (Hohehot), Jining No. 8 Middle School, Jining No. 5 Middle School and others; its residential projects include: Youxin Garden, Fenghua Garden, Modern Chinese Town, Sunshine Tower, Yixin Recreational Garden, Ordos Phoenix New Town and Hongyuan Class-1, Junggar South Garden Block B, Wuhai Sunshine View Garden, Inner Mongolia Higher Court Galaxy Apartment and Haidong Apartment etc.

The core philosophy of "design creating values" penetrates our works, therefore, the foregoing design projects provide customers with cost-effective, practical, safe and reliable design products. Our constant quality guidelines are "quality based, technology leading, sustainable development; progress powered, originality sourced, famous brand building". Henceforth, Zhuyou will persist its own design philosophy, and work hard to build up first class design brands, with passions and dreams!

法定代表人：郭建忠 / 地址：内蒙古呼和浩特市新城区财神庙街29号 / 邮箱：zhuyou666@126.com / 传真：+86-471-6900455

内蒙古敕勒川文化旅游产业园区大剧院

建筑面积：16 000平方米
项目地点：内蒙古

鄂尔多斯准格尔旗大路新区综合服务中心

建筑面积：4 200平方米
项目地点：内蒙古 鄂尔多斯
建筑高度：12.3米

内蒙古敕勒川文化旅游产业园多媒体博物馆

建筑面积：26 000平方米
项目地点：内蒙古

巴彦淖尔市临河区金川学校

建筑面积：17 298平方米　项目地点：内蒙古　乌海
结构类型：框架结构　建筑高度：16.15米（四层）
建设情况：已建成

2010第七届中国人居典范建筑规划设计方案竞赛中被评为最佳建筑设计方案金奖

乌海市滨河学校（现改名为乌海市第九中学）

建筑面积：11 197.55平方米　项目地点：内蒙古　乌海
结构类型：框架结构　建筑高度：16.70米（四层）
建设情况：在建

2010第七届中国人居典范建筑规划设计方案竞赛中被评为最佳建筑设计方案金奖

乌海市第二中学

建筑面积：10 654.74平方米　项目地点：内蒙古　乌海
结构类型：框架结构　建筑高度：23.95米（六层）
建设情况：已建成

2010第七届中国人居典范建筑规划设计方案竞赛中被评为最佳建筑设计方案金奖

锡林浩特市西乌旗
白音华工人文体中心

建筑面积：4 800平方米
项目地点：内蒙古　锡林浩特
建成时间：2010年

呼和浩特市华侨新村住宅小区

建筑面积：235 417平方米
项目地点：内蒙古　呼和浩特
结构形式：剪立墙

呼和浩特市电影制片厂
地块城市设计

建筑面积：270 000平方米
项目地点：内蒙古　呼和浩特

内蒙古自治区党委政府
联合接访中心

建筑面积：10 558.4平方米
项目地点：内蒙古　呼和浩特
项目状况：在建

呼和浩特市金利达酒店

建筑面积：11 430平方米
项目地点：内蒙古　呼和浩特
项目状况：在建

锡林浩特市西乌旗白音华
医院门诊楼

建筑面积：14 820平方米
项目地点：内蒙古　锡林浩特
项目状况：已建成

内蒙古景城建筑设计有限责任公司成立于2010年10月，是具有建筑行业（建筑工程）乙级资质的民营设计公司，注册资金300万元，位于内蒙古呼和浩特市新华大街中银城市广场A座7层。主营建筑装饰工程设计、建筑幕墙工程设计、轻型钢结构工程设计、建筑智能化系统设计、照明工程设计和消防设施工程设计等。公司现拥有国家一级注册建筑工程师2人、国家一级注册结构工程师4人、国家一级注册设备工程师4人、中高级职称20人、助理工程师及设计人员若干，共计50人。办公场所面积800平方米。

公司在从事各种类型的大中型民用与工业建筑设计的过程中，坚持“以精心树精品，以诚信铸辉煌”的经营理念，成立至今设计作品遍及内蒙古自治区。公司严守“诚立德、德立业”的企业理念，致力于不断“创新思维、与时俱进”，对内发挥团队精神，对外广泛开展团队合作，积极向国内外同行学习。在积极开展业务的同时积累了一定的设计经验，一些项目赢得了社会各界的广泛赞扬。公司将继续秉承“以人为本、以诚为根、以和为贵、以信为先”的原则一如既往地开拓进取、锐意创新，向更远大的目标迈进。

地址：内蒙古呼和浩特市新华东街中银城市广场A座707
邮编：010050
电话：+86-471-5254302 / 5254305 / 5254301
传真：+86-471-5254302
邮箱：nmgjcjz@163.com
网址：www.nmgjcjz.com

Inner Mongolia Landscape City Architectural Design Co., Ltd. which was established in October 2010, is a private design company with construction industry (Construction) Grade B qualification, registered capital of 300 million yuan, and is located in Floor 7, Tower A, Zhongyin City Plaza, Xinhua Street, Hohhot, Inner Mongolia. The company engages in architectural decoration engineering design, architectural curtain wall engineering design, light steel structure engineering design, building intelligent systems design, lighting design and fire facilities engineering design. At present, the company has the total of 50 people, including 2 First Class registered building construction engineers of China, 4 First Class registered structural engineers of China, 4 are the First Class registered equipment engineers of China. There are 20 persons with senior and intermediate grade titles, a number of assistant engineer and designers. It has 800 square meters office space.
In the process of engaging in various types of large, medium-sized civil and industrial architectural design , the company insists on the business philosophy of "concentration establishes quality, brilliance cast in good faith", since its foundation the design works throughout Inner Mongolia Autonomous Region. The company sticks to the corporate philosophy of morality originates from honesty, business originates from morality, commits to continuous "innovative thinking, advancing with the times", and emphasizes internal teamwork, and external cooperation with partners, actively learning from domestic and foreign counterparts. Active in business, the company has accumulated some design experience, whose projects have won the wide praise from the society. The company will continue to uphold the principle of "people-oriented, sincerity rooted, harmony as the best, honesty first", always take the pioneering spirit and innovation, move on to the more ambitious goal.

Add: Room 707, Tower A, Zhongyin City Plaza, Xinhua Street, Hohhot, Inner Mongolia
P.C.: 010050
Tel: +86-471-5254302 / 5254305 / 5254301
Fax: +86-471-5254302
E-mail: nmgjcjz@163.com
Http://www.nmgjcjz.com

1

3

2

1–4 蒙古族文化苑

工程地点：内蒙古 鄂尔多斯
项目规模：约100 000平方米
使用功能：集展览、演出会议、餐饮于一体的多功能组合体
设计理念：草原文化原生态

4

5–7 呼伦贝尔讲寺

工程地点：内蒙古 呼伦贝尔
建筑面积：17 011平方米
（我国目前规模最大的仿宋式古建）

8 锡林郭勒盟东苏旗博物馆
（方案设计）

建设地点：内蒙古 锡林郭勒盟
总占地面积：6 947.77平方米
总建筑面积：10 793.1平方米
使用功能：礼堂、展览馆、影剧院

9 呼伦贝尔景观湖（环境设计）

建设地点：内蒙古 呼伦贝尔
总占地面积：300 000平方米

10 乌海市星云大酒店（设计方案）

建设地点：内蒙古 乌海
项目面积：29 220平方米

11 呼和浩特市金川湿地度假中心
（方案设计）

建设地点：内蒙古 呼和浩特
项目面积：约600 000平方米

12 呼和浩特文化大厦（施工图设计）

工程地点：内蒙古 呼和浩特
建筑面积：26 144平方米
项目规模：18层、27层、25层高层住宅建筑

2010—2011

主要建筑设计作品

Main Architectural Design Works

Annual Review of Chinese Architectural Design Works

四川省建筑设计院创立于1953年，是城市建设和开发领域的大型建筑设计咨询机构。半个多世纪以来，已形成较为成熟的三大业务板块，即：设计咨询、工程勘察和项目管理。先后有约220余项工程项目荣获国家级、省（部）级、市级优秀设计奖，优秀勘察奖，优秀标准设计奖，重大科技成果及科技进步奖，工程涉及海内外许多国家和地区，在国内、国际享有良好的社会信誉。

在发展历程中，本院始终将人才视为组织发展的源动力，秉持着"尊重、合作、创新、活力、分享"的团队宗旨，已培养、成就了三代设计人。目前专业技术团队已达到近900人，其中有高、中级工程师420余人，四川勘察设计大师7人和省突出贡献专家及享受政府特殊津贴专家13人。

面对未来，本院将"以设计咨询业务为龙头，发展成为西部一流、国内先进，提供全面技术解决方案的大型现代工程设计咨询集团"作为企业发展的愿景，并一如既往地秉承"质量为本、铸就经典"的设计理念，繁荣建筑创作，关注"城市·人·自然"的和谐发展，为城市建设事业而不断努力！

Sichuan Provincial Architectural Design Institute, a large architectural design and consultation institution engaged in urban construction and development, was founded in 1953. After over half a century, the institute has already established three main business fields: design and consultation, engineering investigation, and project management. Over 220 engineering projects have won outstanding design prizes, excellent investigation prizes, outstanding standing design prizes, and significant scientific achievement and progress prizes of national, provincial (ministry), and municipal levels. The projects cover many countries and regions and have earned good domestic and international reputation.

During the development, our institute has always been considering talents as the source of development and has trained designers of three generations under the tenet of "respect, cooperation, innovation, vitality, and sharing". At present, our professional team has reached a size of nearly 900 people, including over 420 engineers with medium or high professional title, 7 provincial investigation and design masters, and 13 experts with "prominent contributions" and recipients of "special government allowance".

Making "focusing on the design and consultation business to become a top large-scale modern engineering design and consultation group of China that is capable of providing complete technological solutions" as our corporation vision, our institute will keep following the design philosophy of "creating classics on the base of quality" to promote architectural design, to focus on the harmonious development of city, people, and nature, and to keep working hard for urban construction business.

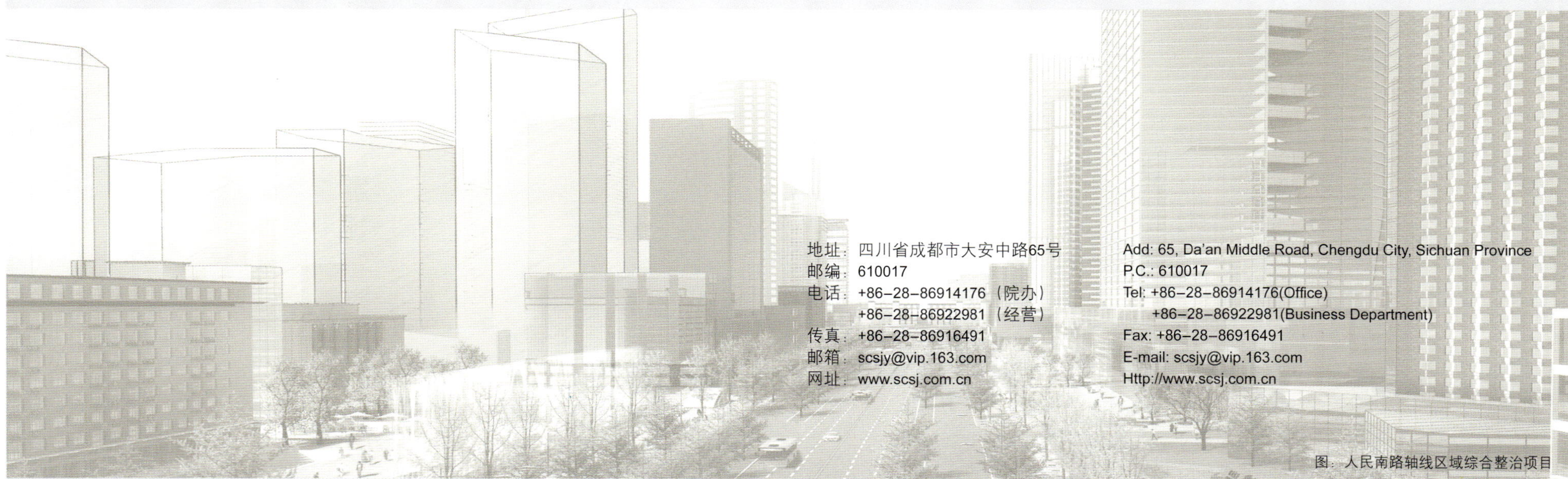

地址：四川省成都市大安中路65号
邮编：610017
电话：+86-28-86914176（院办）
+86-28-86922981（经营）
传真：+86-28-86916491
邮箱：scsjy@vip.163.com
网址：www.scsj.com.cn

Add: 65, Da'an Middle Road, Chengdu City, Sichuan Province
P.C.: 610017
Tel: +86-28-86914176(Office)
+86-28-86922981(Business Department)
Fax: +86-28-86916491
E-mail: scsjy@vip.163.com
Http://www.scsj.com.cn

图：人民南路轴线区域综合整治项目

“二轴四片”人民南路轴线区域综合整治

总建筑面积：全长5.5千米
设计时间：2010年

The Comprehensive Treatment of Renmin Nanlu (Rd.)

Gross Floor Area: 5.5 km long
Design Time: 2010

项目介绍

人民南路全长5.5千米，北起成都市天府广场南至天府立交，是成都市的城市中轴线。成都市人民南路区域综合整治是成都市灾后重建的重点工程，是1958年人民南路建成50年以来最完整的一次改造，是成都市城市建设历史上的一次重大事件。其涉及范围之广（沿线涉及青羊、锦江、武侯三个区）、历时之久（2年多时间）、系统之巨（包括地铁、地面交通、市政、景观、建筑、夜景、店招、街道家具等方面）是前所未有的。正是因为其城市中轴线地位和城市发展密切的关系，本轮设计工作的实施，成都市人民南路已经成为成都市的城市名片，受到了省内外各方面的好评，实现了群众、专家、领导三满意，并被评为2011年度“四川最美街道”第一名。

Project Introduction

Renmin Nanlu (Rd.), which is 5.5 km long from the Tianfu Square in the north to the Tianfu Flyover in the south, is the axis of Chengdu City. The comprehensive treatment of Renmin Nanlu is a key project of the post-disaster reconstruction of Chengdu City, which is the most complete reconstruction since the completion of this road in 1958 and also an important event in the urban construction history of Chengdu City. The wide area (the Qingyang, Jinjiang, and Wuhou District are covered), the long project time (more than 2 years), and the wide coverage (including subway, aboveground traffic, municipal works, landscape, buildings, night scene, signboards, street furniture and more) are all unprecedented. After this project, Renmin Nanlu has become the business card of Chengdu City and was honored as the “most beautiful street in Sichuan” in 2011.

四川省投资集团调度中心

总建筑面积：9.5万平方米
设计时间：2008年

项目介绍

川投调度中心占地2公顷，甲级写字楼，地上建筑面积6.9万平方米，地下2.6万平方米。建筑设计成双塔造型，以化解庞大的建筑体量，双塔一南一北相向布置，北塔21层，南塔23层，总高98米，中间用15层高中庭连接。造型采用异型平面及柔和且富有流动感的曲线，塔楼从下向上不断放大，给人蓬勃向上的感觉。建筑采用蓝灰色全玻璃幕墙，基座外挂不规则变化的咖啡色百叶，形成独特的肌理。门厅东西贯通，15层高的中庭气势恢宏，天桥纵横其间，富于动感。在四层嵌入一个两层高的报告厅，其顶面设计成空中花园，为办公区人员提供休闲场所。中庭采用单索幕墙，精巧的受力构件，形成非常通透的视觉效果。本项目与蔡德勤建筑设计公司合作完成。

The Control Center of Sichuan Provincial Investment Group

Gross Floor Area: 95,000 m^2
Design Time: 2008

Project Introduction

The control center of Sichuan Provincial Investment Group covers an area of 2 ha. It is a class A office building with an above ground floor area of 69,000 m^2 and an underground floor area of 26,000 m^2. The building is designed to have two towers to scatter the huge building volume. The two towers are north-south oriented; the north tower has 21 floors and the south one has 23 with a height of 98 m; the middle part which connects the two towers has 15 floors. A typical surface and soft and fluctuating curves are adopted in the outline. The towers are growing larger from the bottoms to the tops, giving a feeling of thriving. Grey-blue all-glass curtain walls are adopted; irregular louvers are attached to the bases to make unique texture. The hallway connects the east and the west; accompanied by the vivid platform bridge, the 15-floor central body looks very grand. A two-floor lecture hall is implanted into the fourth floor, the top of which is designed to be a garden in the air as a leisure site. The central body uses single-layer cable curtain walls and delicate stressed frames to achieve a very transparent visual effect. This project is accomplished together with Cai Deqin Architectural Design Company.

锦里二期——水岸锦里

总建筑面积：8 467.2平方米
设计时间：2009年
获奖情况：天府杯景观大赛金奖

项目介绍

锦里以国家重点文物保护单位“武侯祠”为依托，营造最富传统韵味、最具市井生活魅力、最有商业气息的旅游商业步行街，被评为“全国十大城市商业步行街”，有“成都版清明上河图”的美誉。占地约5公顷的二期工程——“水岸锦里”，以“历史文脉传承延续、传统空间当代诠释、生态景观有机植入”为设计原则，空间布局体现传统西蜀园林“文秀清幽”的特征，通过项目水系景观化处理，表现成都因水而生的城市脉络，展示城市的平民精神和地域文化特征。建筑设计上以历史为支撑、文化作表象，以秦汉、三国精神为灵，取明清近代风貌作其表，融合西蜀民居、民俗为内容，将历史与现代有机结合，打造出一个具有浓重历史底蕴的锦里延伸段。

Jinli Second Term – Waterside Jinli

Gross Floor Area: 8,467.20 m^2
Design Time: 2009
Honor: Gold Medal in Tianfu Cup Landscape Competition

Project Introduction

Jinli is a commercial pedestrian street based on Wuhouci, a national key cultural relic protection unit. With the most traditional features, the most characteristic folk customs, and the most prosperous commercial environment, it has been listed into the "Top 10 Urban Commercial Pedestrian Streets" and honored as the "Chengdu Version of the Riverside Scene at Qingming Festival". The second term project – Waterside Jinli, which covers an area of around 73 mu, is designed in accordance with the principles of "inheriting the ancient culture, expressing modern ideas in traditional spaces, and being integrated with ecological sceneries". The elegant, quiet, and secluded features of traditional Western Sichuan gardens are shown from the space layout; the spirits of civilians and the features of local culture and their relationship with water are shown from the water system made as a part of the whole scene. Histories and cultures of the Qin, Han Dynasty and the Three Kingdoms are all taken into consideration, the styles of the Ming, Qing Dynasty and the recent modern time shape the overall appearance, and the folk customs of Western Sichuan are also merged into the architectural design. With an organic combination of the history and the modern time, an extension of Jinli with profound historical implications comes into being.

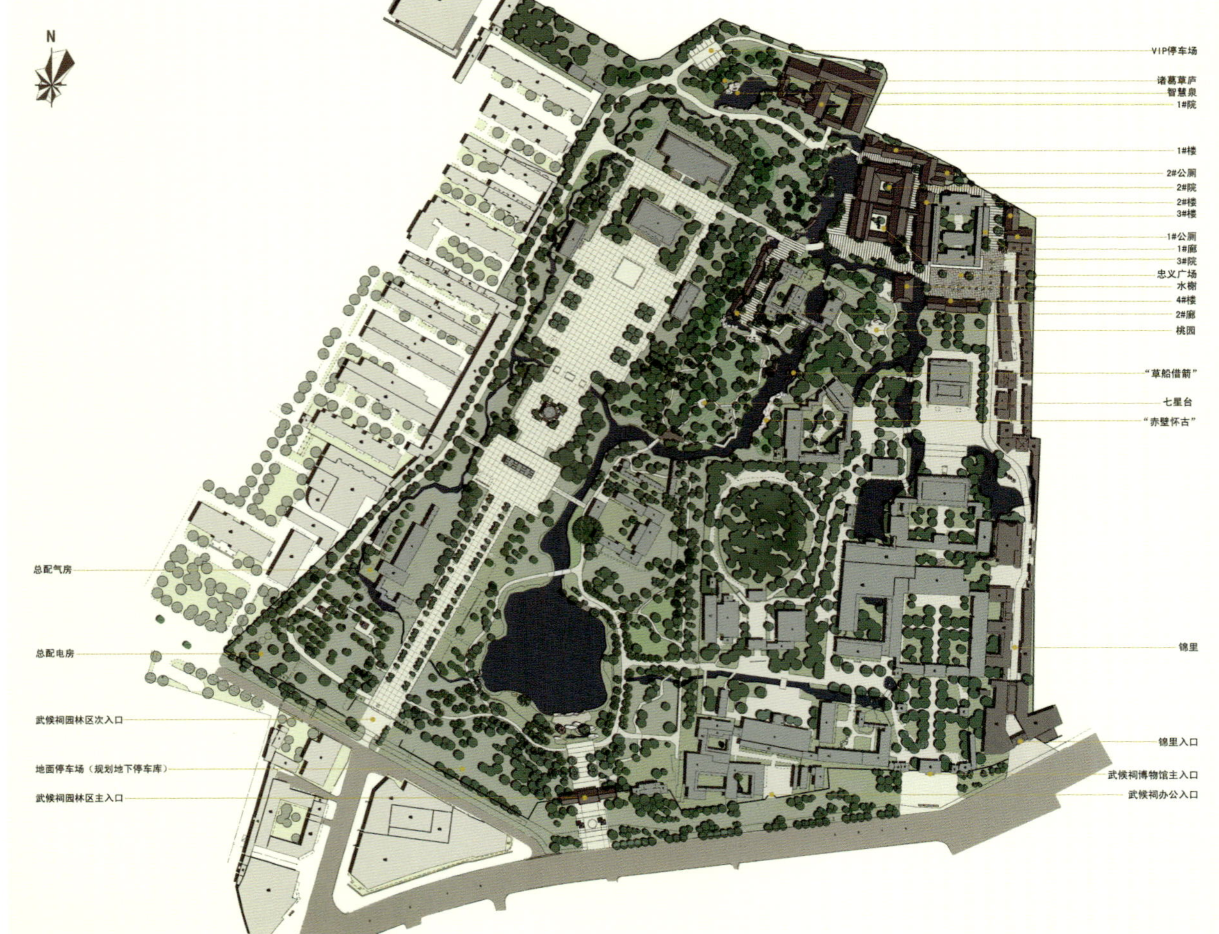

峨眉半山七里坪国际休闲旅游区

Qiliping International Tourist Resort Half Way up Mt. Emei

总建筑面积：6 172.5平方米（一期）
设计时间：2010年

Gross Floor Area: 6,172.5 m^2 (Phase I)
Design Time: 2010

项目介绍

该项目是峨眉山地区最大的度假项目，总用地10平方千米，建设周期5～8年。项目配置有高尔夫、温泉、户外运动和风情小镇。该项目是通过国际投标获得的设计权。

建筑设计注重对地形的结合和利用，注重对别墅品质的追求，注重对建筑除湿的低技策略的运用，住宅单体设计回归自然主义的理性，重新追求建筑美与自然美的和谐统一，以批判的地域主义风格创造出独特的建筑风情，使居住者的情感回归于宁静与自然。完美和谐的整体格局和精心设计的细节充分体现出对人性的全面关怀。设计始终以＂山地度假＂为出发点，追求使用者有别于城市居家的体验和品位，为人们提供一个＂面佛＂的心灵寓所。

Project Introduction

This project, which covers an area of 10 sq. km. with a construction period of 5-8 years, is the largest tourist resort project in the region of Mt. Emei. It is accompanied by golf, hot spring, outdoor exercises, and featured towns. The designer won this project through an international bidding.

The architectural design focused on the use of the landform, the quality of the villas, and the use of low-tech strategy in dehumidification. The monomer design of residence returned to naturalistic reason to pursue the harmonious integration of the architectural beauty and the natural beauty and created a unique architectural feature through a critical regionalism style for guests to put their feelings back into peace and nature. The perfect and harmonious overall pattern and the elaborate details fully reflect the great concerns to humanity. The design has been focusing on the concept of “mountain resort” and been trying to provide guests a Buddhism home for soul to gain an experience different from their urban inhabitance.

万达索菲特大酒店

总建筑面积：2.8万平方米
设计时间：1997年

项目介绍

万达索菲特大酒店为成都市较早的五星级酒店，位于成都滨江路北侧，临人民南路锦江大桥，项目布局时考虑与老滨江饭店客层楼的关系，将临城市一侧的广场空间连贯起来，并由此确定了建筑临街一侧的形态。客层楼主体尽量向后退离，避免对前广场形成压抑，同时利于布置公共裙房部分大空间。建筑风格简洁明快，强调凹凸曲面变化，与临水的环境相呼应。

Sofitel Wanda Hotel

Gross Floor Area: 28,000 m^2
Design Time: 1997

Project Introduction

Sofitel Wanda Hotel, located on the north side of Binjiang Lu (Rd.) and adjacent to Jinjiang Bridge on Renmin Nanlu, is among the first five-star hotels in Chengdu. The style of the building is simple and lively, which focuses on the changes of the curved surface to fit in the riverside environment.

成都市春熙路商业步行街城市及环境设计

总建筑面积：约6万平方米
设计时间：2001年
获奖情况：四川省优秀工程设计一等奖

Urban and Environmental Design for Chunxilu Commercial Pedestrian Street of Chengdu City

Gross Floor Area: about 60,000 m^2
Design Time: 2001
Honor: First Prize of Sichuan Excellent Engineering Design Prize

项目介绍

为将春熙路打造成展示成都城市形象的西南第一商业步行街，其改造建设被视为成都市旧城改造建设的示范工程，为此我们制定了四项设计原则：人车分离原则、整体协调原则、地域文化原则、全程控制原则。在街道与广场的设计中，协调历史轴与未来轴上重要节点的整体关系，处理好广场的空间尺度与空间环境以及穿越人流的交通关系，在地面铺装、街道家具、绿化灯具、小品等方面做了细致深入的分析与设计，在获得良好设计效果的同时，有效地与市政改造配合，达到了艺术性与可操作性的合理结合。

Project Introduction

In order to make Chunxilu the first commercial pedestrian street in Southwestern China to show the city image of Chengdu City, we established four principals for this demonstration project of old city reconstruction: separation of people from cars, overall harmony, regional culture, and overall control. In the design of streets and square, we try to coordinate the relationship between axes of the past and the future, to deal with the traffic relationship between the space of the square, the environment, and the stream of people, to make careful and deep analysis on the ground pavement, street furniture, greenbelt lamps, decorations and more, and to coordinate with the municipal reconstruction while maintaining the design effects to achieve a ideal integration of artistic effects and practicability.

四川大学华西临床医学院教学楼

总建筑面积：约29 214平方米
设计时间：2005年
获奖情况：四川省优秀设计一等奖

项目介绍

四川大学华西临床医学院教学楼位于华西校区西区，北临校内主干道，是靠近学校被列为重点保护对象的传统风格建筑。为了协调校园新老建筑的整体风格，项目设计秉持"高效共生"的设计理念，将建筑作为一个有机体，在自我完善的同时与环境协调、沟通与对话，在空间形式与建筑造型上延续华西校园的文脉，突出场所认同感与归宿感。

Teaching Building of Western China Clinical Medical College of Sichuan University

Gross Floor Area: about 29,214 m^2
Design Time: 2005
Honor: First prize of Sichuan Provincial Outstanding Design Prize

Project Introduction

The teaching building of Western China Clinical Medical College of Sichuan University is located at the west zone of the Huaxi campus; it is adjacent to the main stem of the campus and to the traditional style buildings which are key protection objects of the school. To coordinate the styles of the old and new buildings, the design of the project tries to make the building an organic body to be harmonious with the existing environment of the Huaxi campus.

ICT成都国际医学城

总建筑面积：14.74万平方米
设计时间：2011年

ICT Chengdu International Medical Sciences Town

Gross Floor Area: 147,400 m^2
Design Time: 2011

项目介绍

ICT成都国际医学城位于成都西部的温江区永宁镇。建设用地面积约8公顷，建筑面积14.74万平方米。我们将地块设想成为一片具有生命的树叶。植物是有生命的，医院更是与生命的主题密不可分。借树叶的形式对用地进行规划设计，建筑的布局按照树叶的形状及叶脉的形式进行总体平面规划。每栋单体建筑都成为树叶的一个支脉。同时温江素有"金温江"之称，更是鱼凫王国的发源地，因此建筑设计的总体组合更形成了鱼的形态。现代、温暖、和谐、统一的建筑外形使得建筑也充满着生命的活力，能够作为地标级的建筑长期立于规划新区内。

Project Introduction

ICT Chengdu International Medical Sciences Town is located at Yongning Town of Wenjiang District. The area of the land used for construction is about 8 ha and the floor area is 147,400 m^2. We conceived this area as a piece of living leaf. Plants are alive and hospitals are closely related to life. The construction planning was made with a piece of leaf as the reference model: the overall planning of the distribution of the buildings was made according to the shape and veins of a piece of leaf. Every single building becomes a leaf vein. Meanwhile, Wenjiang has been honored as "Golden Wenjiang", which was also the cradle of Yufu Kindom (kindom of fishes and waterfowls). Therefore, the overall architectural design was also made to take the shape of a fish. The modern, warm, harmonious, and united architectural appearances give vivid life to our buildings, which are entitled to be standing in the new planning area as the landmarks.

中国建筑西南设计研究院有限公司

CHINA SOUTHWEST ARCHITECTURAL DESIGN AND RESEARCH INSTITUTE CORP.LTD

中国建筑西南设计研究院有限公司始建于1950年，是中国同行业中成立时间最早的大型甲级建筑设计院之一，隶属世界500强企业中国建筑工程总公司。建院60年来，本院设计完成了近万项工程设计任务，项目遍及我国各省、市、自治区及全球10多个国家和地区，是我国拥有独立涉外经营权并参与众多国外设计任务的大型建筑设计院之一。2004年以来连续四年被亚洲建筑师协会评为“中国十大建筑设计公司”，并获得“全国工程勘察设计百强”企业称号。

目前，本院有各类专业技术人员1 500余人，其中教授级高级建筑师、教授级高级工程师40余人，高级建筑师、高级工程师260余人，国家一级注册建筑师、一级注册结构工程师、注册造价工程师、注册设备工程师、注册电气工程师300余人，培养出中国工程勘察设计大师4人、对国家有突出贡献的中青年专家1人，建设部有突出贡献中青年专家2人，四川省学术和技术带头人3人，四川省有突出贡献专家2人，享受政府特殊津贴23人，四川省工程设计大师16人。

作为中西部最大的建筑设计院和国家基本建设的重要国有骨干企业，本院以“精心设计、服务社会”为己任，坚持以繁荣建筑创作为宗旨，不断完善创新设计理念，力创建筑设计精品，在工程设计和科研方面获国家级、部级和省级以上优秀奖近500项，并取得了国家优秀设计金质奖四项、银质奖三项、铜质奖五项的创优佳绩。60年的设计耕耘，西南院在博览文化建筑、体育建筑、医疗建筑、教育建筑、旅游建筑、居住建筑以及空间结构等设计领域具有独特的设计优势，而严格、规范的ISO9001质量体系认证管理更使本院的设计质量为业界广泛认同。

在加强生产经营、科技创新和内部管理工作的同时，本院认真贯彻企业党建工作和企业文化建设的各项工作，提倡三个文明建设同步发展、共同进步。本院在改革发展中相继获得建设部“八五期间全国工程建设管理先进单位”、“精神文明建设先进单位”、全国优秀勘察设计院、四川省先进单位、“重点工程建设先进单位”、省级“最佳文明单位”以及“国家‘十五’期间科技进步先进集体”等荣誉称号。

Being established in 1950, China Southwest Architectural Design and Research Institute Corp. Ltd, one of the large-scale, Grade A and comprehensive architectural design institutes, is founded among the earliest in the same industry in China. It is subordinated to one of the Fortune Global 500—China State Construction Engineering Corporation. The institute has fulfilled almost 10 thousand design tasks during the 60 years or more since it was founded, which extend all over China and 10 other countries and regions. Boasting independent and foreign related operating authority, it has been awarded by Asia Institute of Architectures as one of “China's Top 10 Architectural Design Firms” for the successive three years since 2004. Moreover, it has been awarded as “one of the Best 100 National Reconnaissance Design Corporations”.

At present, there are more than 1,500 professional and technical personnel in the institute, including over 40 senior architects and engineers with the rank of professor, over 260 senior architects and engineers, over 300 national Class 1 registered architects, structural engineers, cost plan engineers, facilities engineers and electrical engineers; 4 national engineering design masters, 1 national expert with great contribution in middle and young age, 1 expert with great contribution in middle and young age awarded by ministry of construction, 3 provincial experts with great contribution in middle and young age, 2 provincial academic and technical experts, 23 experts who get special allowance from the government and 16 provincial engineering design masters.

As the biggest architectural design institute in middle west of China, the important and backbone state-owned enterprise in the country's capital construction, the institute takes “Elaborate design and service for society” as the task and persists in the aim of prospering architectural creation. We constantly improve design concepts, bring forth new ideas, and take efforts to produce fine works in architectural design. Thus, we have won over 500 items of national, departmental and provincial award in engineering design and scientific research, including 4 gold, 3 silver and 5 bronze awards of National Excellent Design. More than 60 years' efforts in design make the institute show unique design advantages in international fair culture architecture, sports buildings, medical treatment buildings, educational buildings, tourism buildings, residential buildings and spatial structure design and so on. And the strict and standardized authentication management of the ISO9001 quality system makes the design quality of the institute recognized by the whole industry.

In strengthening the production and operation, technological innovation and internal management work, the institute conscientiously implements the Party and the enterprise culture construction work, to promote the simultaneous development of three civilizations and common progress. With the reformation and development, the institute has received from the Ministry of Construction, "The eighth five-year period of national construction management advanced units", "spiritual civilization construction advanced unit", the National Excellent Survey and Design Institute, Sichuan Province Advanced Unit, "Major projects advanced unit", the provincial class "Best civilization Unit" and "National 10th Five-Year period the advanced science and technology group," and other honor titles.

地址：四川省成都市天府大道北段866号
邮编：610042

院办公室：
电话：+86-28-62550866　传真：+86-28-62550900
经营处：
电话：+86-28-62550032　传真：+86-28-62550030
重庆分院：
电话：+86-23-67742572　传真：+86-23-67870494

客户服务：+86-28-62550315

Add: No. 866, North Section of Tianfu Avenue, Chengdu, China
P.C.: 610042

Institute Office:
Tel: +86-28-62550866 Fax: +86-28-62550900
Business Office:
Tel: +86-28-62550032 Fax: +86-28-62550030
Branch Institutes in Chongqing:
Tel: +86-23-67742572 Fax: +86-23-67870494

Customer Service: +86-28-62550315

http://www.xnjz.com

Guangzhou Museum
广州博物馆

总建筑面积：5万平方米　　Total Building Area: 50,000 m^2

设计借鉴传统岭南园林与广府大屋的意象，将地域历史与文化的精髓收纳其中，建筑与展品，由外及内浑然一体。室内室外空间相互渗透，收放开合，历史、文化、建筑、园林交替呈现，营造意境深远的观展氛围。运用建筑形体错落咬合，建筑空间为主，庭院空间从属，相辅相成、相得益彰，变化出一个"入狭而得景广"的壶中天地，令观赏者产生小中见大的空间体验。建筑形体借鉴传统岭南建筑"高墙围院，连房博厦"的意象。

Borrowing the image of the traditional Lingnan garden and the Cantonese big house, the design incorporates the essence of regional history and culture. The building and the displays form a harmonious entity from the outside and inside. Mutual penetration of indoor and outdoor space is achieved, closing and opening together. Alternating presentation of history, culture, architecture and landscape creates a visiting atmosphere full of artistic conception. Build a bilevel-shaped architecture, that is, the building space is taken as the main element and courtyard space is taken as the subordinate space. The two elements complement each other and bring out the best in each other, creating a pot space, that is, the entrance of a place is small, but large sceneries can be gained later. In this way, the visitors can have an experience of seeing big landscape standing in small places. The shape of the building borrows the image of high-walled house and connecting rooms of Lingnan architecture.

The Project of Commercial Buildings in No. 966 of the North Section of Tianfu Avenue in the High-tech Zone, Chengdu.

成都市高新区天府大道北段966号商业用房工程

总建筑面积：16.79万平方米
设计时间：2010—2011年

Total Building Area: 167,900 m^2
Period of Design: 2010–2011

本项目位于天府国际金融中心园区内，是为实现园区功能转型的改扩建项目。

加建建筑为两栋超高层及一栋金融营业厅，包含高端办公及其配套公寓、商业、餐饮、会议、会所、银行营业厅等多种功能，并合理利用地铁资源布置地下商业。在建筑形体及细部设计上，传承原园区标志性建筑群的优雅弧线特征，并在和谐基础上实现突破创新。通过交通组织计算分析，在办公楼筒体中采用亚洲及国内领先的双子高速电梯。为将本项目打造成高完成度产品，设计采用参数化三维建模，以及建筑信息模型协同设计。

本项目按国家绿色二星标准设计。

This project is located in the park of Tianfu International Finance Center, and it is an expansion project to realize the function transformation of the park.

The added buildings are two mega-high-rise buildings and a business lobby of financial business, including high-class office buildings and the matched building such as apartments, commercial buildings, restaurants, convention center, clubhouses and banking hall. Commercial businesses are arranged underground by using the subway underground resources rationally. In architectural form and the detail design, the features of elegant curves used in the landmark buildings in the original park are inherited, and breakthrough and innovation are also achieved on the basis of harmony. Through analysis of traffic calculation, twin high-speed elevators which are leading products in Asia and China are used in the office buildings. To make this project a product of high-degree completion, three-dimensional and parametric modeling is used in the main design process, and BIM is also used in the design.

National green two-star design standard is used in this project.

The New Part of Library of Sichuan Province
四川省图书馆新馆

总建筑面积：5.10万平方米
设计时间：2011年

Total Building Area: 51,000 m^2
Design Time: 2011

四川省图书馆新馆项目，位于成都天府广场的西翼。新馆以四川地区富于特色的汉代石阙为形态原型，用现代的形式语汇再现了古典建筑的神韵。室内台阶式中庭两侧以通高的书墙围合，赋予中庭空间以文化特征。

设计通过屋顶大尺度出檐与立面石材百叶的遮阳作用，减少夏季太阳辐射进入室内。屋面雨水收集系统将雨水汇入建筑周边的浅水池中，既有利于改善局部微气候，又创造出一片动人的水体景观。

The project of the new part of Library of Sichuan Province is located in the west part of Tianfu Square, Chengdu. Adopting the stones from Han Dynasty which are full of Sichuan characteristics as the prototype, the new part of Library of Sichuan Province reproduces the charm of classical architecture by using modern forms. Both sides of the indoor stepped atrium are enclosed by all-high book wall, providing cultural characteristics for the atrium space.

In summer, solar radiation entering into rooms is reduced because of the shading effects of large-scale part out of the roof eaves and stone screen in the façade. The rooftop rainwater connection system collects the rainwater in the shallow water pool around the building, which can not only improve the local micro-climate, but also create a moving body of water landscape.

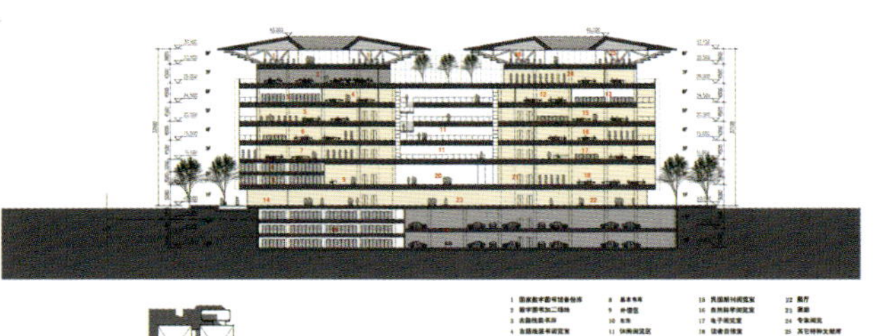

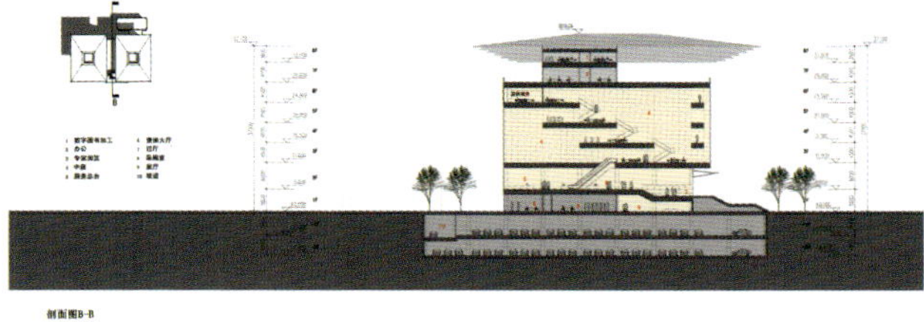

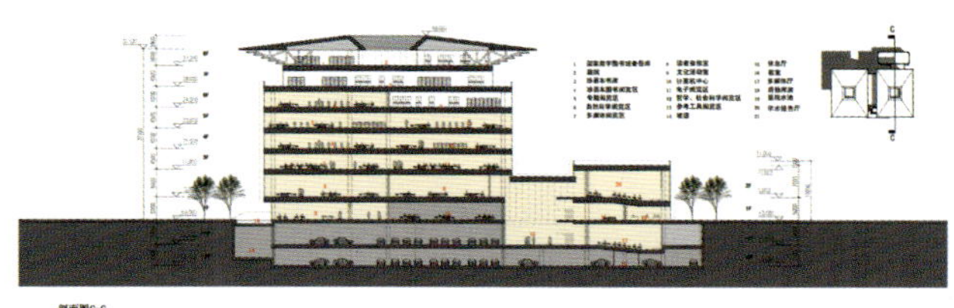

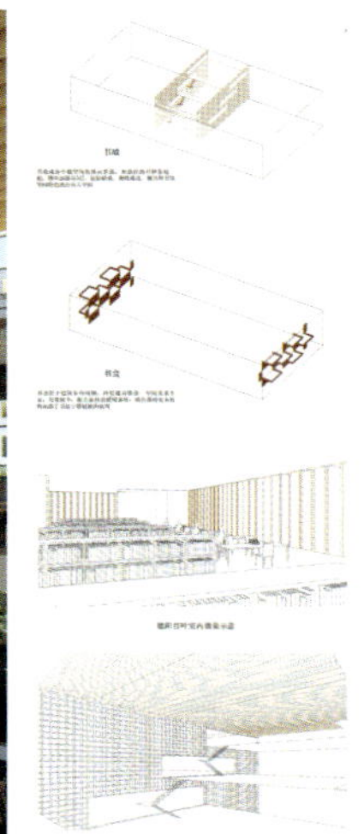

The Eastern Station Area Planning and T3 Terminal Architecture Schematic Design of Chongqing Jiangbei International Airport

重庆江北国际机场东航站区规划及T3航站楼设计

总建筑面积：80万平方米　　Total Building Area: 800,000 m^2

重庆，巴文化发源地，历史源远流长。

重庆，两江汇流之地，气势雄浑壮阔。

T3新航站楼作为彰显城市魅力的门户与窗口，设计构型将“比翼神鸟”与“两江汇流”意向融为一体，寓意城市和谐发展，并凸显对多元文化的兼容并蓄。方案由一个主航站楼和一个卫星厅构成，一方面与空侧站坪区实现高效运行对接，另一方面与陆侧交通换乘中心实现旅客零换乘。T3航站楼内具有功能布局合理、工艺流程简洁、商业配置最优、标志系统清晰等特点，且满足建筑对绿色空间、人文关怀等要求。

Chongqing is the cradle of Ba Culture with a long history, and the confluence of two rivers with forceful magnificent.

As the new landmark of the city, the new T3 terminal highlights the city's charm and fascination. The intention of its unique design is to integrate "Bizart God Bird" with "Confluence of Two Rivers", which means embracing the harmonious development of the city and its cultural diversity. The design is composed of one main Terminal and one Satellite, on one hand, it realizes the goal of efficient docking operation in the airside apron; on the other hand, it makes it possible for zero-transfer passenger in the ground transportation center. The inside of T3 terminal enjoys many advantages, such as reasonable functional layout, simple process, superior commercial configuration, and clear signage system. Moreover, It meets with the requirements of calling on green space, humane care etc.

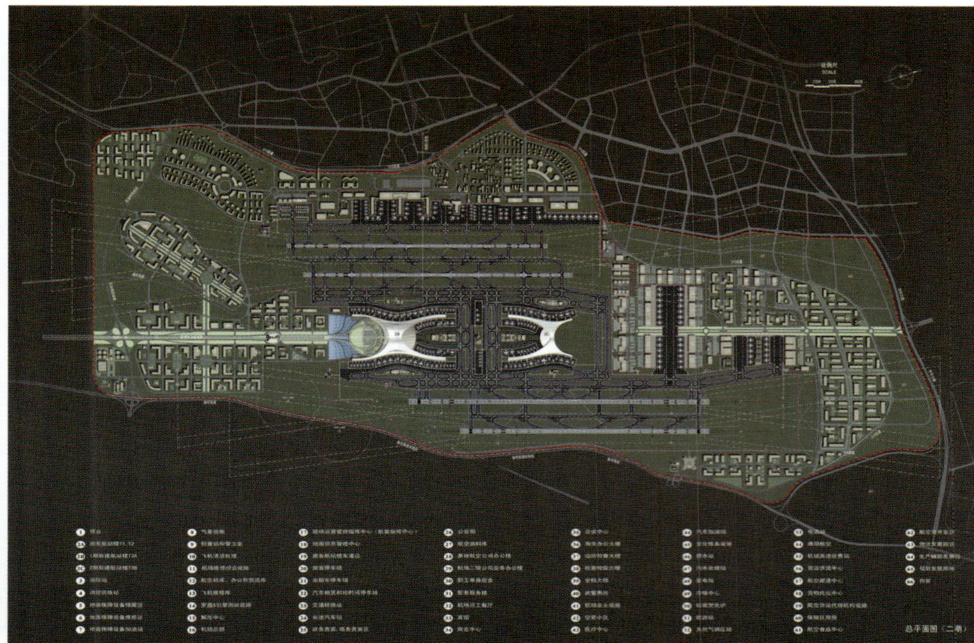

Deyang Special Education School
德阳特殊教育学校

总建筑面积：0.83万平方米　Total Building Area: 8,300 m^2
设计时间：2011年　Design Time: 2011

德阳特殊教育学校主要招收智障和聋哑类儿童。

设计师从儿童画里面找到最基本的形式原型：一个坡屋顶带方窗的房子——这是儿童潜意识里对家的想象。所以建筑造型采用围绕中心庭院的小房子模式，构成了微型的村落。每栋小房子又以天井为核心。所有房间围绕天井布置，天井成为人流、视线和光影汇集的中心。设计师还特意在天井立面上开满大大小小的方窗，提供了看与被看的视觉乐趣。

Deyang Special Education School mainly enrolls mentally retarded and deaf children.
The designers get idea of the most basic form of the building, that is, a house with a rectangular window on the pitched roof. This is the imagination of home from the subconsciousness of children. Therefore, an architectural model is formed, that is, a central courtyard is surrounded by many small houses, and they constitute a miniature village. Courtyards are taken as the core for each small house and every room in the house surrounds the courtyard of this house, making the courtyard the center of people stream, sight line, light and shade. Large and small rectangular windows are designed specially in the courtyard facade by the designers, in this way the visual enjoyment of seeing and being seen is achieved.

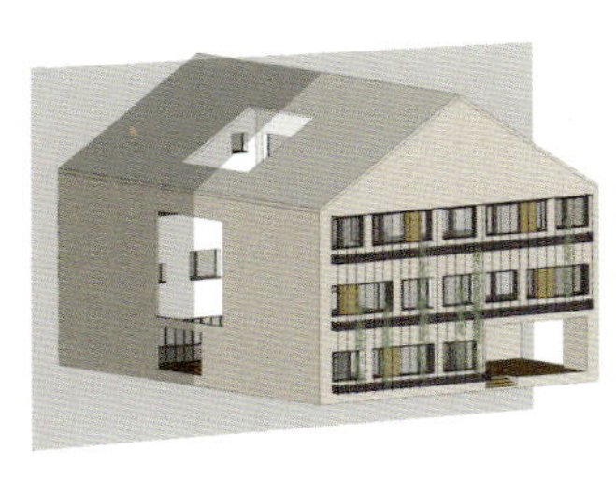

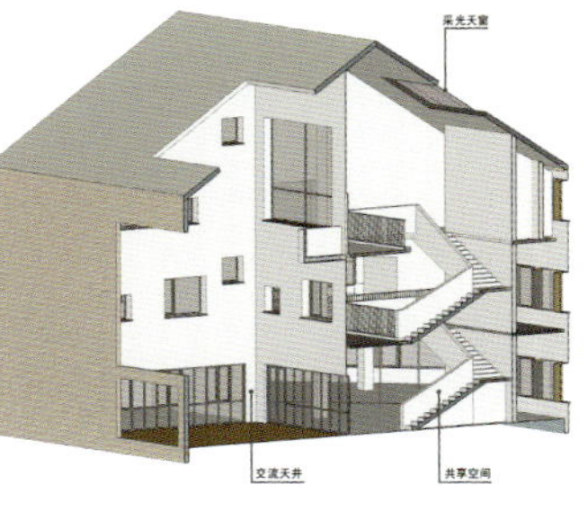

The Design Scope of the Urban Design of the South Island in the Chengdu East New City
成都东部创意新城南岛城市设计

“东部创意新城南岛城市设计”设计范围2平方千米，位于成都市“东部新城文化创意产业综合功能区”起步区域。涵盖创意产业、精品商业、生态住区及商务办公等多种业态。

延续上位规划的指引，以多元化及开放性的设计手法，关注都市空间层次及形态构成，结合轨道及区域交通，营造强烈的可识别性。构建繁荣、创新和活力无穷的创意核心区，为创意产业提供生长及发展平台。

The design scope of the urban design of the South Island in the East New City is two square kilometers, and it is located in the starting area of “Comprehensive District of Cultural and Creative Industry in the East New City, Chengdu”. This project includes diversified commercial format such as creative industry, boutique business, ecological residential area and office buildings.
The design of the project is continuation of the upper planning guidelines. The design idea is open and diversified, which concerns about the level and form of urban space structure and combines rail and regional transportation, creating a strong identifiability. To build a prosperous, creative and dynamic creative district provides a platform of growth and development for the creative industry.

Wolong Nature Reserve, Dujiangyan Giant Panda Center for Disease Control and Prevention
卧龙自然保护区都江堰大熊猫救护与疾病防控中心

总建筑面积：1.17万平方米　Total Building Area: 11,700 m^2
设计时间：2010年　Design Time: 2010

本案系抗震救灾香港援建项目，是全球首个熊猫医院。基于可持续发展策略，以尊重地域文化及满足熊猫疾病防控功能要求为指导原则。

设计提出了川西林盘布局，原生态、地域化的构思理念。在空间组织上，自然山体为空间的主要构成背景，建筑聚落依托山体布置，形成若干开合有序、野趣盎然的“院坝”空间，各建筑聚落参差错落，随机自由。以传统川西民居的穿斗结构，青砖墙为表达元素，辅以现代化的技术、材料表达，形态内敛，整体意象朴素自然。本项目设计申报国家绿色建筑三星标准。

This is a relief aid project assisted by Hong Kong, and it is the first giant panda hospital all around the world. Based on Sustainable Development Strategy, the guidance principals should respect the local culture and meet the requirements of giant panda disease control and prevention .

The design puts forward a concept of respecting the forest & woods layout of West Sichuan, primitive ecology and localization. In spatial organization, the natural mountain is taken as the main component for the space background. Supported by the mountains, the buildings form a number of courtyard space which is full of rustic charm. Each building scatters casually. The design takes the tenon through structure which used in traditional West Sichuan and green walls as the presentation elements, and it is also supported by modern technology and materials. With restrained form, the overall image is simple and natural. The design declares the national green building three-star design standard.

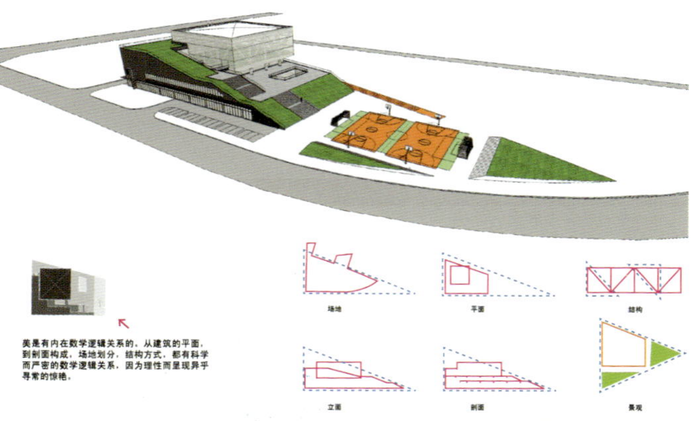

Sports Fitness Center in Sanwayao, Chengdu
成都市三瓦窑片区体育健身中心

总建筑面积：1.2万平方米　Total Building Area: 12,000 m^2
设计时间：2009年　Design Time: 2009

场地的设计用一纵一斜两条景观轴线，打通了阻隔的南北街区，为市民活动带来便捷途径。一、二层设置健身中心，三层为全钢结构的大跨度羽毛球馆。简洁的方形体量用U形玻璃墙包裹，仿佛是漂浮在草坡上的白色灯笼。斜向草坡将人流引入建筑屋顶，斜坡本身是天然的露天看台，与室外运动场相得益彰，为街区创造出浓郁的运动氛围。

One longitudinal landscape axis and one oblique axis are used in the site design to break down the barrier of the north-southern street, which brings a convenient way for public activities.

Fitness centers are installed in level 1 and level 2, and a full-span steel badminton court is installed in level 3. The simple square shape body is wrapped with a U-shaped glass wall, as if it were a white lantern floating on the grass slope. The inclined grass slope guides the people to the building roof. The natural bleacher of the inclined grass slope and the outdoor playground bring out the best in each other, creating a rich sports atmosphere on this block.

Xuzhou Sports Center
徐州市体育中心

总建筑面积：23.60万平方米
Total Building Area: 236,000 m^2

设计项目包括35 000座的体育场、3 000座的综合馆、球类馆，2 000座的游泳跳水馆，约4万平方米的体育宾馆以及中心广场地下商业、休闲娱乐配套设施等综合性体育中心。

规划方案以"山水相依、五省通衢"为规划理念，以象征交通枢纽、极具速度感的中央景观统率全局，将建筑布置在两侧，并利用西北水系形成湖面和建筑前景，在规划中体现徐州山水风光和交通枢纽意向，建设绿色体育中心。

建筑设计结合徐州地域文化，以"玉"和"帛"为构思源泉，将古玉和丝绸的气质融入建筑造型中，表达"以和为贵，和谐发展"的共同愿望，象征徐州从兵家必争之地发展为商家必争之地，和谐共进的积极城市精神。

The design program includes a 35,000-seat stadium, a 3,000-seat complex building, a ball games gym, a 2,000-seat swimming & diving hall, a sports hotel with 40,000 m^2 and the complex sports center, such as the underground businesses of central square, matched leisure and entertainment facilities, etc.

The planning is based on the idea of "landscape dependency, major juncture of five provinces", unifying the overall plan with the central landscape which indicates the transport hub and possesses the sense of speed, laying out the buildings on the two sides. Moreover, it makes use of northwest water system to form the lake and the front view of the building. By this way, the planning shows the landscapes of Xuzhou, the intent as the transport hub and the will of constructing the green sports center.

The architectural design combines the local culture of Xuzhou, using jade and silks as the idea of the source and integrating the architectural modeling with the temperament of jade and silks, which express the common aspiration of accommodative and harmonious development. It also indicates that Xuzhou has been developed from the contested battleground to contested business place, and the positive spirit of harmony and enterprising of the city has been shown.

Sichuan Radio and Television Center
四川省广电中心

总建筑面积：13万平方米
设计时间：2006年

Total Building Area: 130,000 m^2
Design Time: 2006

四川省广电中心在空旷的城市新区环境中，其超高层塔楼成为最具标志性的建筑景观。135米高的主楼造型构成矩形体与直纹曲面并置的有机体，充满几何逻辑且变幻的动态造型给人以极强的视觉冲击力。长约200米的广电长廊空间高达15米，尺度开阔的空间容纳演出、文化展览、旅游参观、休闲等多元化功能，体现现代媒体开放与交流的理念。

Standing in the open and new urban district, the high-rise tower of Sichuan Radio and Television Center becomes the landmark of that area. The shape of the 135 meters high main building is a combination of rectangular body and straight line curved surface, and this geometric, logic and dynamic shape gives a strong visual impact. The length and height of the corridor space of Sichuan Radio and Television Center is 200 meters and 15 meters respectively. The open space can accommodate diversified activities such as performance, cultural exhibition, visiting and entertainment, which reflects the idea of openness and communication adopted by the modern media.

基准方中建筑设计事务所

JI ZHUN FANG ZHONG ARCHITECTURAL DESIGN ASSOCIATES

公司简介 / Introduction

基准方中建筑设计事务所是以创建领先的商业模式，追求市场、技术、经济、文化、管理和谐发展，合伙人性质的大型建筑设计企业。公司成立于2002年12月。

事务所现有各级专业技术人员近1000人，其中四川工程勘察设计大师1人，各专业教授级高级工程师、高级建筑师、高级工程师80余人，建筑师、工程师近200人。拥有国家一级注册建筑师、一级注册结构工程师、注册电气工程师、注册设备工程师等逾100人。

事务所的总部设在四川省成都市，以立足西南、面向全国、走向世界为发展理念。2008年、2009年和2011年，事务所分别在重庆、西安和北京设立了全资分支机构，上海事务所也将在近期开业。事务所计划在未来的三至五年内，有计划地在全国各地和世界范围内设立更多的分支机构，为更广泛的客户提供便利、高效、高质量的订制服务。

基准方中的中、长期发展目标是成为具有高度社会责任感和历史责任感、市场化的、以创造具有持久价值的建设产品为根本目的，具有强大研究、咨询、设计和管理能力，服务于建设的大型综合技术服务公司，通过在设计实践中不断吸收、学习欧美优秀大型设计公司的成功经验和先进的经营模式（理念），努力将自身建设成为国内领先、在国际上有影响的大型综合建筑设计服务企业。

Ji Zhun Fang Zhong Architectural Design Associates (JZFZ), established in December 2002, is a large scale cooperation dedicated to architectural design. Our target is to provide high quality design and whole process service to our customers by means of advanced business simulation and to pursue the accord development of market, technology, economy, culture and management.

JZFZ now has around 1,000 employees, including 1 awarded Sichuan “Design Master”, more than 80 certified senior architects and engineers (equivalent to a FAIA title); about 100 A registered architects and engineers and more than 200 registered architects and structural and mechanical engineers.

The headquarter of JZFZ is located in Chengdu. In 2008, 2009, and 2011, we have established branch offices in Chongqing, Xi’an, and Beijing. With our new branch in Shanghai opening in the near future, a network of architectural and engineering professionals is expected to expand nationwide, even worldwide in 3-5 years, providing a wider range of customers with convenient, efficient, and high quality customized services.

It is the responsibility of JZFZ to create buildings that take social issues, local culture, and environment into consideration. JZFZ will always keeping learning from developed design firms worldwide in order to become a highly motivated team with outstanding design, project managing, consulting, and researching ability, creating long-standing works, remaining advanced at home and impactive abroad.

成都事务所	重庆事务所	西安事务所	北京事务所
西玉龙街6号	高新区新南路164号	南关正街88号	朝阳区广渠东路3号
新世纪广场20F	龙湖水晶国际10F	长安国际中心A座10F	中水电国际大厦7F
邮编：610017	邮编：401147	邮编：710068	邮编：100124
电话：+86-28-86582121	电话：+86-23-67039777	电话：+86-29-87651710	电话：+86-10-57795086
传真：+86-28-86582990	传真：+86-23-67039933	传真：+86-29-87651687	传真：+86-10-57795186
邮箱：KF@jzfz.com.cn	邮箱：JZFZcq@163.com	邮箱：xa@jzfz.com.cn	邮箱：bj@jzfz.com.cn

主要成员 / Partners

钟 明 Zhong Ming

高级建筑师

出生于1961年12月；
1982年8月毕业于同济大学建筑系建筑学专业，工学学士；
1982年8月至1995年先后在中国建筑西北设计研究院和中国建筑西南设计研究院从事建筑设计工作，先后任助理建筑师、建筑师、高级建筑师；
1999年创办四川基准建筑设计有限公司；
2001年10月发起创办成都基准方中建筑设计事务所，现任首席执行官、执行合伙人。

周 颿 Zhou Fan

高级建筑师、国家一级注册建筑师

出生于1968年1月；
1989年毕业于东南大学建筑系建筑学专业，工学学士；
1989年7月至2002年6月在中国建筑西南设计研究院从事建筑设计工作；
2002年7月加入成都基准方中建筑设计事务所，现任总建筑师、总经理、高级合伙人。
四川省土木建筑学会常务理事。

章玉华 Zhang Yuhua

高级工程师、国家一级注册结构工程师

出生于1972年12月；
1994年7月毕业于同济大学工业与民用建筑专业，工学学士；
1994年7月至2002年6月在中国建筑西南设计研究院从事结构设计工作；
2002年7月加入成都基准方中建筑设计事务所，现任执行总经理兼业务经营室主任，高级合伙人。

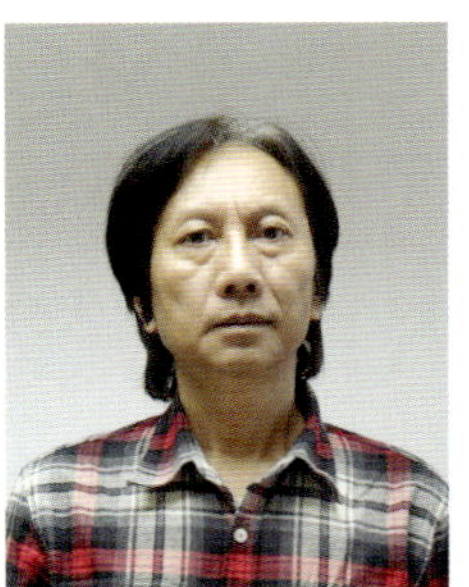

龚 进 Gong Jin

高级建筑师

出生于1958年6月；
1982年7月毕业于重庆建筑工程学院建筑系，工学学士；
1982年8月至2002年7月在中国建筑西南设计研究院从事建筑设计工作；
2002年7月加入成都基准方中建筑设计事务所，现任首席总建筑师，高级合伙人。

龙湖 · 成都北城天街

客户名称：成都龙湖北城置业有限公司
建筑面积：90万平方米
设计时间：2009年

北城天街项目位于成都市金牛区五块石，二环与三环之间，西侧为宝成铁路，东、南、北侧均为市政规划道路，是城北的一个形象制高点。

项目总规划面积约95万平方米，分东西两个地块，含SOHO办公、住宅、商业街、购物商场等业态，对整个片区大环境、城市形象均有一个良好的引导作用。

立面设计时尚大气，在城市重要节点处设置形象口部，在城市形象市场需求方面结合政府对片区的整体规划，做了深刻、细致的思考。

广联·摩根中心

客户名称：成都广联置业有限公司
建筑面积：12.77万平方米
设计时间：2008年

东大街作为未来城市金融中心，对提升城市形象、确保城市区域经济中心地位起着重要的作用，而地铁线的规划更使沿线土地价值日趋珍贵。广联摩根中心项目正位于成都市东大街和海椒市街交界处，三面临街，交通便捷，同时因海椒市街环境优良，成为都市办公和居住的极佳选择。

外形特质是吸引购房者的一大因素，广联摩根中心立面设计以现代简约主义的手法与语言，线条干净纯粹，总体简洁、现代、大气，结合建筑高度给人以视觉冲击。建筑通过体块、材料、虚实的对比来展现建筑独特的现代感。住宅楼采用同样的公建化设计，既保证了项目的整体性，又对东大街金融一条街城市景观作出贡献。色彩是考虑周围环境的影响以及所针对的客户群的审美喜好，选用了白色、灰色、棕色进行搭配，整体给人一种温馨宜人的尺度及色彩感觉。立面设计追求精细化，对开窗比例、尺度的详尽推敲，竖向线条的精心设计体现出建筑师对建筑美的完美诠释。

城投·西环广场

客户名称：成都城投地产有限公司
建筑面积：37.4万平方米
设计时间：2011年

西环广场商业综合体由成都城投地产有限公司、四川同辉实业有限公司共同开发，是一个集大型商业、零售、办公、酒店、公寓以及居住等功能于一体的综合开发项目。

项目用地位于武侯区红牌楼街道辖区，二环路与武侯大道交叉口。用地东北面紧临二环路，东南临武侯大道双楠段，西面为12米规划道路，区域交通十分便利，拥有良好的可达性。

地块周边环境可塑性强，有伊藤、人人乐等大型超市，具有一定的商业基础；同时该区域是一个较为成熟的居住社区，常住人口多、消费能力强，这些外部因素使该项目有条件成为一个商业氛围浓厚的区域中心。项目计划引进王府井百货，与伊藤、人人乐形成有级差的互补业态，呈三角形的商业布局，进一步提升区域的商业价值。

合景泰富·誉峰

客户名称：合景泰富地产控股有限公司
建筑面积：47.5万平方米
设计时间：2010年

在开始方案设计之前，经过充分的市场调研和论证，确定了"城市核心区高档电梯豪宅"的整体定位。围绕这一定位，我们与客户一起，从规划、户型、装潢、景观、配套等方面进行创新研发，力图全面超越成都本地市场对高层豪华电梯公寓的概念认知，树立全新的豪宅开发建设标准，打造"豪宅尖峰之作"。

小区建筑立面采用现代的建筑风格，对建筑中的虚实体量加以整合和变化组合，创造出简洁明快的建筑形象，以一种类公共建筑的立面造型手法去创造具有高品质和高识别性的住宅形象。

广联 · 仁寿五星级酒店

客户名称：成都广联置业有限公司
建筑面积：4.99万平方米
设计时间：2010年

酒店是按照欧洲豪华商务酒店理念设计建造的五星级酒店，是最具浪漫情调及人文关怀的精品酒店。酒店楼高20层，欧陆风格，外观气势宏伟，为仁寿县专业打造的标志性酒店建筑之一。

酒店地处仁寿新区城市公园侧，位居新开发区经济和商业中心，距仁寿黑龙滩风景区15分钟车程，距离成都市约65分钟车程，距眉山市约35分钟车程，是213国道与106省道交汇处的重要交通区位的高星级酒店。

酒店一楼西餐厅设计为可容纳120余宾客同时用餐，中餐厅豪华大气，挑高13余米，拥有600余个餐位，可举办各种大型宴会，气质尊贵、典雅大方。

二楼设计有小型特色中餐厅及40余间行政及豪华包房，环境优雅，格调温馨；三楼康乐部设计有桑拿沐足贵宾房、水疗馆、高标准的恒温游泳池、多器械健身房、休闲娱乐茶坊等服务项目；同时拥有多功能大、中、小型会议室及多媒体会议中心。酒店共设计有豪华套房、行政套房、豪华及高级客房280间（套）。各类客房风格独特，设有独立的 VIP 行政楼层、行政酒廊。各功能区分区明确，流线迅捷。

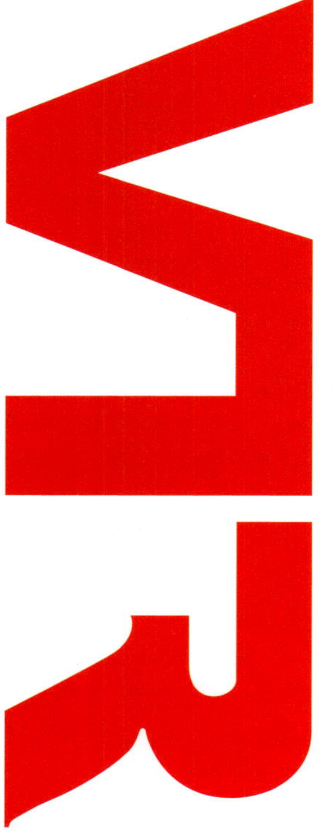

VTR(美国)国际设计研究有限公司(中国机构)

VTR (USA) International Design Institute Ltd. (Chinese Organization)

"建筑设计，反映着每个时代科技、经济与文化的内涵。设计的意义也随着时间而演变，成为一种新的国际语言。"

VTR（美国）国际设计研究有限公司（中国机构）（简称：VTR），是一家拥有建筑工程甲级设计证书、甲级工程咨询证书的专业设计单位，并在2008年通过了国家质量管理GB/T19001—2000 idt ISO 9001:200标准认证。

VTR（美国）国际设计研究有限公司（中国机构）自2002年进入中国以来，一直致力于融合东、西方文化背景和内涵，强调"人"、"从"、"众"的规划设计理念，同时遵从"建筑以人为本"的人性化共性。公司现有员工220人，其中教授级高工11人、高级工程师20人、工程师66人、助理工程师90人、行政后勤人员33人。几年来，VTR完成了一大批城市标志性建筑，如京龙国际城、天赋龙庭、四川省烟草公司阿坝州公司卷烟物流配送中心灾后重建项目、威玻·国际新城、元象·岭郡、荣威云岭、晋阳中学、银桦半岛酒店、远大·林语城、天慧国际总部基地等项目。

继往而拓进，传承而创新，是VTR一直坚持的建筑设计理念。创作是创造的基础，创新是创作的灵魂。我们的团队一直专注把握这一稍纵即逝的设计灵感，并结合个体项目特征，为业主们创作出一个又一个在市场上具有达标性的作品。

"Architectural design reflects the science, technology, economy and culture of the time. The aim of design is changing with the time and is becoming a new international language."

VTR (USA) International Design Institute Ltd (Chinese Organization) (VTR for short) specializes in design with Grade A certificates for construction engineering design and engineering consulting service. The company has passed the certification of the national quality management system of GB/T19001-2000 idt ISO 9001:2000 in 2008.

VTR (USA) International Design Institute Ltd (Chinese Organization) has been devoting itself to combining the cultural settings and connotation between the West and the East since it came into China in 2002, and emphasizes the planning design concept of "unity is strength", and besides follows the humanized general characteristics of "construction on people oriented". There are 220 employees in the company, including 11 professor senior engineers, 20 vice-professor senior engineers, 66 engineers, 90 assistant engineers and 33 administrative logistic employees. In recent years, VTR has completed a lot of projects, such as Jinglong International City, Tianfu Longting, Sichuan Province Tobacco Company Aba Branch Office Logistics Distribution Center Post-earthquake Project for Reconstruction, Weibo-International New town, Yuanxiang • Lingjun, Rongwei Yunling, Jinyang Middle School, Yinhua Peninsula Hotel, Yuanda • Linyu Town and Tianhui International Headquarters Base.

Continuing with the past and opening up the future, and inheriting and innovating are always the construction design concept that VTR is insisting on. Innovation is the soul of creation. Our team always focuses on the fleeting design inspiration and combines the individual project characteristics, to design a work up to standard for the owners in the market.

地址：四川省成都市高新区天府大道北段1480号9号楼3栋5层
首席执行官：谢绍宁
电话：+86–28–86786662
传真：+86–28–86788919
邮箱：vtr@vtr-sc.com
网址：www.vtr-sc.com

Add: 5F, office building 9,3#, Tianfu Avenue North 1480, Gaoxin District, Chengdu, Sichuan
CEO: Xie Shaoning
Tel: +86–28–86786662
Fax: +86–28–86788919
E-mail: vtr@vtr-sc.com
Http://www.vtr-sc.com

京龙国际城

占地面积：35 293.74平方米
建筑面积：37 792.1平方米
设计时间：2006年
建设地点：四川 资阳
容 积 率：6.11

Jinglong International City

Occupancy Area: 35,293.74 m^2
Building Area: 37,792.1 m^2
Design Time: 2006
Construction Site: Ziyang, Sichuan
Floor Area Ratio: 6.11

天赋龙庭

占地面积：42 141.96平方米
建筑面积：约26万余平方米
设计时间：2011年
建设地点：四川 成都
容 积 率：4.49

Tianfu Longting

Occupancy Area: 42,141.96 m²
Building Area: about 260,000 m²
Design Time: 2011
Construction Site: Chengdu, Sichuan
Floor area ratio: 4.49

四川省烟草公司阿坝州公司卷烟物流配送中心灾后重建项目

占地面积：14 570平方米
建筑面积：19 841.8平方米
设计时间：2010年
建设地点：四川 阿坝
竣工时间：在建
容 积 率：1.06

Sichuan Province Tobacco Company Aba Branch Office Logistics Distribution Center Post-earthquake Project for Reconstruction

Occupancy Area: 14,570 m²
Building Area: 19,841.8 m²
Design Time: 2010
Construction Site: Aba, Sichuan
Completion Time: Construction in progress
Floor Area Ratio: 1.06

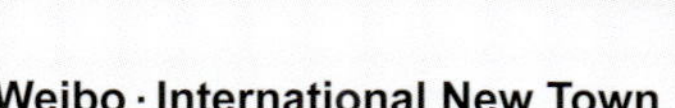

威玻·国际新城

占地面积：293 079.5平方米
建筑面积：1 000 060.5平方米
设计时间：2011年
建设地点：四川 威远
容 积 率：3.46

Weibo·International New Town

Occupancy Area: 293,079.5 m²
Building Area: 1,000,060.5 m²
Design Time: 2011
Construction Site: Weiyuan, Sichuan
Floor Area Ratio: 3.46

元象・岭郡

占地面积：46 744.43平方米
建筑面积：184 833.60平方米
设计时间：2010年
建设地点：四川 安岳
竣工时间：2012年
容 积 率：2.8

Yuanxiang • Lingjun

Occupancy Area: 46,744.43 m^2
Building Area: 184,833.60 m^2
Design Time: 2010
Construction Site: Anyue, Sichuan
Completion Time: 2012
Floor Area Ratio: 2.8

荣威云岭

占地面积：27 593.35平方米
建筑面积：98 591平方米
设计时间：2010年
建设地点：四川 威远
竣工时间：在建
容 积 率：3.3

Rongwei Yunling

Occupancy Area: 27,593.35 m^2
Building Area: 98,591 m^2
Design Time: 2010
Construction Site: Weiyuan, Sichuan
Completion Time: Construction in progress
Floor Area Ratio: 3.3

晋阳中学

占地面积：44 558平方米
建筑面积：26 154.62平方米
设计时间：2009年
建设地点：四川 成都
竣工时间：2010年12月一期工程已竣工
容 积 率：0.56

Jinyang Middle School

Occupancy Area: 44,558 m^2
Building Area: 26,154.62 m^2
Design Time: 2009
Construction Site: Chengdu, Sichuan
Completion Time: The first phase project having been completed in December 2010
Floor Area Ratio: 0.56

银桦半岛酒店

占地面积：36 347.2平方米
设计时间：2009年
建设地点：四川 威远
竣工时间：2011年3月主体封顶，约2011年12月竣工
容 积 率：0.49

Yinhua Peninsula Hotel

Occupancy Area: 36,347.2 m^2
Design Time: 2009
Construction Site: Weiyuan, Sichuan
Completion Time: Project main cap in March 2011; completed in about December 2011
Floor Area Ratio: 0.49

远大·林语城

建筑面积：72 351平方米
设计时间：2009年
建设地点：四川 成都

Yuanda • Linyu Town

Building Area: 72,351 m^2
Design Time: 2009
Construction Site: Chengdu, Sichuan

天慧国际总部基地

占地面积：120 382.99平方米
建筑面积：449 726.52平方米
设计时间：2011年
建设地点：四川 成都
容 积 率：3.0

Tianhui International Headquarters Base

Occupancy Area: 120,382.99 m^2
Building Area: 449,726.52 m^2
Design Time: 2011
Construction Site: Chengdu, Sichuan
Floor Area Ratio: 3.0

四川成都瑞康建筑设计院

Sichuan Chengdu RVKΛNG Architecture Design Institute

四川成都瑞康建筑设计院(以下简称"瑞康"或"设计院")位于成都华阳，成立于1997年11月，具有独立法人资格、甲级设计资质和丙级城乡规划资质，是集规划设计、研发于一体的民营建筑设计院。

经过十多年的积累与发展，瑞康已具有较强的综合实力。现有国家一级注册建筑师4名，国家一级注册结构工程师5名，国家注册设备工程师2名，高级工程师15名、中级工程师13名。

瑞康始终秉承"创新产品设计、创造客户价值"的宗旨，坚持"科学、精细、高效、客户至上"的原则，为客户提供完美的项目整体解决方案，是房产综合解决方案的供应商。设计领域涵盖各类住宅、高档别墅、商业建筑、园林景观、市政工程等，拥有强势的地产产品综合策划能力、大项目的掌控能力、产品与市场的结合能力，在方案中坚持追求项目的综合价值最大化，设计与施工中实行无缝连接，全程式优质服务、为客户着想体现在产品设计的每个环节中。

瑞康设计院汲取众家之长，加强与其他学术和设计团体的交流与合作，完善质量控制体系，探索创新机制，营造团结和谐、积极向上的企业文化和氛围，不断增强迎接经济、技术及市场变化的能力，以优秀的设计作品回报社会。

瑞康为有志之士在设计、规划、开发、营销、管理等领域提供持续学习、全面提升、发挥所长、超越拓展的无限空间。随着技术实力及实践经验的不断提升，瑞康已跻身于西南一流建筑设计院的行列，企业整体发展更上一层楼。

Sichuan Chengdu RVKΛNG Architecture Design Institute (Ruikang, or Institute) is located in Huayang, Chengdu. Founded in Nov. 1997, it is an independent legal entity with Grade-A Design Qualification and Class-C Town and Country Planning Qualification, as well as a private architectural design institute integrated with planning & design and research & development.

Upon years of accumulation and development, Ruikang has become a very competitive institute. It has 4 Class-1 registered architects, 5 Class-1 registered structural engineers, 2 registered equipment engineers, 15 senior engineers and 13 middle engineers.

Persisting in the purpose of "Innovating product design, Creating value" and the principle of "Scientific, Delicate, Efficient and Customer comes first", Ruikang provides perfect overall project solution for customers, and becomes a provider of property overall solution. Its design projects include all kinds of residential communities, high-class villas, commercial buildings, garden landscape and municipal administration constructions etc. With a strong ability of comprehensive planning of property products, the controlling ability of big projects, integration ability of products and markets, Ruikang insists on pursuing the maximum comprehensive value of projects, carrying out seamless connection between design and construction and providing overall high-quality services, while "Customer comes first" is shown in the whole process of product design.

Ruikang grows up by learning from others. It emphasizes communication and cooperation with other academic and design institutions, improving quality control system, exploring innovation mechanism, creating an united, harmonious and active enterprise culture and environment, constantly strengthening the ability of embracing economic, technological and market changes, and reciprocates the society with excellent design works.

In the fields of design, planning, development, marketing and management, Ruikang provides adequate space for talents to learn, to promote and develop comprehensively, to excel and expand.

府河音乐花园

占地约45公顷，建筑面积40余万平方米。2000年被评为成都市住宅样板小区，2001年获建设部颁发的“景观环境示范楼盘”、成都市颁发的“成都十大金奖楼盘”。

Fuhe River Music Garden

The land area is over 45 ha, and the floor area is more than 400,000 m^2. It was selected as Model Community of Chengdu in 2000. In 2001, it was selected as the “Model Building for Landscape Environment” issued by MOHURD and won the “Top 10 Gold Award Building of Chengdu” issued by the Municipal Government of Chengdu.

美茵河谷

融入欧洲田园风情的现代建筑典雅而灵动，营造出通透、疏朗的居住感受。

Meiyin Valley

The modern in European country style is classic and bright, creating a clear and comfortable living environment.

森宇音乐花园

占地约33公顷，建筑面积23万平方米。2002年该楼盘开创了四川大盘时代新纪元，森宇集团总经理也因此被评为“2002年中国住宅推动力人物”之一。

Senyu Music Garden

The land area is over 33 ha, and the floor area is 230,000 m^2. In 2002, the community opened a new era of Sichuan Big Community. The General Manager of Senyu Group is therefore voted as one of the “2002 Chinese Residential Community Promoters”.

维也纳森林别墅

占地约40公顷，建筑面积18万平方米；俯瞰高尔夫球场，拥有133余万平方米原生松林和近万平方米内湖。荣获“2004年四川首届房地产项目五十强别墅十强第一名”、“四川首届房地产核心影响力推介年会十大核心名盘”荣誉称号。

Vienna Forest Villa

The land area is over 40 ha, and the floor area is 180,000 m^2; overlooking the golf court, it has 1,333,333 m^2 of original pine forest and 10,000 m^2 area of internal lake. It wins the “No.1 of Top 50 Villas of the First Property Project of Sichuan, 2004” and “Top 10 Key Famous Community of the First Property Key Influence Promoting Annual Meeting of Sichuan”.

翰林上岛

占地约5.7公顷，建筑面积22.7万平方米，处于航空港黄金地段，荣获2007年成都市房管局第五届房地产“金芙蓉杯”奖，2008年度成都市住宅销售面积十强楼盘、2008年成都传媒集团房地产年度区域推动力楼盘、2008年度明星楼盘。森宇集团因其良好的销售业绩而荣获2008年度成都市销售十强企业。

Han Lin Shang Dao

The land area is 56,667 m^2, and the floor area is 227,000 m^2. Located in the golden area of airport, it won the Fifth Property Gold Lotus Cup hosted by Bureau of Housing Management of Chengdu (2007), the Top 10 Community for Residence Sales Area of Chengdu (2008), Property Annual Regional Promoting Community of Chengdu Media Group (2008) and Star Community (2008). Due to its excellent sales performance, Senyu Group was selected as Top 10 Enterprises for Sales in 2008.

南湖度假风景区

总占地约400公顷，核心景区106余万平方米，是以府河、江安河、内湖（南湖）、森林、湿地等自然资源为依托，以河滨休闲区、欧洲街、占地约20万平方米的大型游乐场及大型体育运动功能区等大型人造景观为支撑的都市商务休闲旅游景区。

South Lake Resort Scenic Spot

The total land area is over 400 ha, in which, the central scenic area is 1,066,667 m^2. Surrounded by Fuhe River, Jiang'an River, internal river (South Lake), forest, wetland and other natural resources, it is an urban business leisure tourist site supported by riverside leisure area, the Europe street, large-scale playground taking up an area of 200,000 m^2, large-scale spots area and other large-scale artificial landscapes.

Rvkang 四川成都瑞康建筑设计院

南湖国际社区

建筑面积150余万平方米，社区每个户型的外部景观和内部空间相互渗透，最大限度地达到人与环境的交流沟通，建筑定向采用了最经久不衰的装饰艺术风格，赋予了建筑高贵典雅的艺术气质。该项目荣获"第二届中国西部航都二十大经典楼盘"和"第六届金芙蓉杯(2008)成都地产年度楼盘金奖"荣誉称号。

South Lake International Community

The floor area is over 1,500,000 m^2. Exterior landscape and interior space of each household infiltrate into each other, realizing the maximum communication between human and nature. The architectural design adopts the classic decoration art style, giving the building an elegant and noble artistic look. The project was selected as "Top 20 Classic Community of the 2nd Western China-Chengdu" and "Chengdu Property Annual Gold Award of the 6th Gold Lotus Cup (2008)".

河心公园商业街

建筑面积约4万公顷，1.5千米长的原味欧洲商务休闲街，受到全国众多开发商及设计院关注。

Business Street of Hexin Park

The floor area is over 40,000 m^2. It is a 1.5 km long, European-style business & leisure street, which has attracted attention from many developers and design institutes of China.

Rvkang
四川成都瑞康建筑设计院

成都梦幻岛体验公园

占地20公顷集娱乐、探险、刺激于一体,为亚洲科技含量最先进的游乐场之一。乐园精心打造的商业街和主题餐饮、主题商品结合天然河景，加上互动喷泉、雕塑景观，为游客营造一个忘掉烦恼和压力、动静皆有的梦幻天堂。

Dream Island Park, Chengdu

The land area is 20 ha Integrated with entertainment, adventure and excitement, it is one of the most technologically-advanced playgrounds of Asia. Specially-built business street, theme restaurants and theme products are combined with natural river-view, interactive fountains and sculpture landscapes to provide a tension-free, quiet yet active dream heaven for tourists.

秦皇岛龙屿游艇会

占地面积约38公顷，建筑面积约70万平方米。

Longyu Yacht Club, Qinhuangdao

The land area is 38 ha, and the floor area is about 700,000 m^2.

秦皇岛观山海

占地面积6公顷，建筑面积10万平方米；迄今为秦皇岛山海关区销售最好楼盘之一。

Guan Shan Hai Community, Qinhuangdao

The land area is 6 ha, and the floor area is 100,000 m^2; till now, it's still the best selling community of Shanhaiguan District, Qinhuangdao.

秦皇岛观海

占地面积约13公顷，建筑面积20余万平方米。

Guan Hai, Qinhuangdao

The land area is over 13 ha, and the floor area is over 200,000 m^2.

Rvkang
四川成都瑞康建筑设计院

四川贵嘉建筑设计有限公司

SICHUAN GUIJIA ARCHITECTURAL DESIGN CO., LTD.

四川贵嘉建筑设计有限公司是经国家建设部批准成立的具有建筑工程甲级资质、风景园林工程设计专项甲级资质和城乡规划编制乙级资质的综合性设计机构。公司设有建筑、结构、给排水、电气、暖通、风景园林、城乡规划及建筑经济等专业，提供从城乡规划、前期策划、方案设计、初步设计到施工图设计的全过程建筑设计服务。

公司始终秉承“追求卓越、为社会服务、为业主及员工创造最大化价值”的经营宗旨，发扬“务实求真、精心设计”的企业精神，竭诚为各界用户服务，努力打造优质的建筑设计，为创建美好城乡家园，促进社会经济发展而努力。

地址：四川省成都市锦江工业园区三色路209号
火炬动力港A区2栋8楼
电话：+86-28-85925198 +86-28-85925330
传真：+86-28-85925077
邮箱：gjsjgs@126.con
网址：Http://www.gjjzsj.com

SICHUAN GUIJIA ARCHITECTURAL DESIGN CO., LTD. is an A-class qualification integrated corporation which is permitted to build up by Ministry of Housing and Urban-Rural Development of the People's Republic of China, further more in possession of A-class qualification concerning building engineering, landscape architecture, B-class qualification on urban planning. Our group sets comprehensive department not only in construction, architectural structure, watersupply&drainage, electrical engineering, and heating & ventilating, but also in landscape architecture, urban planning and engineering budget. We offer architectural services from conceptual through developed designs and construction documents.

Based on maximizing the value for the staff, client and society as the operation tenet, and realistic and pragmatic design as the enterprise spirit, the company is dedicated to building a good home and promoting social development with high-quality design.

Add: 8 Floor Building 2 Section A Sanse Road Jinjiang Industrial Park Habor Chengdu, Sichuan
Tel: +86-28-85925198 / +86-28-85925330
Fax: +86-28-85925077
E-mail: gjsjgs@126.com
Http://www.gjjzsj.com

西藏军区军史馆，占地面积约15 000平方米，总建筑面积6 000平方米。该项目为纪念西藏和平解放60周年而建，项目设计将西藏军区军史馆与民族教育馆相结合。设计围绕地形中部队进藏时种植的已有60年历史的钻天杨为中轴，将军史馆和民族教育馆设计在中轴两侧，通过连廊将两馆有机地联结在一起。

该项目设计严肃、大气，充分体现了西藏军区部队的优良传统，增进了民族团结。

成都市成华区职业技术学校，位于成华区龙潭寺，占地面积20万平方米，总建筑面积17公顷。招收学生7 000人，为职业培训和学历教育学校。该项目于2007年10月建设，2009年9月招生开学。

成都市熊猫体育公园，位于成都市成华区龙潭寺，项目占地约11.3公顷，为全民健身体育练习及比赛场地。项目设羽毛球练习馆1座、400米标准跑道田径场1个、6片网球场、6片篮球场。小球馆设计采用薄壳钢筋混凝土与网架结构相结合，整体设计流畅、现代，不失为一座标志性建筑。

Sichuan Yugao Architectural Design Co., Ltd.

四川域高建筑设计有限公司

四川域高建筑设计有限公司是国家建设部批准的甲级建筑设计单位，是具有建筑设计、规划设计、风景园林设计、装饰工程设计、市政工程设计、岩土工程勘察等资质的综合性勘察设计单位。是一个社会责任心强、充满创意与朝气的设计创作团队。

公司现有职工200人，其中享受国家政府特殊津贴1人、教授级高级工程师6人、高级工程师20人、工程师54人、国家一级注册建筑师7人、二级注册建筑师10人、国家一级注册结构工程师11人、二级注册结构工程师9人、国家注册岩土工程师3人、注册公用设备工程师（电气）2人、专业技术人员占全院职工总数的80%。是一支经过专业培训、取得职业资格证书并拥有大型重点工程设计实践经验的专业队伍，具有专业配套、人员结构合理的特点，保证在设计上为业主提供优质服务。

公司自设立以来，始终坚持“重合同、守信誉”的原则，精心设计，不断完善创新设计理念，科学管理，追求一流，力创建筑设计精品。设计成果得到相关部门、业主和社会的充分肯定。已通过ISO9001：2000质量管理体系认证。

该公司长期发展目标是成为具有高度社会责任感和历史责任感、彻底市场化的、以创造具有持久价值的产品为根本目的，具有强大设计、研究、咨询和管理能力的大型综合技术服务公司。

Sichuan Yugao Architectural Design Co., Ltd. which has Grade A qualification certificate of architectural design issued by the Ministry of Housing and Urban-Rural Development, is a comprehensive exploration and design company including architectural design, planning and design, landscape design, decoration engineering, municipal engineering, geotechnical engineering investigation and other qualification. It is a creative design team with strong sense of social responsibility, creativity and vitality.

At present, composed of 200 staff, 1 with special government allowances , 6 professor level senior engineers, 20 senior engineers, 54 engineers,7 first grade registered architects of China, 10 second grade registered architects,11 first grade registered structural engineers, 9 second grade structural engineers , 3 certified geotechnical engineers, 2 registered public facility engineer (electrical) , technical staff account for 80% in the total number of staff in the company, it is a professional team with technology training, professional credentials and practical experience of large and important engineering design, providing high quality services on designing for the owners with the characteristics of professional support and reasonable structure.

Since its establishment, always adhering to the "order the contract, keep the promise" principle, it has continuously improved the innovative design concept, scientifically managed the company, pursued good quality, to create the fine architectural design. Design results were fully affirmed by the relevant authorities, landlords and the community. The company has passed the ISC9001-2000 quality management system certification.

Its long term goal is to become a large-scale integrated technology services company with highly social responsibility and sense of history, a thorough market-oriented manner, creating products with lasting value as the fundamental purpose, and the strong capability of design, research, consulting and management.

总公司
地址：四川省成都市金牛区九里堤星辰路64号1幢4层
电话：+86-28-87609659
传真：+86-28-87624833
邮箱：tj@scygdesign.com
Hr@scygdesign.com
网址：www.scygdesign.com
邮编：610031

重庆分公司
地址：重庆市渝北区新南路166号2栋1单元22-1234号
电话：+86-23-63088122
邮编：401147

贵阳分公司
地址：贵州省贵阳市高新区科技产业园G-2-17-53都匀路
电话：+86-28-62013776
邮编：550001

Company
Add: F4, Tower I, No.64 Xingchen Read, Jiulidi, Jinniu District, Chengdu, Sichuan
Tel: +86-28-87609659
Fax: +86-28-87624833
E-mail: tj@scygdesign.com
Hr@scygdesign.com
Http: //www. scygdesign.com
P.C.: 610031

Chongqing Branch
Add: 22-1234, Unit1, Tower 2, No. 166 New South Road, Yuzhong District, Chongqing
Tel: +86-23-63088122
P.C.: 401147

Guiyang Branch
Add: Guiyang Hi-tech Zone Science and Technology Industrial Park, Guiyang, Guizhou
Tel:+86-28-62013776
P.C.: 550001

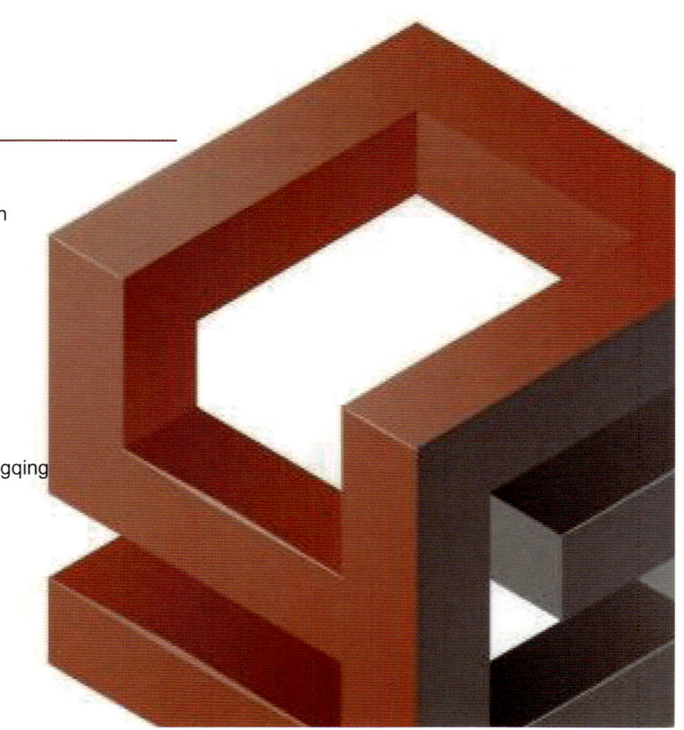

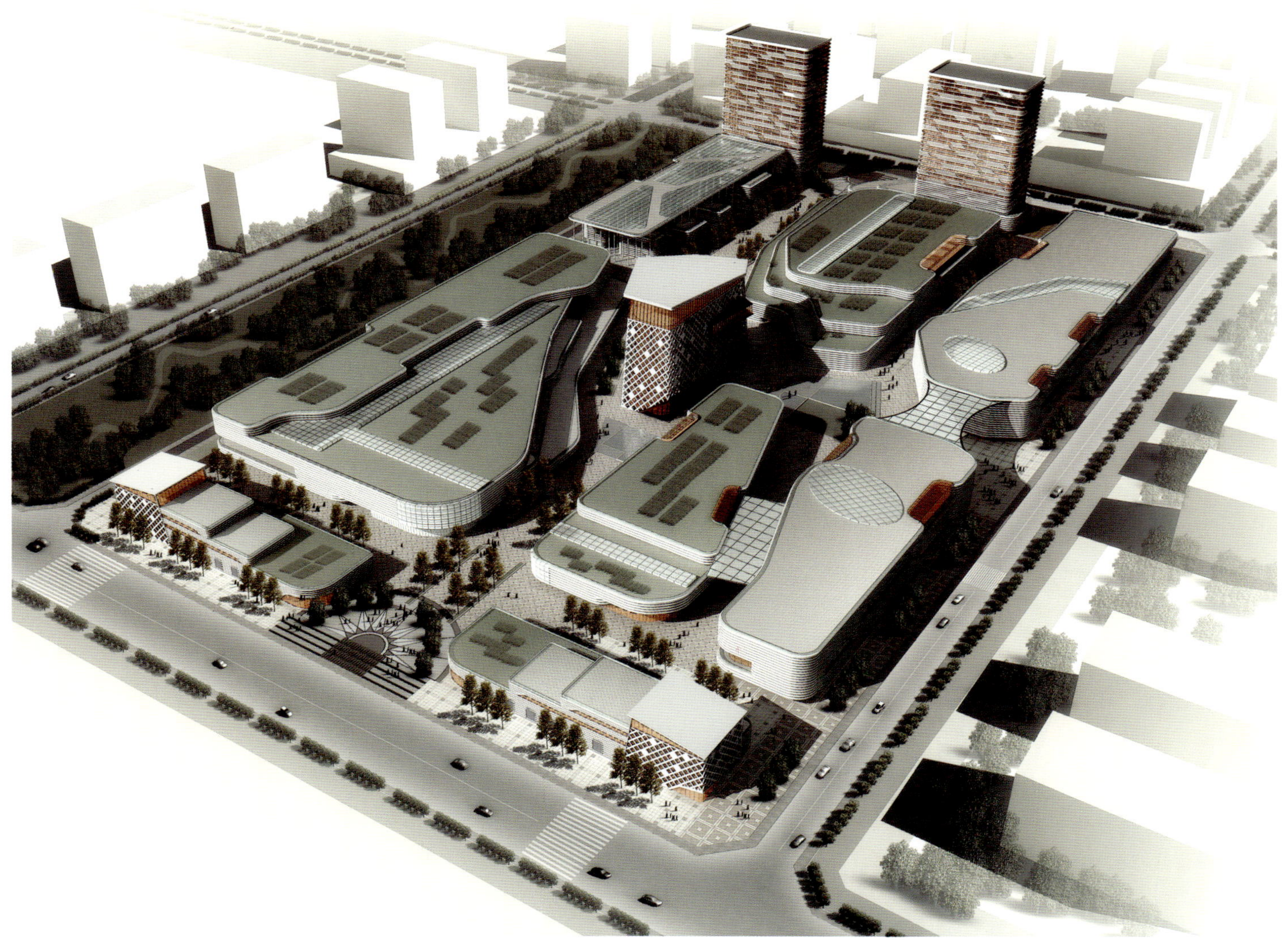

ARCHITECTURAL DESIGN WORKS
建筑设计作品
Public Buildings
公共建筑

百年国际家居展贸城
建设地点：四川 成都
项目规模：39万平方米

ARCHITECTURAL DESIGN WORKS
建筑设计作品
Residential Buildings
住宅建筑

月光湖

建设地点：四川 成都
项目规模：30万平方米

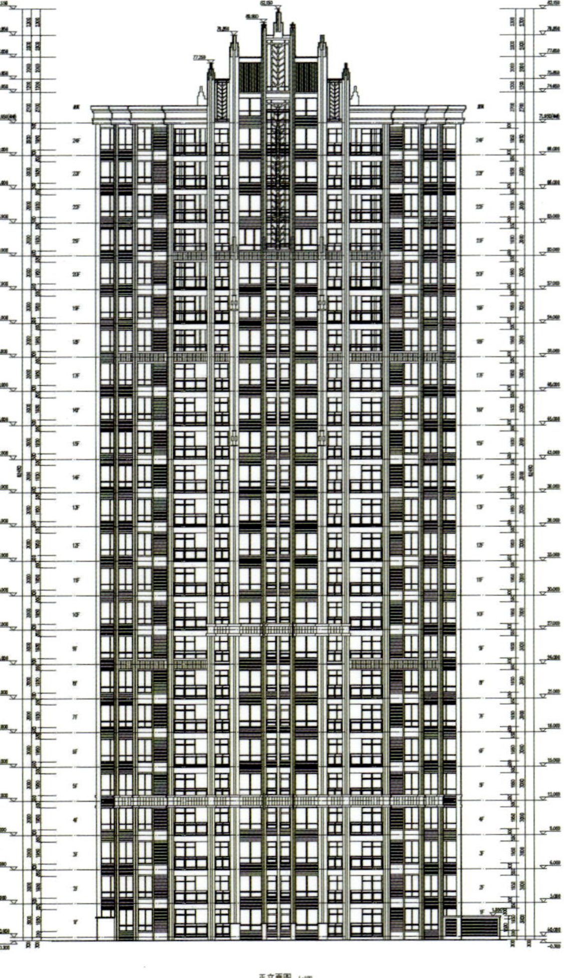

ARCHITECTURAL DESIGN WORKS
建筑设计作品
Public Buildings
公共建筑

国宾领域壹号
建设地点：四川 成都
项目规模：3万平方米

ARCHITECTURAL DESIGN WORKS
建筑设计作品
Public Buildings
公共建筑

润和国际商务酒店
建设地点：四川 成都
项目规模：3万平方米

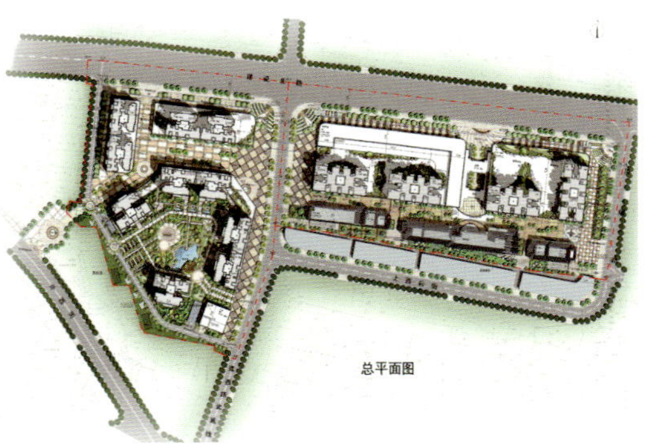

ARCHITECTURAL DESIGN WORKS
建筑设计作品
Public Buildings
公共建筑

达高国际城市综合体
建设地点：四川 资阳
项目规模：28万平方米

ARCHITECTURAL DESIGN WORKS
建筑设计作品
Public Buildings
公共建筑

盛大国际城市综合楼
建设地点：四川 成都
项目规模：30万平方米

ARCHITECTURAL DESIGN WORKS
建筑设计作品
Residential Buildings
住宅建筑

金色阳光
建设地点：山东 金乡
项目规模：30万平方米

ARCHITECTURAL DESIGN WORKS
建筑设计作品
Residential Buildings
住宅建筑

润禾花园
建设地点：四川 成都
项目规模：27万平方米

ARCHITECTURAL DESIGN WORKS
建筑设计作品
Residential Buildings
住宅建筑

北新润苑
建设地点：四川 成都
项目规模：20万平方米

ARCHITECTURAL DESIGN WORKS
建筑设计作品
Public Buildings
公共建筑

上上创业大厦
建设地点：河南 郑州
项目规模：10万平方米

ARCHITECTURAL DESIGN WORKS
建筑设计作品
Residential Buildings
住宅建筑

南湖国际
建设地点：四川 成都
项目规模：25万平方米

ARCHITECTURAL DESIGN WORKS
建筑设计作品
Public Buildings
公共建筑

羊西线城市综合体
建设地点：四川 成都
项目规模：28万平方米

山鼎建筑
Cendes Architecture

Cendes

Cendes Architecture Design Corporation Ltd.

山鼎建筑设计咨询有限公司

山鼎建筑（Cendes）拥有国际背景的公司决策层和国际化项目管理运营体系，是一家提供专业技术服务和整体解决方案的国际专业集团公司。山鼎建筑以国际智慧融合当地经验，秉承“筑就城市理想”的理念，其各地运营公司的400多名国内外专业人士，在建筑、规划、环境、能源等领域为政府及私人机构提供专业咨询和综合工程设计服务。山鼎建筑拥有各地的丰富经验和本土知识，通过不断的技术创新和项目实践，持之以恒地改善并维护各地的建筑品质、城市环境和社会环境的可持续发展。

山鼎建筑（Cendes）各地运营公司：

· 北京山鼎建筑工程设计有限公司
· 上海山鼎建筑工程设计有限公司
· 四川山鼎建筑工程设计咨询有限公司（建设部甲级资质）
· 西安山鼎建筑工程设计咨询有限公司

山鼎建筑以全面和超越的视野、整合的解决方案，应对各个领域日益复杂的挑战，为公司品牌创造持久发展的影响力。

Cendes Architecture has an international background of the company decision-making and international project management operating system, is an international professional group providing professional technical services and integrated solutions. Cendes integrates the local experience with international wisdom, adhering to the "Creating Ideas Realizing Dreams" concept, its local operating companies of more than 400 domestic and foreign professionals provide professional comprehensive engineering consulting and design services for the government and the private sector in architecture, planning, environment, energy and other areas. Cendes has extensive experience and local indigenous knowledge, through constant technological innovation and project practices, consistently improves and maintains the sustainable development architecture quality, urban environment and social environment.

Cendes Architecture contains:

• Beijing Cendes Architecture Engineering Design Co., Ltd.
• Shanghai Cendes Architecture Engineering Design Co., Ltd.
• Sichuan Cendes Architecture Engineering Design Consulting Co., Ltd. (Ministry of Construction Grade A)
• Xi'an Cendes Architecture Engineering Design Consulting Co., Ltd.

Cendes Architecture has the comprehensive and prospective vision, integrated solutions to deal with increasingly complex challenges in various fields, creating a lasting development impact for the company brand.

龙湖长桥郡

项目地点：四川 成都
项目规模：469 882平方米
设计时间：2007年
竣工时间：——
项目业主：龙湖房地产
项目参数：容积率（0.51）
建筑密度（20.14%）
机动车车位（1 053辆）
总户数（480户）

设计说明

项目位于成都市新津县，拥有独一无二的生态环境。在设计中根据项目的地理环境特点，运用水岸、绿地生态环境和现代北美别墅高档住宅的规划模式，始终强调别墅区的生态社区感，有意营造人与自然的交流空间，强调人与环境的关系，使得别墅成为环境的一部分。将会所布置在基地临道路的中部，提高了配套设施的商业化，以弥补周边商业空白，内容包括会所、健身、休闲、超市、餐饮等。

别墅建筑采用传统庭院的布局。每幢别墅既有前园，又有内庭和私家后花园。在总体规划和建筑功能的设计上，参照北美式别墅建筑的现代风格。根据规划结构形成组团式别墅布置模式，每个组团内部建筑摆放及朝向布置均经过精心考量，根据不同区域、不同的景观条件，布置不同的别墅类型，且不拘于形式，做到建筑与环境的高度和谐。

尚都服饰广场

项目地点：四川 成都
项目规模：104 000平方米
设计时间：2004年
竣工时间：2007年
项目业主：成都通生房地产有限公司
项目参数：容积率（10.08）
建筑密度（74.29%）
建筑高度（67.9米）
机动车车位（280辆）

设计说明

本项目着重于商业氛围的营造。广场空间、街道线性空间与建筑本身的结合，有效地引导人流进入商场且起到了调节大量人流的功能；建筑内部精心设计了一条商业流线，室内商业中庭不受室外环境影响而又可以享受阳光和景观，创造出更为舒适的、以人流动线为设计重点的休闲购物环境。

由于本案位于繁华的春熙路商圈，周边商场、写字楼林立，故造型和色彩上的设计反其道而行之，以传统的方式在细节处理上使本案明显有别于周边建筑，同时注重外立面的整体效果。

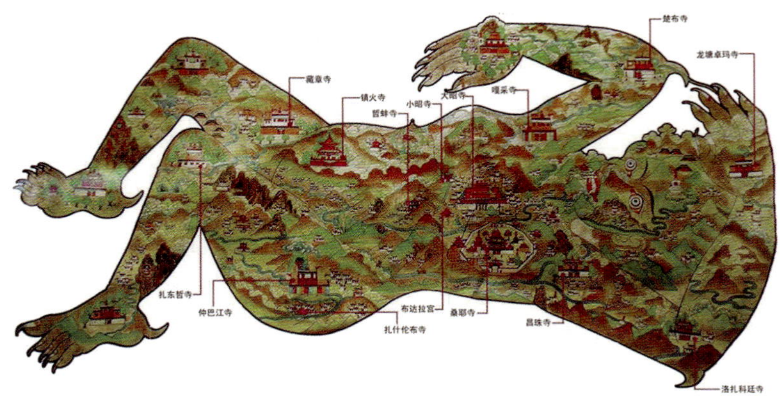

西藏镇魔图
Legend map of Tibet

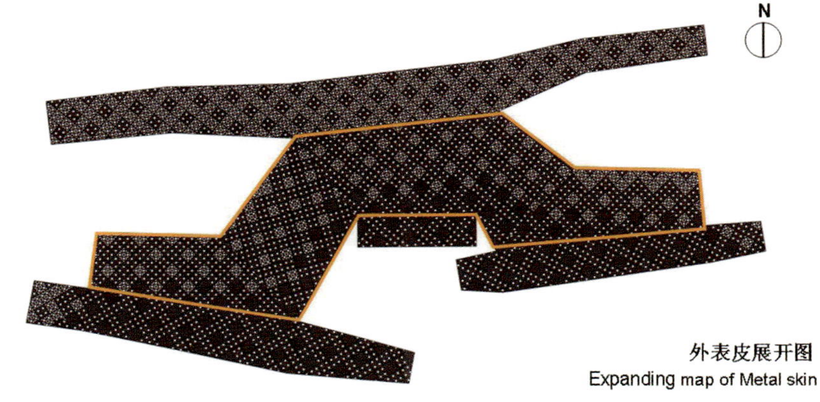

外表皮展开图
Expanding map of Metal skin

西藏自然科学博物馆

项目地点：西藏 拉萨
项目规模：33 000平方米
设计时间：2008年
竣工时间：2011年
项目业主：西藏自治区科技厅
项目参数：容积率（0.24）
建筑密度（9.4%）
建筑高度（42米）
项目殊荣：英国建筑科学院BREEAM认证

设计说明

建筑设计灵感来自于西藏自然意境和藏族文化。从藏族以星辰变迁制定藏历的传统可以得知，自古以来自然科学就一直在指导藏族人民的生活和祭礼，故取项目的案名为“星河”。立面纹案通过屋顶和玻璃外墙，将西藏充沛的光影诗歌演绎在入口大厅和公共空间，使建筑在遮阳和节能方面得到平衡，并充满人文气息和科技含量。纯白色的建筑与远处皑皑白雪的山顶遥相呼应，犹如洁白的哈达飘落在拉萨河的怀抱中。西藏自然科学博物馆是向公众展示人文和科技的空间，整个建筑呈带状布置，功能完善的展厅沿建筑北侧布置，而南侧为中庭和公共廊道。南立面高30米、宽120米，屋顶进深达17米，采用多折面单层钢网格结构体系，使主立面无柱体系在超限结构体系方面承受了巨大挑战。

淄博SM城市广场

淄博SM城市广场是SM商业集团(东南亚排名第一的商业开发集团和世界500强企业)自20世纪90年代进入中国市场后，第一个在北方地区的商业项目。基地位于淄博市的核心区，西临孝妇河，沿河有明代的城墙遗址，赋予了项目一定的历史和文化内涵，后期开发将结合城墙遗址打造出一个有特色的休闲商业和景观区域。

淄博SM城市广场也是在国内为数不多的由外资独资开发、完全自持和自营的商业中心。因完全采用国际商业标准和模式，设计理念上有多处突破了中国现有的商业和建筑规范。

淄博SM城市广场是一栋全长300多米、标准层近3万平方米、总建筑面积约15万平方米的大型纯商业中心。功能上主要由百货主力店、大型超市、精品商业街、电影城和多个餐饮中心组成。在停车场的设计上采用了立体停车场和地下停车场相结合的模式，顾客可在最近的商业区停车。

在平面布置上，淄博SM采用了SM城市广场一贯使用的、典型美式商业商场的布局——以一条商业内街和大型中庭为主轴贯穿整个建筑。沿商业街布置垂直交通体系来组织人流，并和绝大多数美式商业商场一样采用透明顶盖，使内街在全中央空调的环境里能享受到自然光。该项目总平面布局和商业内街的设计打破了以往惯用的直线手法，而采用流线处理，使整个建筑内外空间都充满了活力。由于是一栋多层建筑，从高处俯视淄博SM城市广场，建筑物宛如一只飞翔的海鸥。

在立面设计上，采用大体量现代建筑的设计手法，凸显地标性建筑的气质。外立面造型采用同色渐变金属板和玻璃为主要材料，局部设有LED广告位，使建筑简约大气而又不失细部刻画，商业气息浓烈。

淄博SM城市广场在结构和机电专业的设计方面具有独特性和专业性。考虑到建成后商业运营的变化和空间舒适度，建筑主体采用了15米X15米的大跨度柱网，主梁都为预应力结构。结构设计上，为满足建筑对中庭空间不设柱的要求，自动扶梯平台的挑梁长度设为8米，且对梁高也有所限制。机电专业设计方面对功能针对性和后期的商业运营可变性都有很高的要求。在现代商业中心的设计上，对照明、通风和室温等设计进行了综合考虑，使空间达到最佳舒适度。

锦绣森邻

项目地点：四川 成都
项目规模：220 000平方米
设计时间：2004年
竣工时间：2005年
项目业主：西藏自治区科技厅
项目参数：容积率（0.24）
建筑密度（9.4%）
建筑高度（42米）
项目殊荣：2005年建设部人居规划设计优秀奖

设计说明

"森"——自然生态的社区景观
"邻"——和谐融洽的邻里生活

规划以4个"森林"岛组成建筑群体，围而不合的退台式花园洋房构成各具特色的邻里单元，精心设计的不同尺度、风格的林带将邻里单元包裹。小高层以点式景观建筑的姿态有机地融入，形成丰富的空间形体。户型以极具创新和人性化的设计在业内赢得广泛的赞誉。

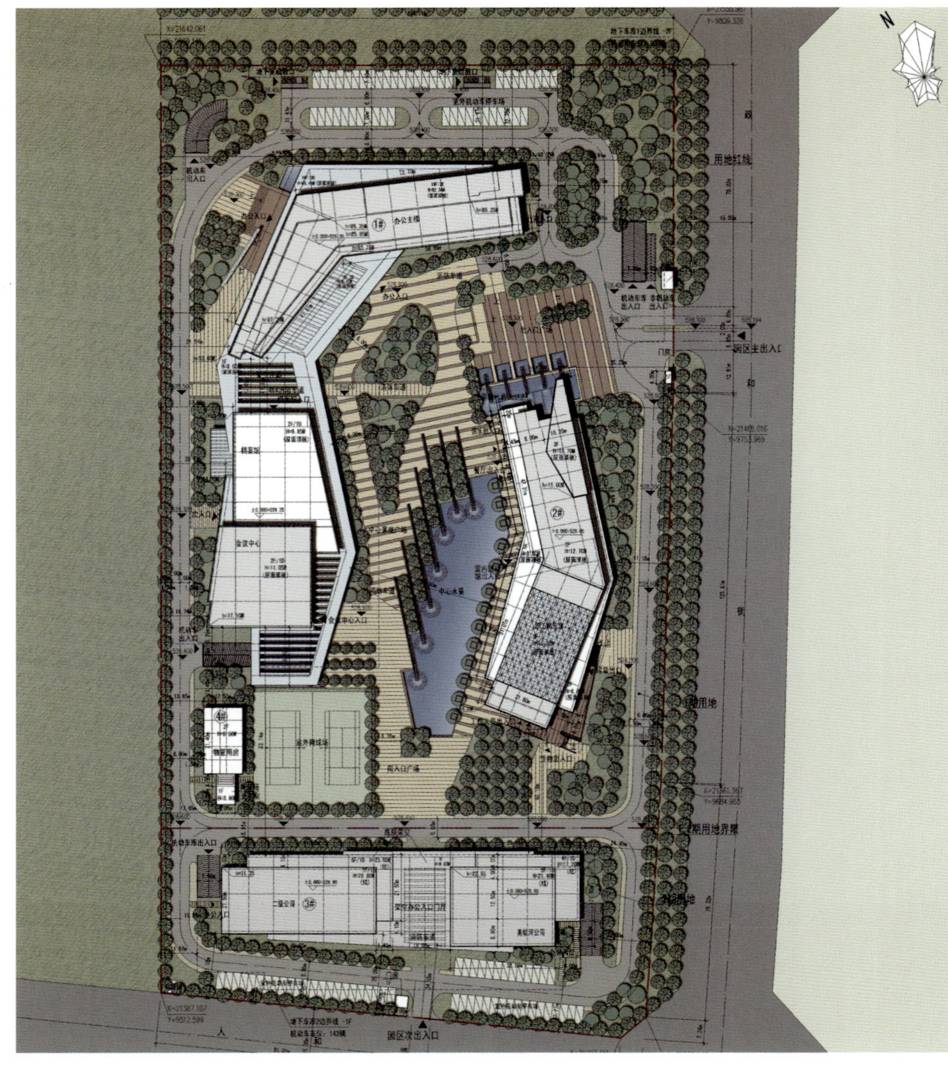

成勘院办公室

项目地点：四川 成都
项目规模：115 000平方米
设计时间：2004年
竣工时间：——
项目业主：中国水电顾问集团成都勘测院
项目参数：容积率（1.52）
建筑密度（23.6%）
建筑高度（89.3米）
机动车车位（1 144辆）

设计说明

作为中国水电顾问集团成都勘测院的新办公区，项目从开始就按高标准、高起点进行打造，并一直受到各界广泛关注。

项目是由办公主楼、档案馆、员工餐厅、员工俱乐部等四幢建筑单体组成的办公群体建筑。总面积逾12万平方米。建筑总平面采用折板围合式布局，形成中央大尺度办公庭院。折板式造型带来了建筑体形的动感，在不同的角度均能体会到全新的视觉体验，使得建筑形象卓尔不群，印象突出。隐喻一本打开的书籍，与项目业主行业特征相吻合，充分体现人文关怀。项目设计充分体现"和谐、生态、高效、创新"的理念，必将成为光华大道上的重要标志。

都城雅颂居

项目地点：四川 成都
用地面积：142 000平方米
总建筑面积：880 000平方米
建筑最高高度：82.2米
项目业主：香港嘉里建设
工程进度：建设中

设计说明

作为嘉里建设在大陆西南地区开发布局的第一项目，它的重要性不言而喻。前后历经四年打造，为业界所广泛关注。

规划方式以点板结合，建筑布局围而不合，通而不透，充分尊重成都当地独特的气候条件和居住习惯。户型设计充分体现"舒适，实用，成熟，大气"，建筑立面设计线条干净利落，造型沉稳低调，色彩含蓄隐忍，充分体现都市高端华宅的独特魅力。

川大科技园

项目地点：四川 成都
项目规模：480 000平方米
设计时间：2008年
竣工时间：2009年
项目业主：川大科技园（南区）开发有限公司
项目参数：容积率（5.5）
建筑密度（34.6%）
建筑高度（200米）
机动车车位（2 916辆）

设计说明

本项目由四川大学科技园（南区）开发有限公司开发建设，地块位于成都市高新区的国际金融服务区域核心区，总用地面积90 346平方米。

项目由四幢高标准甲级写字楼组成。其中，A座100米，B座130米，C座100米、D座200米。高层塔楼两两成组，分为两个办公单元，每组办公单元由底层共享大堂联结，形成互相区分又互为联系的群体关系。

在立面塑造上，以建筑本身的形式来丰富立面的层次，统一且简洁，具有强烈的时代感。在经济、合理的前提下巧妙地解决了玻璃幕墙建筑和普通开窗建筑之间的矛盾。竖向线条的运用，强调建筑的挺拔和隽秀，彰显建筑的独特品质。

成都长源建筑设计研究院有限公司

地址：四川省成都市石马巷8号虹鑫大厦4楼
邮编：610017
电话：+86-28-86957595
传真：+86-28-86953456
网址：www.cd-cy.com.cn

Add: F4 Hongxin Building No.8 Shima Xiang, Chengdu, Sichuan
P.C.: 610017
Tel: +86-28-86957595
Fax: +86-28-86953456
Http://www.cd-cy.com.cn

成都长源建筑设计研究院有限公司是一个涉足项目策划、建筑规划、建筑工程设计的综合性建筑设计研究院。公司成立于20世纪90年代初期，是成都虹鑫投资集团旗下的独立法人企业。主要领导及骨干均来自中国建筑西南设计研究院。

成都长源建筑设计研究院有限公司现为国家甲级资质（A151009731）。现有国家一级注册建筑师6名、国家二级注册建筑师4名、国家一级注册结构工程师5名以及近百名设计技术人员。在城市规划、商业设施、办公建筑以及住宅小区设计等方面有着丰富的实践经验和卓越的设计理念，所参与的工程设计项目涉及各个领域。在十多年的工程设计实践中，设计及竣工面积逾数百万平方米，并与新加坡、加拿大等的国际设计机构有着长期广泛的合作。

Chengdu Changyuan Architecture Design and Research Institute Co., Ltd is a comprehensive architectural design and research institute involving project planning, and architectural planning, construction design. It was established in early 1990s, as an independent legal entity under Chengdu Hongxin Investment Group. Its main leaders and cadres come from China Southwestern Architectural Design Institute.

Now, it holds the national license Class-A (A151009731). It employs 6 Class-A registered architects, 4 Class-B registered architects, 5 Class-A registered structural engineers and nearly one hundred technical designers. It has rich experience and excellent design ideas in the design of urban planning, commercial facilities, office building and residential community etc. It has participated in various fields of project designs. With more than 10 years' practice of construction design, it has a total design and completion area over 1,000,000 m^2, and has long-term cooperation with many international design institutions in Singapore, Canada and other countries.

1

1–2 中房 高新西区极目项目

1-2 Zhongfang Jimu Project, High and New-Tech West District, Chengdu

3 西安曲江 亮丽家园

3 Qujiang Bright Home, Xi'an

2

3

Chengdu Changyuan Architecture Design And Research Institute Co., Ltd